Lecture Notes in Artificial Intelligence 4120

Edited by J. G. Carbonell and J. Siekmann

Subseries of Lecture Notes in Computer Science

T0223719

Jacques Calmet   Tetsuo Ida
Dongming Wang (Eds.)

Artificial Intelligence and Symbolic Computation

8th International Conference, AISC 2006
Beijing, China, September 20-22, 2006
Proceedings

 Springer

Series Editors

Jaime G. Carbonell, Carnegie Mellon University, Pittsburgh, PA, USA
Jörg Siekmann, University of Saarland, Saarbrücken, Germany

Volume Editors

Jacques Calmet
University of Karlsruhe, IAKS
Am Fasanengarten 5, D-76131 Karlsruhe, Germany
E-mail: calmet@ira.uka.de

Tetsuo Ida
University of Tsukuba
Department of Computer Science
Tennoudai 1-1-1, Tsukuba 305-8573, Japan
E-mail: ida@cs.tsukuba.ac.jp

Dongming Wang
Beihang University
School of Science
37 Xueyuan Road, Beijing 100083, China
and Université Pierre et Marie Curie, Paris, France
E-mail: Dongming.Wang@lip6.fr

Library of Congress Control Number: 2006932044

CR Subject Classification (1998): I.2.1-4, I.1, G.1-2, F.4.1

LNCS Sublibrary: SL 7 – Artificial Intelligence

ISSN   0302-9743
ISBN-10   3-540-39728-0 Springer Berlin Heidelberg New York
ISBN-13   978-3-540-39728-1 Springer Berlin Heidelberg New York

Springer is a part of Springer Science+Business Media

springer.com

© Springer-Verlag Berlin Heidelberg 2006
Printed in Germany

Typesetting: Camera-ready by author, data conversion by Scientific Publishing Services, Chennai, India
Printed on acid-free paper   SPIN: 11856290   06/3142   5 4 3 2 1 0

Foreword

AISC 2006, the 8th International Conference on Artificial Intelligence and Symbolic Computation, was held on the campus of Beihang University, China, in the golden autumn of 2006. On behalf of the Organizing Committee and Beihang University, I had the pleasure to welcome the participants of this conference. The AISC series of specialized biennial conferences was founded in 1992 by Jacques Calmet and John Campbell with initial title "Artificial Intelligence and Symbolic Mathematical Computing" (AISMC) and the previous seven conferences in this series were held in Karlsruhe (AISMC-1, 1992), Cambridge (AISMC-2, 1994), Steyr (AISMC-3, 1996), Plattsburgh, NY (AISC 1998), Madrid (AISC 2000), Marseille (AISC 2002), and Hagenberg, Austria (AISC 2004).

Artificial intelligence and symbolic computation are two views and approaches for automating (mathematical) problem solving. They are based on heuristics and mathematical algorithmics, respectively, and each of them can be applied to the other. The AISC series of conferences has not only provided a lively forum for researchers to exchange ideas and views and to present work and new findings, but has also stimulated the development of theoretical insights and results, practical methods and algorithms, and advanced tools of software technology and system design in the interaction of the two fields and research communities, meanwhile leading to a broad spectrum of applications by the combined problem solving power of the two fields.

The success of the AISC series has benefited from the contributions of many people over the last decade. For AISC 2006, the Program Committee and, in particular, its Chair Tetsuo Ida deserve special credits: it is their expertise and tireless effort that made an excellent scientific program. I am most grateful to the four distinguished invited speakers, Arjeh M. Cohen, Heisuke Hironaka, William McCune, and Wen-tsün Wu, whose participation and speeches definitely helped make AISC 2006 a unique and high-level scientific event. The AISC 2006 General Chair, Dongming Wang, and Local Arrangements Chair, Shilong Ma, together with their organization team made considerable effort on many aspects to ensure the conference was successful. I would like to thank all the above-mentioned individuals, other Organizing Committee Members, referees, authors, participants, our staff members and students, and all those who have contributed to the organization and success of AISC 2006.

September 2006

Wei Li
Honorary Chair
AISC 2006
President of Beihang University

Preface

This volume contains invited presentations and contributed papers accepted for AISC 2006, the 8th International Conference on Artificial Intelligence and Symbolic Computation held during September 20–22, 2006 in Beijing, China. The conference took place on the campus of Beihang University and was organized by the School of Science and the School of Computer Science and Engineering, Beihang University.

In the AISC 2006 call for papers, original research contributions in the fields of artificial intelligence (AI) and symbolic computation (SC), and in particular in the fields where AI and SC interact were solicited. In response to the call, 39 papers were submitted. This volume contains 18 contributed papers, selected by the Program Committee on the basis of their relevance to the themes of AISC and the quality of the research expounded in the papers. The program of the conference featured 5 invited talks, out of which 4 presentations are included in the proceedings.

The papers in this volume cover a broad spectrum of AI and SC. The papers may be characterized by key words such as theorem proving, constraint solving/satisfaction, term rewriting, deductive system, operator calculus, quantifier elimination, triangular set, and mathematical knowledge management. Despite the breadth of the papers, we can observe their mathematical aspect in common.

For 15 years since the conception of AISMC (AI and Symbolic Mathematical Computing), changed to AISC in 1998, the mathematical aspect has remained as the common profile of the conferences. We see challenges as the problems that we face become more complex with the rapid development of computer technologies and the transformation of our society to an e-society. Such problems are tackled by using mathematical tools and concepts, which become highly sophisticated. The interaction of AI and SC bound by mathematics will become more relevant in problem solving. We hope that this unique feature of the conference will remain and gather momentum for further expansion of AISC.

We would like to express our thanks to the Program Committee members and external reviewers for their efforts in realizing this high-quality conference. Our thanks are also due to Wei Li, President of Beihang University, and the Local Arrangements Committee chaired by Shilong Ma for making the conference such a success.

We acknowledge the support of EasyChair for administering paper submissions, paper reviews and the production of the proceedings. With its support, the whole preparation process for the AISC 2006 program and proceedings was streamlined.

September 2006

Jacques Calmet
Tetsuo Ida
Dongming Wang

Organization

AISC 2006, the 8th International Conference on Artificial Intelligence and Symbolic Computation, was held at Beihang University, Beijing, September 20–22, 2006. The School of Science and the School of Computer Science and Engineering, Beihang University were responsible for the organization and local arrangements of the conference. AISC 2006 was also sponsored by the State Key Laboratory of Software Development Environment and the Key Laboratory of Mathematics, Informatics and Behavioral Semantics of the Ministry of Education of China.

Conference Direction

Honorary Chair	Wei Li (Beihang University, China)
General Chair	Dongming Wang (Beihang University, China and UPMC–CNRS, France)
Program Chair	Tetsuo Ida (University of Tsukuba, Japan)
Local Chair	Shilong Ma (Beihang University, China)

Program Committee

Luigia Carlucci Aiello	(Università di Roma "La Sapienza," Italy)
Michael Beeson	(San Jose State University, USA)
Bruno Buchberger	(RISC-Linz, Austria)
Jacques Calmet	(University of Karlsruhe, Germany)
John Campbell	(University College London, UK)
William M. Farmer	(McMaster University, Canada)
Martin Charles Golumbic	(University of Haifa, Israel)
Thérèse Hardin	(Université Pierre et Marie Curie – LIP6, France)
Hoon Hong	(North Carolina State University, USA)
Joxan Jaffar	(National University of Singapore, Singapore)
Deepak Kapur	(University of New Mexico, USA)
Michael Kohlhase	(International University Bremen, Germany)
Steve Linton	(University of St Andrews, UK)
Salvador Lucas	(Technical University of Valencia, Spain)
Aart Middeldorp	(University of Innsbruck, Austria)
Eric Monfroy	(UTFSM, Chile and LINA, France)
Jochen Pfalzgraf	(University of Salzburg, Austria)
Zbigniew W. Ras	(University of North Carolina, Charlotte, USA)
Eugenio Roanes-Lozano	(Universidad Complutense de Madrid, Spain)
Masahiko Sato	(Kyoto University, Japan)

Carsten Schürmann	(Yale University, USA)
Jörg Siekmann	(Universität des Saarlandes, DFKI, Germany)
Carolyn Talcott	(SRI International, USA)
Dongming Wang	(Beihang University, China and UPMC–CNRS, France)
Stephen M. Watt	(University of Western Ontario, Canada)
Jian Zhang	(Chinese Academy of Sciences, China)

Local Arrangements

Xiaoyu Chen	(Beihang University, China)
Li Ma	(Beihang University, China)

External Reviewers

Enrique Alba	Tony Lambert
Jesús Almendros	Ziming Li
Jamie Andrews	Ana Marco
Philippe Aubry	Mircea Marin
Carlos Castro	Marc Moreno Maza
Arthur Chtcherba	Immanuel Normann
Abram Connelly	Vicki Powers
Robert M. Corless	Maria Cristina Riff
Marco Costanti	Renaud Rioboo
Andreas Dolzmann	Miguel A. Salido
Frederic Goualard	Frédéric Saubion
Martin Henz	Ashish Tiwari
Sandy Huerter	Sergey P. Tsarev
Stephan Kepser	Christian Vogt
Ilias Kotsireas	Min Wu
Temur Kutsia	Yuanlin Zhang

Table of Contents

Constraint Satisfaction/Solving

Mathematical Knowledge Management

Interactive Mathematical Documents

Arjeh M. Cohen

Technische Universiteit Eindhoven
a.m.cohen@tue.nl
http://www.win.tue.nl/~amc/

Abstract. Being so well structured, mathematics lends itself well to interesting interactive presentation. Now that computer algebra packages have come of age, their integration into documents presenting the mathematics as natural as possible is a new challenge.

XML techniques enable us to record mathematics in a Bourbaki like structure and to present it in a natural fashion. In this vein, at the Technische Universiteit Eindhoven, we have built a software environment, called MathDox, for playing, creating, and editing interactive mathematical documents. Computer algebra systems function as services to be called by the document player. Specific applications are

- the build-up of context, providing information about the variables and notions involved in the document;
- a package providing an argument why two given graphs are isomorphic or —the more interesting case— non-isomorphic;
- an exercise repository from which exercises can be played in various natural languages (generated from the same source).

Parts of the work have been carried out within European projects like LeActiveMath, MONET, OpenMATH, and WebALT.

J. Calmet, T. Ida, and D. Wang (Eds.): AISC 2006, LNAI 4120, p. 1, 2006.
© Springer-Verlag Berlin Heidelberg 2006

Algebra and Geometry

Interaction Between "Equations" and "Shapes"

Heisuke Hironaka

Japan Association for Mathematical Sciences, Coop Olympia 506, Jingumae 6-35-3, Shibuya-ku, Tokyo 150-0001, Japan
`as6h-hrnk@asahi-net.or.jp`

Abstract. It is an elementary fact, even known to high schoolers, that there are two different ways of showing "shapes", the one by *photography* and the other by *drawing*. Mathematically, they correspond to the one by giving *equations* which show algebraic relations among coordinates and the other by using *parameters*, of which coordinates are expressed as functions. Think of a circle of radius one, on one hand expressed by an equation $x^2 + y^2 = 1$ and on the other by $x = \cos t, y = \sin t$ with a *parameter* $t, 0 \leq t \leq 2\pi$. As is well known, if the equation is cubic with no *singularity*, then we have a parametric presentation using elliptic functions.

The correspondence between *equational presentation* and *parametric presentation* becomes complex but interesting in the case of *many variables and presence of singularities*. I will present how the correspondence can be processed in general.

Theoretically there are two technical approaches.

1. Newton Polygons and Toroidal Geometry
   For a plane curve defined by $f(x, y) = 0$, which is for simplicity assumed to have only one local branch of multiplicity m at the origin, we choose a suitable coordinate system and look at the Newton polygon which is one segment of slope $-1/\delta$. (The coordinate is chosen to make δ the biggest possible.) Then we get a graded $\mathbb{R}[x, y]$-algebra

$$\wp(f) = \sum_{i=1}^{\infty} \{x^i y^j \mid j + i/\delta \geq k/m\} \mathbb{R}[x, y] T^k$$

   where T is a dummy variable whose exponent indicates the degree of homogeneous part of the graded algebra. It is toroidal. I will show that this algebra can be generalized with meaningful geometric properties to all the cases of many variables. Thus the technique of *toroidal geometry* is applicable.

2. Ideals and Tropical Geometry
   Think of an expression of ratio $\frac{J}{b}$ where the numerator J is an ideal (or system of equations) and the denominator b is a positive integer. The first equivalence is $\frac{J^k}{kb} = \frac{J}{b}$ for all positive integers k. We introduce more geometrically meaningful equivalence and discuss natural operations on it in connection with the language of the *tropical* geometry.

J. Calmet, T. Ida, and D. Wang (Eds.): AISC 2006, LNAI 4120, p. 2, 2006.

An Inductive Inference System and Its Rationality

Wei Li

State Key Laboratory of Software Development Environment
Beihang University, Beijing 100083, P.R. China
liwei@nlsde.buaa.edu.cn

Abstract. An inductive inference system **I** is introduced to discover new laws about the generality of the theory for a given model describing knowledge of a specific domain. The system **I** is defined in the first order language and consists of the universal inductive rule, the refutation revision rule and the basic sentence expansion rule. A rule of **I** can be applied to a theory and a given instance depending on their logical relation, and generates a new version of the theory. When the instances are taken one after another, a version sequence will be generated. The rationality of the system **I** is demonstrated by the following three properties of the generated version sequence: the convergency, commutativity, and independency. The rationality of the system **I** is formally proved by constructing a procedure GUINA which generates such version sequences.

Keywords: Belief, induction, refutation, inference, rationality.

1 Motivation

Inductive reasonings are the most frequently invoked and most effective means to discover new laws in scientific investigations. The purpose of this paper is to construct formally an inductive inference system **I** in the first order language, to present formally the rationality of the inductive inference systems, and to prove that the system introduced is rational. In order to achieve this goal, we begin by pointing out the main characteristics of inductive inference.

Inductive inference is a kind of mechanism to discover the new laws or propose new conjecture from individuals, and it is used in the evolutionary process of forming a theory of knowledge for a specific domain. Using the terminology of the first order language, inductive inference is used to find those sentences containing quantifier $\forall$ which are true for a given model **M**.

The basic idea of inductive inference was given by Aristotle about 2000 years ago. In the chapter "Nous" of his masterpiece "Organon", he wrote *"The induction is a passage from individual instances to generalities"* [1]. In the first order language, it could be described by the following rule:

$$P(c) \Longrightarrow \forall x P(x)$$

J. Calmet, T. Ida, and D. Wang (Eds.): AISC 2006, LNAI 4120, pp. 3–17, 2006.

where c is a constant symbol, $P(c)$ denotes an individual instance, $\forall x P(x)$ denotes a general law (generalities) called inductive conclusion, and $\Longrightarrow$ denotes inductive inference relation (passage).

It is obvious that the above inductive conclusion $P(x)$ is just a conjecture about generality, or a belief [2] about generality, and the rule does not have the soundness as the deductive inference usually does. For example, let Johnson be a constant symbol and $W(x)$ be a unary predicate which is interpreted as "x is a white swan". An application of the above rule could be

$$W[\text{Johnson}] \Longrightarrow \forall x W(x).$$

It means that from the instance "Johnson is a white swan", we induce a universal proposition "every swan is white". It is obvious that the inductive conclusion is not true because there exists some swans which are not white.

Thus, the truth of an inductive conclusion of the above rule depends on whether there is a counterexample, or say, whether it is refuted by some facts. If it is refuted, then it is false; however if an inductive conclusion has never met a refutation, then it will be accepted. Therefore, we need to verify each inductive conclusion in a model (in the first order sense) which describes some knowledge in a specific domain. By all means, inductions and refutations are two indispensable parts of inductive inferences that supplement each other, and an inductive inference systems must contain a rule for refutation revision.

Let us use Γ to denote our initial theory, or a set of beliefs. After each application of the inductive inference rule, the following two cases would arise. For the first case, we obtain a new law for generality. Namely, Γ will be expanded, or a new version will be generalized. For the second case, a universal belief of Γ encounters a refutation by facts. Namely, Γ should be removed, and a revised version is generated.

After inductive inference rules and refutation revision rules are alternately invoked, a sequence of versions of the set of beliefs Γ is generated as follows:

$$\Gamma_1, \ \Gamma_2, \ \ldots, \ \Gamma_n, \ \ldots$$

where Γ_n denotes the n-th version of Γ.

The above versions in the sequence describe the evolution of the versions of the set of beliefs Γ by applying inductive inference rules. From the view point of the version sequence, the key point of rationality of an inductive inference system should be as follows: For any given model describing knowledge of a specific domain, there exists a version sequence starting from an initial belief (conjecture) and every version is generated by applying the rules of this inductive inference system, such that the version sequence is convergent and the interpretation of its limit containing all of the laws about generalities of the knowledge of that specific domain.

In this paper, we will formalize the evolution of knowledge as outlined above in the terminology of the first order language.

2 A Formal Language of Beliefs

In order to avoid the syntactical details, in this paper, the first order language is chosen to be a formal language to describe sets of beliefs (or knowledge bases) [3]. Briefly, a first order language $\mathcal{L}$ has two sets of symbol strings. They are the set of terms and the set of beliefs. The terms are used to describe the constants and functions used in knowledge bases. The set of terms are defined inductively on the set of variable symbols $\mathbf{V} : \{x, y, z, \ldots\}$, the set of function symbols $\mathbf{F} : \{f, g, h, \ldots\}$, and the set of constants symbols $\mathbf{C} : \{a, b, c, \ldots\}$ using the following BNF like definition:

$$t ::= c \mid x \mid f(t_1, t_2, \ldots, t_n).$$

where t_1, t_2, ..., t_n are terms.

The beliefs of $\mathcal{L}$ are used to describe laws, rules, principles or axioms in the knowledge bases. The beliefs of $\mathcal{L}$ are represented by φ, χ, ψ, and are defined on the set of predicate symbols $\mathbf{P} : \{P, Q, R, \ldots\}$, quality symbol $\doteq$, the set of logical connective symbols including: $\neg$, $\wedge$, $\vee$, $\supset$, and the set of the quantifiers $\forall$ and $\exists$. The set of beliefs $\mathcal{L}$ is defined inductively as follows:

$$\varphi ::= t_1 \doteq t_2 \mid P(t_1, t_2, \cdots, t_n) \mid \neg\psi \mid \varphi \wedge \psi \mid \varphi \vee \psi \mid \varphi \supset \psi \mid \forall x \varphi \mid \exists x \varphi.$$

A belief is called closed belief if there occurs no free variable. A belief is called atomic if it is a closed predicate.

In this paper, the concept of consistency of the first order language are employed for the set of beliefs [3].

Definition 1 (Base of belief set). *A finite consistent set Γ of the beliefs is called a base of belief set, or called a base for short. The beliefs contained in Γ are called the laws of the base.*

A model $\mathbf{M}$ is a pair $\langle M, I \rangle$, where M is called the domain of $\mathbf{M}$, which is a nonempty set; I is called the interpretation of $\mathbf{M}$, which is a mapping from $\mathcal{L}$ to M. The form $\mathbf{M} \models \varphi$ means that for the given domain M and the interpretation I, φ is true in M. $\mathbf{M} \models \Gamma$ mean that for every $\varphi \in \Gamma$, $\mathbf{M} \models \varphi$.

Finally, φ is called a logical consequence of Γ and is written as $\Gamma \models \varphi$, if and only if for every $\mathbf{M}$, if $\mathbf{M} \models \Gamma$ holds, then $\mathbf{M} \models \varphi$ holds.

The individual instances and counterexamples are described by the predicates or negation of predicates of $\mathcal{L}$ without free variables defined formally as below.

Definition 2 (Herbrand universe). *Let $\mathcal{L}$ be a first order language. The set H of terms in $\mathcal{L}$ is defined as follows:*

1. *If c is a constant symbol, then $c \in H$.*
2. *If f is an n-ary function symbol and terms t_1, ..., $t_n \in H$, then $f(t_1, \cdots, t_n) \in H$ as well.*

H is called Herbrand universe or term universe of $\mathcal{L}$, and elements in H are called Herbrand terms or basic terms.

Definition 3 (Complete set of basic sentences for model M). *Let* **M** *be a model of* $\mathcal{L}$, *and* H *be Herbrand universe of* $\mathcal{L}$. *Let*

$$\Omega_0 = \{\varphi \mid \varphi \text{ is a constant predicate } P \text{ and } P \text{ is true under } \mathbf{M}, \text{ or } \varphi \text{ is } \neg P$$
$$P \text{ is a constant predicate and } \neg P \text{ is true under } \mathbf{M}\},$$

$$\Omega_{n+1} = \Omega_n \cup \{\varphi[t_1, \ldots, t_n] \mid t_j \in H, \ \varphi \text{ is } P[t_1, \ldots, t_n] \text{ and } P[t_1, \ldots, t_n]$$
$$\text{is true under } \mathbf{M}, \text{ or } \varphi[t_1, \ldots, t_n] \text{ is } \neg P[t_1, \ldots, t_n] \text{ and } \neg P[t_1, \ldots, t_n]$$
$$\text{is true under } \mathbf{M}\},$$

$$\Omega_{\mathbf{M}} = \bigcup_{i=1}^{\infty} \Omega_i.$$

$\Omega_{\mathbf{M}}$ *is called a complete set of basic sentences in* $\mathcal{L}$ *for model* **M**.

$\Omega_{\mathbf{M}}$ is countable, and it is also called a complete sequence of base sentences for model **M** of $\mathcal{L}$. It is interpreted as the set of all positive examples and counterexamples.

A Gentzen style logical inference system **G** is used in this paper, which is a modified version of the inference system given in [3]. It is used for the formal deductive reasoning of the beliefs. The system **G** is built on sequent. A sequent is formed as $\Gamma \vdash \Delta$, where Γ and Δ could be any finite sets of beliefs, Γ is called antecedent of the sequent and Δ is called succedent of the sequent, and $\vdash$ denotes the deductive relation [3, 5]. The system **G** consists of axioms, the rules for logical connectives and the rules for quantifiers [3, 5].

$Cn(\Gamma)$ is the set of all logical consequences of the base of belief set Γ.

$Th(\mathbf{M})$ is the set of all sentences of $\mathcal{L}$ which are true in **M**.

Finally, it should be assumed that two beliefs α and β are treated as the same belief if and only if $\alpha \equiv \beta$, that is $(\alpha \supset \beta) \wedge (\beta \supset \alpha)$, is a tautology.

3   Refutation by Facts

As mentioned in the first section, a law for generality of a base of a belief set would be rejected by a counterexample which contradicts this law. This phenomenon can be defined in a model-theoretic way as below [4].

Definition 4 (Refutation by facts). *Let* Γ *be a base and* $\Gamma \models \varphi$. *A model* **M** *is called a refutation by facts of* φ *if and only if* $\mathbf{M} \models \neg\varphi$ *holds. Let*

$$\Gamma_{M(\varphi)} := \{\psi \mid \psi \in \Gamma, \ \mathbf{M} \models \psi \text{ and } \mathbf{M} \models \neg\varphi\}.$$

M *is further called an* ideal *refutation by facts of* φ *if and only if* $\Gamma_{M(\varphi)}$ *is maximal in the sense that there does not exist another refutation by facts* **M**′ *of* φ, *such that* $\Gamma_{M(\varphi)} \subset \Gamma_{M'(\varphi)}$.

The above definition describes the following situation that $\Gamma \vdash \varphi$ holds, but we have found a counterexample **M** that makes $\neg\varphi$ true. $\Gamma_{M(\varphi)}$ is a subset of Γ which does not contradict $\neg\varphi$. The refutation by facts meets the intuition that whether a base is accepted, depends only on whether its logical consequences agree with the facts. The ideal refutation by facts meets the Occam's razor, which says: *Entities are not to be multiplied beyond necessity* [6]. Here, it means that if a logical consequence φ deduced from a base Γ is rejected by facts or counterexamples, then the maximal subsets of the base which are consistent with $\neg\varphi$ must be retained and are assumed to be true in the current stage of the developing process of the base, but the rest of laws contained in the base Γ must be removed because they lead to the refutation by facts.

In the rest of the paper, we consider ideal refutation by facts only, and simply call them refutation by facts. Sometimes, we say that $\neg\varphi$ is a refutation by facts of Γ. It means that $\Gamma \vdash \varphi$ holds and that there is an ideal refutation by facts $\mathbf{M}$ which satisfies $\mathbf{M} \models \neg\varphi$.

Definition 5 (Maximal contraction). *Let $\Gamma \vdash \varphi$ and $\Lambda \subset \Gamma$. Λ is called a maximal contraction of Γ by $\neg\varphi$ if it is a maximal subset of Γ and is consistent with $\neg\varphi$.*

Lemma 1. *If $\Gamma \vdash \varphi$ holds and Λ is a maximal contraction of Γ by $\neg\varphi$, then there exists a maximal refutation by facts $\mathbf{M}$ of φ and $\mathbf{M} \models \Lambda$ holds, and for any χ, if $\chi \in \Gamma - \Lambda$, then $\mathbf{M} \models \neg\chi$ holds.*

Proof. The proof is immediate from Definitions 4 and 5.   □

The maximal contraction given here is a proof-theoretic concepts, and can be viewed as a special kind, but most useful contraction given by AGM in [1]. The refutation by facts is a corresponding model-theoretic concept of the maximal contractions.

4   Inductive Inference System I

This section will introduce the inductive inference system $\mathbf{I}$ including the universal inductive rule, the refutation revision rule and the instance expansion rule. For simplicity, in this paper we prescribe that the language $\mathcal{L}$ only contains unary predicates.

The key to construct an inductive inference rule is to describe "the passage from individual instances to generalities" formally. In this paper, we use the following fraction to represent inductive inference:

$$\frac{condition(\Gamma,\ P[t],\ \Omega_{\mathbf{M}})}{\Gamma \Longrightarrow \Gamma'}.$$

where P is an unary predicate. $P[t/x]$ denote that term t is used to substitute free variable x occurring in P [3]. If t does not contain free variables, then $P[t/x]$ can be written as $P[t]$ [3]. Here since $t \in \Omega_{\mathbf{M}}$, t does not contain any free variable.

The bases Γ and Γ' in the denominator of the fraction are the versions before and after applying inductive inference rules respectively. The $condition(\Gamma, P[t], \Omega_{\mathbf{M}})$ in the numerator of the fraction represents the relations between the current version Γ and the basic sentence $P[t] \in \Omega_{\mathbf{M}}$. The fraction can be interpreted as:

if $condition(\Gamma, P[t], \Omega_{\mathbf{M}})$ holds, then a new version Γ' should be induced from the version Γ. The role of $condition(\Gamma, P[t], \Omega_{\mathbf{M}})$ is to make the new version Γ' generated after the induction not only consistent with itself but also consistent with the current version Γ.

Definition 6 (Preempt consistent relation). *Let Γ be a base, $P[t]$ and $\neg P[t']$ be basic sentences, $t, t' \in H$ be basic terms. Let $P[t]$ and Γ be consistent. If there does not exist a basic term $t' \in H$ such that $\neg P[t'] \in \Gamma$, then $P[t]$ is preemptly consistent with Γ and is denoted as $P[t] \bowtie \Gamma$; otherwise, $P[t]$ is not preemptly consistent with Γ and is denoted as $P[t] \not\bowtie \Gamma$.*

Lemma 2. *If the current version Γ is finite, then the preempt consistency between the basic sentence $P[t]$ and Γ is decidable.*

Proof. Because $P[t]$ is an atomic sentence or the negation of an atomic sentence and Γ is finite, the consistency of Γ and $P[t]$ is decidable. □

In the next step, we examine whether Γ contains atomic sentences in a form like $\neg P[t']$, which is also decidable. Having defined the preempt relation between the current version and basic sentence, we now introduce the inductive inference system **I**. In the following rules, we suppose that **M** is a model describing the knowledge for specific domains, and $P[t] \in \Omega_{\mathbf{M}}$ is a basic sentence.

Definition 7 (Universal inductive rule)

$$\frac{P[t] \bowtie \Gamma}{\Gamma \Longrightarrow_i \forall x P(x), \Gamma, P[t]}.$$

The universal inductive rule is a formal rule that induces universal sentences from individual basic sentences. This rule shows the following: under the assumption that the current version is Γ and $P[t]$ is a basic sentence, if $P[t]$ and Γ are preempt consistent, i.e., there does not exist another basic term t' such that $\neg P[t'] \in \Gamma$ holds, then we can induce a universal sentence $\forall x P(x)$ from $P[t]$. The new version generated after the induction is $\{\forall x P(x), \Gamma, P[t]\}$. The subscript i of $\Longrightarrow_i$ in the denominator of the rule represents that this inference is a universal inductive inference.

It is not difficult for readers to find that the universal inductive inference rule can not retain the soundness of inferences. Therefore the following refutation revision rule is indispensable as a supplement to correct errors in case that they happen. The refutation revision rule has the following form:

Definition 8 (Refutation revision rule)

$$\frac{\Gamma \vdash \neg P[t]}{\Gamma \Longrightarrow_r R(\Gamma, P[t])}.$$

This rule shows that when the formal conclusion $\neg P[t]$ of the current version Γ meets the refutation of the basic sentence $P[t]$, the refutation revision rule should be applied to generate a new version $R(\Gamma, P[t])$, which is a maximal contraction of Γ with respect to $P[t]$. The subscript r of $\Longrightarrow_r$ in the denominator of the rule denotes its difference from the inductive inference $\Longrightarrow_i$.

Definition 9 (Basic sentence expansion rule)

$$\frac{P[t] \not\Vdash \Gamma}{\Gamma \Longrightarrow_e P[t], \Gamma}.$$

This rule shows that the current version Γ is not preempt consistent with the basic sentence $P[t]$, i.e., there exists another basic sentence $\neg P[t'] \in \Gamma$. Thus we have to accept $P[t]$ as a new axiom of Γ but we cannot use the universal inductive rule to induce $\forall x P(x)$. Thus the new version is $P[t], \Gamma$. The subscript e of $\Longrightarrow_e$ represents the basic sentence expansion inference.

The universal induction, refutation and basic sentence expansion are all basic ingredients of evolutions of the base. Without causing confusions in the context, this paper uses $\Longrightarrow$ to represent the inductive, refutation and basic sentence expansion inferences, all of which are evolutionary relations.

5   Sequence of Bases

Definition 10 (Sequence of bases). *If for every natural number $n > 0$, Γ_n is a base, then*

$$\Gamma_1, \Gamma_2, \ldots, \Gamma_n, \ldots$$

is a sequence of bases, or sequence for short, and is written as $\{\Gamma_n\}$.

If for every natural number $n > 0$, $\Gamma_n \subseteq \Gamma_{n+1}$ (or $\Gamma_n \supseteq \Gamma_{n+1}$), the sequence is a monotonically increasing (or decreasing) sequence.

A sequence $\{\Gamma_n\}$ is called a version sequence, if for any given $n > 0$, Γ_{n+1} is obtained by applying a rule of the system $\mathbf{I}$ for a belief $\varphi_n \in \Omega_{\mathbf{M}}$ of a given model $\mathbf{M}$.

The definition of limit of sequences introduced in [5] is given below.

Definition 11 (Limit of sequence). *Let $\{\Gamma_n\}$ is a sequence of base. The set of beliefs*

$$\{\Gamma_n\}^* = \bigcap_{n=1}^{\infty} \bigcup_{m=n}^{\infty} \Gamma_m$$

is called the upper limit of $\{\Gamma_n\}$. The set of beliefs

$$\{\Gamma_n\}_* = \bigcup_{n=1}^{\infty} \bigcap_{m=n}^{\infty} \Gamma_m$$

is called lower limit of $\{\Gamma_n\}$.

If the set of beliefs $\{\Gamma_n\}_*$ is consistent, and $\{\Gamma_n\}_* = \{\Gamma_n\}^*$ holds, the sequence $\{\Gamma_n\}$ is convergent, and its limit is its upper (lower) limit, and written as

$$\lim_{n \to \infty} \Gamma_n.$$

From the definition, it is easy to see that $\varphi \in \{\Gamma_n\}^*$ holds, if and only if there exists a countable natural numbers sequence $\{k_n\}$ such that $\varphi \in \Gamma_{k_n}$ holds; $\varphi \in \{\Gamma_n\}_*$ holds, if and only if there exists natural number $N > 0$ such that if $m > N$, $\varphi \in \Gamma_m$ holds.

Lemma 3. $\{Cn(\Gamma_n)\}_* = Cn(\{Cn(\Gamma_n)\}_*)$.

Proof. We need to prove that $Cn(\{Cn(\Gamma_n)\}_*) \subseteq \{Cn(\Gamma_n)\}_*$ holds. Let $\varphi \in Cn(\{Cn(\Gamma_n)\}_*)$ such that $\{Cn(\Gamma_n)\}_* \vdash \varphi$ holds. According to the compactness of first order languages, there exists a finite set of beliefs $\{\varphi_{n_1}, \ldots, \varphi_{n_k}\} \subseteq \{Cn(\Gamma_n)\}_*$, and $\{\varphi_{n_1}, \ldots, \varphi_{n_k}\} \vdash \varphi$ is provable. Thus, there exists $N > 0$, such that if $n > N$ holds, $\varphi_{n_i} \in Cn(\Gamma_n)$, where $i = 1, \ldots, k$, holds. This proves that for every $n > N$, $Cn(\Gamma_n) \vdash \varphi$ holds. Thus, $\varphi \in \{Cn(\Gamma_n)\}_*$ is proved. $\square$

6   Procedure GUINA

In the rest of the paper, we will discuss the rationality of the inductive inference system **I**. The purpose of the inductive inferences is to gradually induce all the true propositions of **M**, which is realized through examining individual instances of Ω_M and modifying the current version of a base by adding new beliefs or deleting wrong ones. In this way, a version sequence is eventually generated and the rationality of inductive inferences is naturally described by the following three properties of the generated version sequence.

Definition 12 (Convergency, Cn-Commutativity and Independency). *Let **M** be a model and $\{\Gamma_n\}$ be a version sequence generated by the system **I** for Ω_M.*

1. *$\{\Gamma_n\}$ is **M**-convergent, if $\{\Gamma_n\}$ is convergent and*

$$\lim_{n \to \infty} Cn(\Gamma_n) = Th(\mathbf{M}).$$

2. *$\{\Gamma_n\}$ is Cn-commute, if*

$$\lim_{n \to \infty} Cn(\Gamma_n) = Cn(\lim_{n \to \infty} \Gamma_n).$$

3. *$\{\Gamma_n\}$ keeps independency, if every Γ_n is independent and $\{\Gamma_n\}$ is convergent then $\lim_{n \to \infty} \Gamma_n$ is independent.*

In order to generate a version sequence satisfying the above three properties, we need to design a procedure that starts from a given initial conjecture and inputs all elements of Ω_M. For each element of Ω_M, the procedure applies appropriate inductive inference rules to generate a new version. All these new versions form a version sequence. If this sequence satisfies the above three properties, then the inductive inference system **I** is rational. In this section we show such a procedure can be constructed and we call it GUINA which is a revised version given in [7]. Its basic design strategy is as follows.

The body of procedure GUINA is mainly a loop which calls the sub-procedure GUINA*. In each call of GUINA*, the current version Γ_n and basic instance $P_n[t]$ in Ω_M are the inputs. A new version Γ_{n+1} will be output corresponding to the following cases:

1. $\Gamma_n \vdash P_n[t]$ is provable. The input basic sentence is a logical conclusion of the current version Γ_n. In this case it is unnecessary to use the inductive rules. The output of GUINA* is $\Gamma_{n+1} := \Gamma_n$, $\Theta_{n+1} := \{P_n[t]\} \cup \Theta_n$ and $\Delta_{n+1} := \Delta_n$.

2. $\Gamma_n \vdash \neg P_n[t]$ is provable. Since $P_n[t]$ belongs to Ω_M, it must be accepted. This shows that the logical conclusion $\neg P_n[t]$ of Γ_n meets a refutation by $P_n[t]$. In this case, the refutation revision rule should be applied. The new version Γ_{n+1} is constructed in two steps. In the first step, the union of a maximal contraction of Γ_n containing Δ_n and $\{P_n[t]\}$ is formed. In the second step, every element of Δ_n and Θ_n is checked. If it is not a logical consequence then it is added to Γ_{n+1}, otherwise it is left off.

3. Neither $\Gamma_n \vdash P_n[t]$ nor $\Gamma_n \vdash \neg P_n[t]$ is provable. There are two cases:
   (a) $P_n[t] \not\bowtie \Gamma_n$ holds. This shows that $P_n[t]$ must be accepted, but for some t', $\neg P[t']$ already exists in Γ_n. Under such circumstances, only the basic sentence expansion rule can be applied. Thus $\Gamma_{n+1} := \{P_n[t]\} \cup \Gamma_n$, $\Delta_{n+1} := \{P_n[t]\} \cup \Delta_n$ and $\Theta_{n+1} := \Theta_n$.
   (b) $P_n[t] \bowtie \Gamma_n$ holds. This shows that $P_n[t]$ must be accepted, but for any t', $\neg P[t']$ does not exist in Γ_n. In this case, the universal inductive rule should be applied. $\Gamma_{n+1} := \{\forall x P_n(x)\} \cup \Gamma_n$, $\Delta_{n+1} := \{P_n[t]\} \cup \Delta_n$ and $\Theta_{n+1} := \Theta_n$.

The procedure GUINA can be specified formally as below.

Definition 13 (GUINA). *Let* **M** *be a model of knowledge for some specific domains whose complete sets of basic sentences* Ω_M *is* $\{P_i[t]\}$.

procedure *GUINA(Γ: theory; $\{P_n[t]\}$: a sequence of base-sentences);*
Γ_n: *theory;*
Θ_n, Θ_{n+1}: *theory;*
Δ_n, Δ_{n+1}: *theory;*

procedure *GUINA* (Γ_n: theory; $P_n[t]$: base-sentence;* **var** *Γ_{n+1}: theory);*
begin
    if $\Gamma_n \vdash P_n[t]$ **then**
      begin
        $\Gamma_{n+1} := \Gamma_n$;
        $\Theta_{n+1} := \Theta_n \cup \{P_n[t]\}$;
        $\Delta_{n+1} := \Delta_n$
      end
    else if $\Gamma_n \vdash \neg P_n[t]$ **then**
      begin
        $\Gamma_{n+1} := \{P_n[t]\} \cup R(\Gamma_n, P_n[t])$;
        loop for every $\psi_i \in \Delta_n \cup \Theta_n$
            if $\Gamma_{n+1} \vdash \psi_i$ **then skip**

```
                else Γ_{n+1} := Γ_{n+1} ∪ {ψ_i}
            end loop
            Θ_{n+1} := Θ_n;
            Δ_{n+1} := Δ_n
        end
    else if P_n[t] ⋈̸ Γ_n then
        begin
            Γ_{n+1} := Γ_n ∪ {P_n[t]};
            Θ_{n+1} := Θ_n;
            Δ_{n+1} := Δ_n
        end
    else
        begin
            Γ_{n+1} := Γ_n ∪ {∀xP(x)};
            Θ_{n+1} := Θ_n;
            Δ_{n+1} := Δ_n ∪ {P_n[t]}
        end
end

begin
    Γ_n := Γ;
    Θ_n := ∅;  Θ_{n+1} := ∅;
    Δ_n := ∅;  Δ_{n+1} := ∅;
    loop
        GUINA*(Γ_n, P_n[t], Γ_{n+1});
        print Γ_{n+1}
    end loop
end
```

The $R(\Gamma_n, P_n[t])$ in the procedure is a maximal contraction of Γ_n with respect to $P_n[t]$.

In GUINA, Δ is a buffer. It is used to store the basic sentences serving as new axioms before the n-th version. The role of Δ is to ensure that all new axioms accepted should be added to the selected maximal contraction when the refutation revision rule is applied.

Θ is also a buffer. It is used to store the inputted basic sentence P_m which is a logical consequece of version Γ_m, $m < n$ during the formations of the first n versions. The role of Θ is to ensure that all basic sentences are not lost when the refutation revision rule is applied. In addition, the initial states of Δ and Θ are $\emptyset$.

Lemma 4. *If the initial base input of GUINA is an empty base, then the conditions $\Gamma_n \vdash P_n[t]$ and $P_n[t] \bowtie \Gamma_n$ occurring in GUINA are both decidable.*

Proof. The proof is omitted. □

It should be metioned that if the initial base of GUINA is a finite base of beliefs, then the conditions such as $\Gamma_n \vdash P_n[t]$ and $P_n[t] \bowtie \Gamma_n$ occurring in **if** and **while** statements may not be decidable. Thus the procedure GUINA will not be a

procedure defined in a programming language such as C or Pascal, it should be
called procedure scheme.

7 Convergency

In this section we prove that the output version sequence of GUINA is convergent.

Theorem 1 (Convergency). *Let* $\mathbf{M}$ *be a model of* $\mathcal{L}$ *and* Γ *be a base of* $\mathcal{L}$.
$\{\Gamma_n\}$ *is the output sequence of GUINA by taking a complete sequence of basic
sentences* $\Omega_{\mathbf{M}}$ *and initial base* Γ *as inputs, then sequence* $\{Cn(\Gamma_n)\}$ *is convergent
and*

$$\lim_{n \to \infty} Cn(\Gamma_n) = Th(\mathbf{M})$$

holds.

Proof. In what follows we prove this theorem in two steps:

1. We first prove that $Th(\mathbf{M}) \subseteq \{Cn(\Gamma_n)\}_*$ holds. It suffices to prove that for
   any belief φ, if $\varphi \in Th(\mathbf{M})$, then $\varphi \in \{Cn(\Gamma_n)\}_*$. We prove by induction on
   the structure of φ:
   (a) φ is an atomic belief. Since $\varphi \in Th(\mathbf{M})$, $\varphi \in \Omega_{\mathbf{M}}$. Let $\varphi = P_N$. By
       the definition of GUINA, we know that P_N is a logical consequence, a
       new instance or a refutation by facts of Γ_N. But in whichever cases,
       we always have $P_N \in Cn(\Gamma_{N+1})$ holds. According to the designs of the
       buffer sets Δ and Θ, we know that when $n > N$, $P_N \in Cn(\Gamma_n)$. That is,
       $\varphi \in \{Cn(\Gamma_n)\}_*$.
   (b) φ is the negation of an atomic belief. It is a negative instance. We can
       just assume that $\varphi = \neg P_N$ and $\neg P_N \in \Omega_{\mathbf{M}}$. By the definition of GUINA
       and using the same proof as in (a), we know that $\varphi \in \{Cn(\Gamma_n)\}_*$.
   (c) φ is $\alpha \vee \beta$. According to the meaning of $\vee$, we know that at least one
       of $\alpha \in Th(\mathbf{M})$ and $\beta \in Th(\mathbf{M})$ holds. Assume the first one holds. By
       the induction hypothesis, we know that $\alpha \in \{Cn(\Gamma_n)\}_*$. Then according
       to the $\vee$ right rule of system $\mathbf{G}$, we have $\alpha \vee \beta \in Cn(\{Cn(\Gamma_n)\}_*)$. By
       Lemma 3, this is $\varphi \in \{Cn(\Gamma_n)\}_*$.
   (d) Similarly we can prove the cases when φ is $\alpha \wedge \beta$ or $\alpha \supset \beta$.
   (e) φ is $\exists x \alpha(x)$ and $\varphi \in Th(\mathbf{M})$. By the meaning of $\exists$, there exists a $t \in H$
       such that $\alpha(t) \in Th(\mathbf{M})$ is true. By the induction hypothesis, $\alpha[t] \in
       \{Cn(\Gamma_n)\}_*$ is true. Then according to the $\exists$ right rule of system $\mathbf{G}$,
       $\exists x \alpha(x) \in Cn(\{Cn(\Gamma_n)\}_*)$. By Lemma 3, that is $\varphi \in \{Cn(\Gamma_n)\}_*$.
   (f) φ is $\neg\beta$ and $\varphi \in Th(\mathbf{M})$. Since the proof in the case of β being an atomic
       belief has been given in (b), we can just assume that β is not an atomic
       belief. Hence β can only be: $\psi \wedge \chi$, $\psi \vee \chi$, $\neg\psi$, $\psi \supset \chi$, $\forall x \psi$ and $\exists x \psi$ with
       ψ and χ being beliefs. Thus $\neg\beta$ can be listed by the following table:

β	$\psi \wedge \chi$	$\psi \vee \chi$	$\neg\psi$	$\psi \supset \chi$	$\forall x \psi$	$\exists x \psi$
$\neg\beta$	$\neg\psi \vee \neg\chi$	$\neg\psi \wedge \neg\chi$	ψ	$\psi \wedge \neg\chi$	$\exists x \neg\psi$	$\forall x \neg\psi$

By the method used in (b) to (e), we can prove that each item in the second row of the proof table belongs to $\{Cn(\Gamma_n)\}_*$. Thus $\varphi \in \{Cn(\Gamma_n)\}_*$. Thus we have proven $Th(\mathbf{M}) \subseteq \{Cn(\Gamma_n)\}_*$.

2. Next we prove that $\{Cn(\Gamma_n)\}^* \subseteq Th(\mathbf{M})$ holds. Suppose there exists a belief φ such that $\varphi \in \{Cn(\Gamma_n)\}^*$ and $\varphi \notin Th(\mathbf{M})$ both hold. $Th(\mathbf{M})$ being complete indicates that $\neg\varphi \in Th(\mathbf{M})$. Since $Th(\mathbf{M}) \subseteq \{Cn(\Gamma_n)\}_*$, there must exists an N such that for $m > N$, $\neg\varphi \in Cn(\Gamma_m)$. Furthermore, since $\varphi \in \{Cn(\Gamma_n)\}^*$, there exist $n_1, \ldots, n_k, \ldots$ such that $\varphi \in Cn(\Gamma_{n_k})$ holds for any natural number k. Thus when $n_k > N$, both φ and $\neg\varphi$ belong to $Cn(\Gamma_{n_k})$. This contradicts the consistency of Γ_{n_k}. Hence $\varphi \in Th(\mathbf{M})$.

These two steps proved that $\{Cn(\Gamma_n)\}^* \subseteq Th(\mathbf{M}) \subseteq \{Cn(\Gamma_n)\}_*$ holds. Thus $\{Cn(\Gamma_n)\}_* = \{Cn(\Gamma_n)\}^* = Th(\mathbf{M})$ holds. The theorem is proved. $\square$

Theorem 1 shows that for any given model $\mathbf{M}$ describing knowledge of specific domains the GUINA procedure, starting from a given initial base, improves this base by generating its new version through examining only one basic sentence of $\Omega_\mathbf{M}$ each time. If this basic sentence is a logical conclusion of the current version, then we regard it as an evidence of recognizing and accepting the current version; if this basic sentence is a refutation by facts, then we need to revise the current version and the new version generated should be a maximal contraction of the original version plus all the basic sentences that were accepted previously but were not logical consequence of the selected maximal contraction; if this basic sentence is a new basic sentence for the current version, then we either use the universal inductive rule to generalize this instance to a universal sentence or only expand the version by adding this basic sentence. As long as all basic sentences of $\Omega_\mathbf{M}$ are examined, the logical closure of the versions output by GUINA will gradually approach all the true sentences of the model $\mathbf{M}$ as a whole. This is the convergency of the inductive inference system $\mathbf{I}$, and can be called the convergency of the procedure GUINA.

8   Commutativity

This section will prove that the version sequence output by the GUINA procedure possesses commutativity between the operator of limit and logical closure.

Theorem 2 (Commutativity). *Let $\mathbf{M}$ be a model of $\mathcal{L}$ and Γ be a base of $\mathcal{L}$. If $\{\Gamma_n\}$ is the output sequence of GUINA by taking a complete sequence of basic sentences $\Omega_\mathbf{M}$ and initial base Γ as inputs, then*

$$\lim_{n \to \infty} Cn(\Gamma_n) = Cn(\lim_{n \to \infty} \Gamma_n)$$

holds.

Proof. Since we have already proved by Theorem 1 that $\lim_{n \to \infty} Cn(\Gamma_n) = Th(\mathbf{M})$ holds, it suffices to prove that

$$\{Cn(\Gamma_n)\}_* \subseteq Cn(\{\Gamma_n\}_*) \subseteq Cn(\{\Gamma_n\}^*) \subseteq \{Cn(\Gamma_n)\}^*$$

holds. In fact, this theorem can be proved in two steps, i.e., to prove that both $\{Cn(\Gamma_n)\}_* \subseteq Cn(\{\Gamma_n\}_*)$ and $Cn(\{\Gamma_n\}^*) \subseteq \{Cn(\Gamma_n)\}^*$ hold.

1. We first prove that $Cn(\{\Gamma_n\}^*) \subseteq \{Cn(\Gamma_n)\}^*$. For any $\varphi \in Cn(\{\Gamma_n\}^*)$, $\{\Gamma_n\}^* \vdash \varphi$ is provable. According to the compactness theorem of system $\mathbf{G}$, there exists a sequence

$$\{\varphi_{n_1}, \ldots, \varphi_{n_k}\} \in \{\Gamma_n\}^*$$

such that

$$\{\varphi_{n_1}, \ldots, \varphi_{n_k}\} \vdash \varphi$$

is provable. By the definition of $\{\Gamma_n\}^*$, $\varphi_{n_i} \in \{\Gamma_n\}^*, i = 1, \ldots, k$, which shows that there exists a subsequence of Γ_n:

$$\Gamma_{n_{i1}}, \ldots, \Gamma_{n_{ij}}, \ldots \qquad j \text{ is natural number.}$$

φ_{n_i} is an element of $\Gamma_{n_{ij}}$ in this sequence and thus is an element of $Cn(\Gamma_{n_{ij}})$. Hence $\varphi_{n_i} \in \{Cn(\Gamma_n)\}^*$, i.e.,

$$\{\varphi_{n_1}, \ldots, \varphi_{n_k}\} \subset \{Cn(\Gamma_n)\}^*$$

holds. According to Theorem 1, $\{Cn(\Gamma_n)\}^* = Th(\mathbf{M})$. Thus $\{Cn(\Gamma_n)\}^*$ is the logical closure. Hence

$$\varphi \in Cn(\varphi_{n_1}, \ldots, \varphi_{n_k}) \subset \{Cn(\Gamma_n)\}^*.$$

2. Then we prove that $\{Cn(\Gamma_n)\}_* \subseteq Cn(\{\Gamma_n\}_*)$. For any $\varphi \in \{Cn(\Gamma_n)\}_*$, there exists a $N > 0$ such that for any $n > N$, $\varphi \in Cn(\Gamma_n)$ holds. This is for any $n > N$, $\Gamma_n \vdash \varphi$ is provable. Thus $\{\Gamma_n\}_* \vdash \varphi$ is also provable, i.e. $\varphi \in Cn(\{\Gamma_n\}_*)$ holds. Hence $\{Cn(\Gamma_n)\}_* \subseteq Cn(\{\Gamma_n\}_*)$. □

This theorem shows that the logical closure of $\lim_{n \to \infty} \Gamma_n$ equals to the limit of $Cn(\Gamma_n)$; therefore in any stage $n > 0$ of investigation, it is enough to consider the finite base Γ_n.

9   Independency

If the initial base is an empty set, then we can prove that GUINA procedure retains its independency.

Theorem 3 (Independency). *Let $\mathbf{M}$ be a model of $\mathcal{L}$ and Γ be a base of $\mathcal{L}$. $\{\Gamma_n\}$ is the output sequence of GUINA by taking a complete sequence of basic sentences $\Omega_\mathbf{M}$ and initial base Γ as inputs. If $\Gamma = \emptyset$, then for any $n > 0$, Γ_n is independent, and $\lim_{n \to \infty} \Gamma_n$ is also independent.*

Proof. The proof can be done in two steps.

I. Prove that for any $n > 0$, Γ_n is independent. Let $\Omega_\mathbf{M}$ be $\{P_1, \ldots, P_n, \ldots\}$. According to the condition given in the theorem, $\Gamma_1 = \emptyset$. Suppose that Γ_n is independent, By the definition of GUINA, there are only the following four possible cases:

1. $\Gamma_n \vdash P_n$ is provable. In this case $\Gamma_{n+1} = \Gamma_n$. Thus Γ_{n+1} is independent.
2. P_n is a refutation by facts of Γ_n. In this case GUINA selects a maximal subset Λ of Γ_n that is consistent with P_n. Because Γ_n is independent, Λ is also independent. By the definition of GUINA, Γ_{n+1} can be generated in two steps: First, we need to combine P_n with Λ. Since the basic sentence P_n is a new axiom of Λ, $\Lambda \cup \{P_n\}$ is still independent. Secondly, GUINA needs to examine the elements in Θ_n and Δ_n individually and then make a union between those lost basic sentences P_{n_j} during the selection of Λ. Since P_{n_j} are independent of $\Lambda \cup \{P_n\}$, each time when P_{n_j} is added, the new version obtained is still independent. Thus Γ_{n+1} is independent.
3. Neither $\Gamma_n \vdash P_n$ nor $\Gamma_n \vdash \neg P_n$ is provable and $P_n \bowtie \Gamma_n$ holds. By the definition of GUINA, P_n must be the first instance of a predicate P facing Γ_n. In this case $\Gamma_{n+1} = \Gamma_n \cup \{\forall x P(x)\}$. Thus Γ_{n+1} is independent.
4. Neither $\Gamma_n \vdash P_n$ nor $\Gamma_n \vdash \neg P_n$ is provable and $P_n \not\bowtie \Gamma_n$ holds. By the definition of GUINA, $\Gamma_{n+1} = \Gamma_n \cup \{P_n\}$. In this case $P_n \notin Th(\Gamma_n)$ but $P_n \in Th(\Gamma_{n+1})$. Thus Γ_{n+1} is independent.

The above four cases prove that if Γ_n is independent, then Γ_{n+1} is still independent. By the induction hypothesis, for any $n > 0$, Γ_n is independent.

II. Since for any $n > 0$, Γ_n is independent and $\{\Gamma_n\}$ is convergent, for any $\varphi \in \{\Gamma_n\}_*$, there exists an N such that $\varphi \in \Gamma_n$ and $\Gamma_n - \{\varphi\} \vdash \varphi$ is not provable. So

$$(\bigcap_{n=N}^{\infty} (\Gamma_n - \{\varphi\})) \vdash \varphi \text{ is not provable, and } \bigcap_{n=N}^{\infty} \Gamma_n \vdash \varphi \text{ is provable.}$$

Therefore $(\{\Gamma_n\}_* - \{\varphi\} \vdash \varphi$ is not provable, and $\{\Gamma_n\}_* \vdash \varphi$ is provable. This shows $\{\Gamma_n\}_*$ is independent. Since $\lim_{n \to \infty} \Gamma_n = \{\Gamma_n\}_*$, $\lim_{n \to \infty} \Gamma_n$ is independent. $\square$

10   Conclusion

In this paper, we have introduced the inductive inference system **I**. It is used to discover the new laws about generality of the base of beliefs for a given model **M** describing knowledge of a specific domain. The system **I** is defined in the first order language and consists of the universal inductive rule, the refutation revision rule, and the basic sentence expansion rule. A rule of **I** should be applied to a base of beliefs and a given basic sentence, and generates a new version of the base depending on their logical relation. When the basic sentences of $\Omega_{\mathbf{M}}$ are taken one after another, a version sequence will be generated. The rationality of the system **I** is exhibited by the convergency, commutativity and independency of the generated version sequences. Finally, we have proved formally the rationality of the system **I** by constructing the procedure GUINA which inputs all basic sentences of $\Omega_{\mathbf{M}}$, and generates the version sequences satisfying these properties.

Acknowledgements. The author would like to take this opportunity to thank Professor Shaokui Mo for his philosophical reasoning about inductive inference, which influences the author deeply. His thanks also go to Jie Luo and Shengming Ma for useful discussions, and proof readings of the paper.

References

1. "The Basic Works of Aristotle" by McKeon, 1941, page 198
2. Gärdenfors, P., Belief Revision: An Introduction, Belief Revision Edited by Gärdenfors, Cambridge Tracts in Computer Science, Cambridge University Press, 1992.
3. Gallier, J. H., Logic for Computer Science, foundations of automatic theorem proving. John Wiley & Sons, 1987, 147-158, 162-163, 197-217.
4. Li, W., A Logical Framework for the Evolution of Specifications. ESOP'94, LNCS 788, Springer-Verlag, 1994.
5. Paulson, L., Logic and Computations, Cambridge University Press, 1987, 38-50.
6. Flew, A., A Dictionary of Philosophy, Pan Books Ltd., 1979.
7. Li, W., Inductive Processes: A Logical Theory for Inductive Inference, Science in China, Series A, Vol. 38, No. 9, Sept. Supp. 1995.

Semantic Guidance for Saturation Provers

William McCune*

Argonne National Laboratory, Illinois, USA
McCune@mcs.anl.gov
http://www.mcs.anl.gov/~mccune/

Abstract. We use finite interpretations to guide searches in first-order
and equational theorem provers. The interpretations are carefully chosen
and based on expert knowledge of the problem area of the conjecture.
The method has been implemented in the Prover9 system, and equational
examples are given the areas of lattice theory, Boolean algebras, and
modular ortholattices.

1   Introduction

Automated deduction methods for first-order and equational problems have fo-
cused mostly on completeness-preserving restrictions on inference rules, effective
term orderings, and special-purpose inference and simplification rules. Less attention has been paid to ordering the search. In saturation systems that use the
given-clause algorithm or one if its variants, the order of the search is determined
mostly by selection of the given clauses. The selection is usually by a weighting function that considers syntactic properties of clauses such as length, depth,
occurrences of particular symbols, and patterns of symbols.

We propose to use semantic criteria in addition to syntactic weighting functions to select the given clauses, and we view this method as a form of semantic
guidance for theorem provers. The semantic criteria are finite interpretations of
the language of the problem.

Interpretations have been used previously for restricting the application of
inference rules, for example, a semantic rule may require that one of the parents
be false in the interpretation [9, 6, 11]. However, semantic inference rules are
frequently incompatible, from both theoretical and practical points of view, with
other important methods such as simplification and restrictions based on term
orderings.

Semantic guidance with finite interpretations has been used previously, most
notably in the recent versions of SCOTT series of provers. In MSCOTT [2],
Hodgson and Slaney use semantic guidance with multiple interpretations that
are generated automatically and updated during the search. In Son of SCOTT
[10], Slaney, Binas, and Price introduced the notion of *soft constraints* that allows the use of just one interpretation that partially models some subset of the

---

* Supported by the Mathematical, Information, and Computational Sciences Division
  subprogram of the Office of Advanced Scientific Computing Research, Office of Science, U.S. Department of Energy, under Contract W-31-109-ENG-38.

J. Calmet, T. Ida, and D. Wang (Eds.): AISC 2006, LNAI 4120, pp. 18–24, 2006.

derived clauses. As in MSCOTT, Son of SCOTT automatically generates the interpretation and updates it during the search. In both systems, the interpretations are small, usually with at most 4 elements.

In this project we are studying the use of larger interpretations that are carefully chosen by the user and fixed throughout the search. The interpretations are intended to be based on expert knowledge on the theory of the problem and on closely related theories. We give several examples of using semantic guidance on difficult equational problems and touch on several ideas for choosing good interpretations.

2   Semantic Strategies and Choice of Interpretations

The roots of semantic strategies for automated theorem proving are in the set of support strategy, introduced by Wos et al in 1965 [14]. The primary motivation for the set of support strategy rests on the assumption that many conjectures have the form *theory, hypotheses* $\Rightarrow$ *conclusion*. The idea is that when searching for a proof, one should avoid exploring the theory and focus instead on the hypotheses and the conclusion. The set of support strategy is a semantic restriction strategy, and (assuming that the set of support consists of the hypotheses and conclusion) an arbitrary model of the theory, in which the hypotheses or the denial of the conclusion are false, is used to prove completeness. The effect is that all lines of reasoning must start with the hypotheses or the denial of the conclusion.

Semantic restriction strategies based on other general interpretations (e.g., positive resolution), or on arbitrary explicit interpretations were developed later, most notably by Slagle in 1967 [9]. These rules generally require that one of the parents for each binary inference be false in the interpretation.

The motivation for the present work on semantic guidance with carefully selected interpretations is similar to the motivation for the set of support strategy. If the conjecture has the form *theory, hypotheses* $\Rightarrow$ *conclusion*, we wish to focus the search on lines of reasoning that connect the hypotheses to the conclusion. If the theory is true in the guiding interpretation, and the hypotheses and conclusion are false, we believe that lines of reasoning consisting mostly of false clauses will be valuable in connecting the hypotheses to the conclusion.

Because we propose to use semantics to guide rather than restrict the search, valuable consequences of the theory (e.g., lemmas true in the interpretation) can be easily derived, and these may help with the "false" lines of reasoning.

If the conjecture has no obvious hypotheses that are separate from the theory, the interpretation should falsify some part of the theory (a model of the theory that falsifies the conclusion gives a counterexample). In particular, we believe that the interpretation should be a model of a slightly weakened theory, in which the conclusion is false. For example, if the goal is to prove that a theory is associative, one might wish to use a nonassociative interpretation that satisfies many other properties of the theory. If one is unfamiliar with or unsure of the theory, one can use a nonassociative interpretation that satisfies properties of closely related theories.

3   Implementation in Prover9

Prover9 [5] is Otter's [4] successor, and it is similar to Otter in many important ways. In particular, it uses the *given-clause algorithm* (the so-called *Otter loop*), in which weighting functions are used to select the next given clause, that is, the next path to explore. Ordinarily (without semantic guidance), Prover9 cycles through two functions: selecting the oldest clause (to provide a *breadth*-first component) and selecting the lightest clause (to provide a *best*-first component). The ratio of the two functions is determined by parameters.

For semantic guidance, Prover9 accepts one or more finite interpretations in the format produced by Mace4 [3]. Each clause that is retained (input or derived) is evaluated in the interpretations. The clause is marked as *true* if it is true in all of the interpretations; otherwise is it marked as *false*. An exception is that if evaluation of a clause in an interpretation would be too expensive (determined by a parameter that considers the number of variables in the clause and the size of the interpretation), the evaluation is skipped and the clause is marked with the default value *true*. The mark is used when (and only when) selecting given clauses.

When using semantic guidance, Prover9 cycles through three methods when selecting the next given clause: (1) the oldest clause, (2) the lightest *true* clause, and (3) the lightest *false* clause. The ratio of the three methods is determined by parameters. (Son of SCOTT [10] uses a similar 3-way ratio.) We use the notation $A : B : C$ to mean A rounds of selecting the given clause by age, B rounds selecting *true* clauses of lowest weight, and C rounds selecting *false* clauses of lowest weight, and so on. If a *false* clause is called for and none is available, a *true* clause is substituted, and vice versa.

4   Examples

The theorems cited here were first proved with Otter by using various search strategies, with substantial interaction from the user. All of the examples are equational theorems, and a paramodulation inference rule with simplification (demodulation) was used, similar to unfailing Knuth-Bendix completion.

The interpretations for the semantic guidance were produced by Mace4 (quickly) after the user had determined the desired properties for the interpretations. Although the Prover9 implementation can handle multiple interpretations, each of the examples uses just one.

The Prover9 jobs used ratio $1 : 1 : 4$ (*age:true:false*) for selecting the given clauses. The weighting function for the *true* and *false* components was simply the number of symbols in the clause.

As is frequently done with Prover9, limits were set on the size of equations for several of the searches; these limits substantially improve the performance of

Prover9. The term-ordering method and symbol precedence/weights are usually very important in equational problems, but they are not so important here, because these examples have so few symbols. We used the lexicographic path ordering (LPO) with the default symbol precedence.

Finally, Prover9 was directed to introduce a new constant when it deduced that a constant satisfying some property exists. For example, if $f(x, f(x, x)) = f(y, f(y, y))$ was derived, the equation $f(x, f(x, x)) = c$, for a new constant c was inferred, with c added to the interpretation in such a way that $f(x, f(x, x)) = c$ is true.

Waldmeister [1] is usually assumed to be the fastest automatic prover for equational logic. We ran each example with version 704 (July 2004) of Waldmeister in its automatic mode with a time limit of four hours, and the results are given below with each example. Comparison between provers on a small number of examples is usually not meaningful; the purpose of the Waldmeister jobs is simply to give another measure of the difficulty of these examples.

4.1 Lattice Theory Identities

This example arose in a project to find weak Huntington identities, that is, lattice identities that force a uniquely complemented lattice to be Boolean [8]. The following two identities (among many others) were proved to be Huntington identities, and we then looked at the problem of whether one is weaker than the other.

$$(x \wedge y) \vee (x \wedge z) = x \wedge ((y \wedge (x \vee z)) \vee (z \wedge (x \vee y))) \tag{H82}$$

$$x \wedge (y \vee (x \wedge z)) = x \wedge (y \vee (z \wedge ((x \wedge (y \vee z)) \vee (y \wedge z)))) \tag{H2}$$

Let LT be an equational basis for lattice theory in terms of meet and join. Mace4 easily finds a counterexample to LT, H2 $\Rightarrow$ H82. The statement LT, H82 $\Rightarrow$ H2 is a theorem and is the focus of this example.

This theorem has the form *theory, hypotheses $\Rightarrow$ conclusion*, and a natural choice for a guiding interpretation is model of the theory that falsifies the hypothesis. Mace4 easily finds a lattice of size 6 satisfying those constraints, and also shows that there is none smaller and none other of size 6. By using semantic guidance with that lattice, Prover9 proved the theorem in 10 seconds; without semantic guidance, Prover9 proved it in about one hour. Waldmeister proved the theorem in about 5 minutes.

4.2 Boolean Algebra Basis

This example is on Veroff's 2-basis for Boolean algebra in terms of the Sheffer stroke [13].[1] Consider the following equations, where f is the Sheffer stroke.

$$f(x, y) = f(y, x) \tag{C}$$

$$f(f(x, y), f(x, f(y, z))) = x \tag{V}$$

$$f(f(f(y, y), x), f(f(z, z), x)) = f(f(x, f(y, z)), f(x, f(y, z))) \tag{S}$$

---

[1] The Sheffer stroke, which can be interpreted as the *not-and* or NAND operation, is sufficient to represent Boolean algebra.

The pair C,V is a basis for Boolean algebra, and S is a member of Sheffer's original basis. The theorem for this example is C,V $\Rightarrow$ S. This statement does not have the form *theory, hypotheses* $\Rightarrow$ *conclusion* for any nontrivial and well-understood theory, and it is not so obvious where to look for a guiding interpretation.

When faced with the conjecture, we know that the goal is to prove a property of Boolean algebra. We use an interpretation that is close to, but not, a Boolean algebra. Consider the chain of varieties *ortholattices (OL), orthomodular lattices (OML), modular ortholattices (MOL), Boolean algebras (BA)*. (See [7] for Sheffer stroke as well as standard bases for these varieties.) Mace4 can be used to find the smallest MOL that is not a BA. It has size 6, and there is exactly one of size 6. With that interpretation as a guide, Prover9 proved the theorem in about 4 minutes. A similar search without semantic guidance produced a proof in about 6.5 minutes. Waldmeister took about 8 minutes to prove the theorem.

4.3   Modular Ortholattice Single Axiom

The two theorems in this section are on a single axiom for modular ortholattices in terms of the Sheffer stroke [7]. The second theorem is quite difficult; it was first proved by Veroff using the proof sketches method [12], and it was first proved automatically (given a helpful interpretation) with the semantic guidance described here. Consider the following equations, all in terms of the Sheffer stroke.

$$f(f(y,x), f(f(f(x,x),z), f(f(f(f(f(x,y),z),z),x), f(x,u)))) = x \quad \text{(MOL)}$$
$$f(x, f(f(y,z), f(y,z))) = f(y, f(f(x,z), f(x,z))) \quad \text{(A)}$$
$$f(x, f(y, f(x, f(z,z)))) = f(x, f(z, f(x, f(y,y)))) \quad \text{(M)}$$

The equation MOL is a single axiom for modular ortholattices, A is an associativity property, and M is a modularity property. The two theorems in focus are MOL $\Rightarrow$ A and MOL $\Rightarrow$ M. Neither suggests an obvious interpretation for guidance.

In the rest of this section, the term *associative* refers to the operations meet and join when defined in terms of the Sheffer stroke. As in the preceding example, we considered the chain of varieties OL–OML–MOL–BA.

For the first theorem (the associativity property), we chose the smallest nonassociative interpretation that satisfies several MOL (modular ortholattice) properties; it has size 8. With that interpretation as guidance, Prover9 proved the theorem in about 8.5 minutes. Without semantic guidance, Prover9 took about the same amount of time to prove the theorem, but required a much larger search, generating 57% more clauses. Waldmeister did not find a proof of the theorem within the time limit of 4 hours.

For the second theorem (the modularity property), we chose the smallest nonmodular orthomodular lattice, which has size 10. The motivation for this choice is similar to that for the Boolean Algebra 2-basis example, that is, to use an interpretation that is very close to, but not, an algebra in the variety corresponding to the goal of the conjecture. With the interpretation as guidance,

Prover9 proved the theorem in about 3.8 hours. Without semantic guidance, Prover9 failed to prove the theorem within 6 hours. Waldmeister proved the theorem in about 3.3 hours.

5 Remarks

The Mace4 jobs that find the interpretations and the Prover9 jobs that find the proofs for the examples can be found on the Web at the following location.

  http://www.mcs.anl.gov/~mccune/papers/semantic-strategy/

When Prover9 selects a given clause, it is printed to the output file, including any *false* marks. The *false* marks are shown as well in any proofs that are printed. These notations help the user to analyse the effects of semantic guidance.

The output files on the Web show the following.

- The *false* clauses that are selected as given clauses are, for the most part, heavier than the *true* clauses. When the *false* clauses are the same weight as the *true* clauses, the *false* clauses have much higher ID numbers. Both of these properties indicate that the *false* clauses would not be selected so soon without semantic guidance.
- Most of the proofs have a preponderance of *false* clauses, especially near the ends of the proofs. The *true* clauses do, however, play very important roles as lemmas of the theory, suggesting that semantic *restriction* strategies that eliminate many *true* clauses may be less useful than semantic guidance strategies.

Comparisons of the Prover9 jobs with and without semantic guidance indicate that semantic guidance may be more helpful in examples where the interpretations are more obvious or are more carefully chosen. The lattice theory example, which immediately suggests a lattice interpretation falsifying the hypothesis, shows a great improvement. The Boolean Algebra 2-basis and the MOL modularity examples, in which the interpretation is close to the theory corresponding to the goal, show a substantial improvement. The MOL associativity example, which uses a rather ad hoc interpretation which was not carefully chosen, shows only a small improvement.

Future work includes more experimentation with various interpretations, experimentation on non-equational and non-Horn problems, and automating the selection of effective interpretations for semantic guidance.

References

1. T. Hillenbrand, A. Buch, R. Vogt, and B. Löchner. Waldmeister. *J. Automated Reasoning*, 18(2):265–270, 1997.
2. K. Hodgson and J. Slaney. TPTP, CASC and the development of a semantically guided theorem prover. *AI Communications*, 15:135–146, 2002.

3. W. McCune. Mace4 Reference Manual and Guide. Tech. Memo ANL/MCS-TM-264, Mathematics and Computer Science Division, Argonne National Laboratory, Argonne, IL, August 2003.

4. W. McCune. Otter 3.3 Reference Manual. Tech. Memo ANL/MCS-TM-263, Mathematics and Computer Science Division, Argonne National Laboratory, Argonne, IL, August 2003.

5. W. McCune. Prover9. http://www.mcs.anl.gov/~mccune/prover9/, 2005.

6. W. McCune and L. Henschen. Experiments with semantic paramodulation. *J. Automated Reasoning*, 1(3):231–261, 1984.

7. W. McCune, R. Padmanabhan, M. A. Rose, and R. Veroff. Automated discovery of single axioms for ortholattices. *Algebra Universalis*, 52:541–549, 2005.

8. R. Padmanabhan, W. McCune, and R. Veroff. Lattice laws forcing distributivity under unique complementation. *Houston J. Math*. To appear.

9. J. R. Slagle. Automatic theorem proving with renamable and semantic resolution. *J. ACM*, 14(4):687–697, 1967.

10. J. Slaney, A. Binas, and D. Price. Guiding a theorem prover with soft constraints. In *Proceedings of the 16th European Conference on Artificial Intelligence*, pages 221–225, 2004.

11. J. Slaney, E. Lusk, and W. McCune. SCOTT: Semantically constrained Otter. In A. Bundy, editor, *Proceedings of the 12th International Conference on Automated Deduction, Lecture Notes in Artificial Intelligence, Vol. 814*, pages 764–768. Springer-Verlag, 1994.

12. R. Veroff. Solving open questions and other challenge problems using proof sketches. *J. Automated Reasoning*, 27(2):157–174, 2001.

13. R. Veroff. A shortest 2-basis for Boolean algebra in terms of the Sheffer stroke. *J. Automated Reasoning*, 31(1):1–9, 2003.

14. L. Wos, G. Robinson, and D. Carson. Efficiency and completeness of the set of support strategy in theorem proving. *J. ACM*, 12(4):536–541, 1965.

Labeled @-Calculus:
Formalism for Time-Concerned Human Factors

Tetsuya Mizutani[1], Shigeru Igarashi[2], Yasuwo Ikeda[3], and Masayuki Shio[4]

[1] Department of Computer Science, University of Tsukuba
mizutani@cs.tsukuba.ac.jp
[2] Graduate School of Community Development, Tokiwa University
kindaisi@mail1.accsnet.ne.jp
[3] Department of Computer & Media Science, Saitama Junior College
ikeda@sjc.ac.jp
[4] College of Community Development, Tokiwa University
shio@tokiwa.ac.jp

Abstract. In the recent years it has come to be more and more important to analyze and verify systems which control complex external systems, like railways and airlines, whose serious accidents have been caused
not only by bugs in software or hardware but also by human errors possibly
involved in recognition or decision. This paper deals with an actual traffic
accident widely known as "Shigaraki Kougen Railway accident" caused by
a fatal decision based upon certain incorrect knowledge, as an example of
systems involving human factor. In order to represent and analyze this accident, the formal system @-calculus is generalized into the "labeled" @-
calculus so as to describe time-concerned recognition, knowledge, belief
and decision of humans besides external physical or logical phenomena,
while the word 'time-concerned' is used to express the property not only
time-dependent mathematically but also distinctively featured by subtle
and sophisticated sensitivity, interest or concern in exact time.

Keywords: Labeled @-calculus, time-concerned knowledge and belief,
Shigaraki Kougen Railway accident.

Topics: Artificial intelligence, logic, mathematical knowledge
management.

1 Introduction

In the recent decades, it has come to be more and more important to study formal
theory for systems controlling complex systems, e.g. railways and airlines, whose
serious accidents have been caused not only by bugs in software or hardware but
also by human errors possibly involved in human factors, i.e. recognition or decision by humans. The authors of this paper have published the formalisms [14], [22],
[23] for verification of hybrid systems involving recognition and decision of humans
in a merging car driving situation to avoid collision as an example. They claim that
these formalisms belong to the earliest to verify and analyze concurrent program

J. Calmet, T. Ida, and D. Wang (Eds.): AISC 2006, LNAI 4120, pp. 25–39, 2006.
© Springer-Verlag Berlin Heidelberg 2006

systems controlling continuously changing outer systems including human factor, even though there are many reasoning systems for temporal or time-related logic of knowledge and belief [1], [3], [6], [7], [8], [10].

In this paper, *labeled @-calculus* which is a generalization of *@-calculus* [17] is introduced for representation, analysis and verification of time-concerned recognition, knowledge, belief and decision of humans besides external physical or logical phenomena, while the word 'time-concerned' is used here to express the property not only time-dependent mathematically but also distinctively featured by subtle and sophisticated sensitivity, interest or concern in exact time, like musical performance, traffic control, etc. @-calculus is one of the formal systems based on the first-order natural number theory introduced as a modification of *tense arithmetic* (*TA*) [15] developed on rational time.

A theory for knowledge and belief [16], [20] have been dealt with by one of the authors of this paper, which used to belong to the area of artificial intelligence. This paper establishes a formal theory of time-concerned human factor by the collaboration of [20] and @-calculus, and moreover, the 42 lives lost train accident happened in 1991 widely known as "Shigaraki Kougen Railway accident" by a fatal decision based upon certain incorrect knowledge is analyzed as an example of systems involving human factor.

Traffic control systems, or collision avoidance protocol more specifically, are represented in hybrid automata models and duration calculus, which study includes a thorough verification of a car distance controlling example [4]. The theoretical treatments in that paper can be translated into the original @-calculus [17], besides the formalism in the present paper treats misunderstanding of subjects e.g. drivers or controllers of traffic systems. It must be noted that the 2001 Japan Air Line (JAL) near-miss off Shizuoka injuring about 100 people involved at least 6 subjects: Two air controllers, one of them making an error, two pilots and two units of the computerized Traffic Alert and Collision Avoidance System (TCAS), while one pilot decided to follow air traffic control instead of the aural TCAS Resolution Advisory [2]. Nevertheless, it can be treated in our formalism with less difficulty, presumably, for the reasoning by humans was much simpler than that during Shigaraki accident.

In section 2, the formal system of labeled @-calculus is introduced with its intended meaning. The outline of the accident is shown in section 3, and the formal description of the accident and its analysis is in section 4. Discussions follow.

2   Labeled @-Calculus

2.1   Preliminary Explanation

As a quick example of the formulas of @-calculus, let us consider a 'clock' J that counts every milisecond [ms]. If the clock value happens to be 1000 now, written as either $J = 1000@0$ or $J = 1000$, equivalently, in @-calculus, then it will be $1000 + n$ after n[ms], written as $J = 1000 + n@n$[ms], and vice versa.

Thus $J = 1000 \equiv J = 1000 + n@n$, "@" connecting stronger than usual logical connectives and [ms] understood. Using usual individual variables x, y and z, we have $J = x \equiv J = x + y@y$, or its universal closure $\forall x \forall y (J = x \equiv J = x + y@y)$, whereas $\forall x \forall y \forall z (z = x \equiv z = x + y@y)$ is incorrect. ($\forall x \forall y \forall z (z = x \equiv z = x@y)$ is correct instead.) Hence we treat J as a 'special' constant rather than a variable in order to prevent quantification. J itself and any term containing J must be discriminated from regular terms farther in derivation, substitution especially, to avoid inconsistency.

Incidentally $J = x + y@y \equiv J - y = x@y$ holds in @-calculus since $J = x + y \equiv J - y = x$ in arithmetic, so that $J = x \equiv J - y = x@y$ holds for the above clock. Thus the term $J - y$ expresses the time 'now' looked back, or evaluated, y[ms] later.

J cannot become smaller than the initial 1000, i.e. $\neg \exists x (J < 1000@x)$. In other word, the time when (or the world where) $J < 1000$ would hold can never be reached, which fact will be usually written as $J < 1000@\infty$.

As to the labeled @-calculus, let us think about a misunderstanding: A person, say l, may well expect the train to depart within 1 minute as scheduled whereas another person or controller, say l', knows that it will depart 5 minutes later. If α, which is a defined special constant called a 'spur' formally, corresponds to the trigger to make the train start, then the former is expressed by $\alpha < 1060@l$, or $\alpha < 1$[min]$@l$, [min] indicating the proper conversion of the clock values, and the latter by $\alpha = 5$[min]$@l'$ together with the very fact $\alpha = 5$[min], i.e. $\alpha = 1300$. Incidentally, $(\mathbf{A}@l)\&\mathbf{A}$ can mean "l knows $\mathbf{A}$." in most cases, while it is not the purpose of the present paper to formalize 'knows' or the operator "$*$" in [20].

2.2 Syntax and the Intended Meaning of Labeled @-Calculus

Peano arithmetic (PA) is extended to $PA(\infty)$ called *pseudo-arithmetic* [17], including the *infinity* (∞) and the *minimalization* (μ) listed in Table 1 [1].

The syntax of the labeled ('$N\Sigma$-labeled' in a more general framework) @-calculus that is an extension of $PA(\infty)$ is introduced.

Infinite number of *special*, or '*tense-sensitive*' constants, J, J_1, J_2, ... and labels, l, l_1, l_2, ... $\in \Sigma$, *the set of labels*, are added. The special constants correspond to program variables taking natural number value and possibly ∞, and the change of their values along the natural number time is expressed by the change of local models, or 'worlds', and vice versa. It must be noticed that a program variable is guarded against quantification in the calculus, since each J is not a variable but a constant. A label indicates a *personality* that is a generalization of an *agent* of multi-agent systems, or an *observer* in physics, including notion of subjectivity. Besides the conventional logical operators $\neg$, $\&$ and $\forall$, a new connective "@" is introduced.

---

[1] We will partially follow [25] in which P stands for Peano arithmetic, while an occurrence of the variable x in a formula $\mathbf{A}$ will be indicated explicitly as $\mathbf{A}[x]$, and $\mu x \mathbf{A}[x]$ will designate the least number sasitfying $\mathbf{A}$, in the present paper.

Table 1. Axioms of $PA(\infty)$

N1. $x + 1 \neq 0$.	**N2.** $x < \infty \supset y < \infty \supset x + 1 = y + 1 \supset x = y$.
N3. $x + 0 = x$.	**N4.** $x < \infty \supset y < \infty \supset x + (y + 1) = (x + y) + 1$.
N5. $x \times 0 = 0$.	**N6.** $y < \infty \supset x \times (y + 1) = (x \times y) + x$.
N7. $\neg(x < 0)$.	**N8.** $y < \infty \supset (x < y + 1 \equiv x \leq y)$.

Axioms for ∞:
N4′. $x + \infty = \infty + x = \infty$.
N5′. $0 \times \infty = 0$.   **N6′.** $0 < x \supset x \times \infty = \infty \times x = \infty$.
N7′. $x \leq \infty$.
Mathematical Induction:
N9. $\mathbf{A}[0] \ \& \ \forall x(x < \infty \supset \mathbf{A}[x] \supset \mathbf{A}[x + 1]) \supset \forall x(x < \infty \supset \mathbf{A}[x])$.
Definition of μ (The least number principle):
N9′. $\exists x \mathbf{A}[x] \supset \mathbf{A}[\mu x \mathbf{A}[x]] \& \forall y(\mathbf{A}[y] \supset \mu x \mathbf{A}[x] \leq y)$.

Definition 1. (*Terms and formulas of* ($N\Sigma$-)*labeled @-calculus*)

1. *A term of* $PA(\infty)$ *is a term.*
2. *A special constant is a term.*
3. *If* x *is a variable (of* $PA(\infty)$*) and* **A** *is a formula, then* $\mu x \mathbf{A}[x]$ *is a term.*
4. *If* a *and* b *are terms, then* $a \leq b$ *is a formula.*
5. *If* x *is a variable, and* **A** *and* **B** *are formulas, then* $\neg \mathbf{A}, \mathbf{A} \& \mathbf{B}$ *and* $\forall x \mathbf{A}$ *are formulas.*
6. *If* a *is a term,* **A** *is a formula, and* l *is a label, then* $\mathbf{A}@\langle a, l \rangle$, $\mathbf{A}@a$ *and* $\mathbf{A}@l$ *are formulas.*

It must be noted that the essential difference between the above inductive definition and that of the original @-calculus lies only in the last part. Hence this particular language and calculus will be called '$N\Sigma$-labeled' in accordance with the typical pair (or a tuple in general) $\langle a, l \rangle$ following @, N standing for the natural numbers (see remark 1).

A term of $PA(\infty)$ is called a *pseudo-arithmetic expression* henceforth to discriminate it from general terms, while a, or a[J], in which any special constant occurs is said to be *tense-sensitive*.

A *tense* means the time relative to a reference 'observation time' called *now* or *present*, which is taken to be 0 throughout the calculus, while any term designates some tense, e.g. 1, n, ∞, etc, where ∞ indicates the tense when **false** holds. The value of a tense-sensitive term may change along with tense since so may that of a special constant, while that of a pseudo-arithmetic one does not.

@-mark as a logical symbol is called the *coincidental* operator. A formula of the form $\mathbf{A}@\langle a, l \rangle$ intuitively means that the personality designated by the label l believes at the observation time the fact that **A** holds at the tense designated by a, while $\mathbf{A}@a$ and $\mathbf{A}@l$ that **A** holds at tense a, and that l believes that **A** holds *now*, respectively.

The terms $\mathbf{a} + \mathbf{b}$ and $\mathbf{a} \times \mathbf{b}$ can be defined by the operators of $PA(\infty)$ and μ as follows:

$$\mathbf{a} + \mathbf{b} \Leftrightarrow \mu x \exists y_1 y_2 (\mathbf{a} = y_1 \And \mathbf{b} = y_2 \And x = y_1 + y_2),$$
$$\mathbf{a} \times \mathbf{b} \Leftrightarrow \mu x \exists y_1 y_2 (\mathbf{a} = y_1 \And \mathbf{b} = y_2 \And x = y_1 \times y_2).$$

The primitive logical symbol " ; " called the *futurity* operator of the original @-calculus is treated as an abbreviation defined by

$$\mathbf{a}; \mathbf{b} \Leftrightarrow \mathbf{a} + \mu x (x = \mathbf{b}@\mathbf{a}),$$
$$\mathbf{a}; \mathbf{A} \Leftrightarrow \mu x (\mathbf{a} \le x \And (\mathbf{A}@x)).$$

Thus the futurity operator moves the observation time toward a future time-point, so that a formula, e.g., $J - y = x@y$ can be written as $y; (J - y) = y; x$. $\mathbf{a}; \mathbf{b}$ is the tense of $\mathbf{b}$ observed at the tense designated by $\mathbf{a}$. $m; n$ is $m + n$ for any pair of natural numbers m and n. $\mathbf{a}; \mathbf{A}$ means the least, i.e. 'the earliest', time when $\mathbf{A}$ comes to hold, or 'rises', after (or at) the tense $\mathbf{a}$, which will be called the *ascent* of $\mathbf{A}$ at $\mathbf{a}$.

The precedence over symbols is defined as follows:

$$\times, +, ; , \{=, \le, <\}, @, \neg, \And, \vee, \supset, \equiv$$

$$\text{(strong)} \longleftrightarrow \text{(weak)}.$$

Abbreviations. $\mathbf{A}@\langle \mathbf{a}, l_1 \rangle \And \mathbf{A}@\langle \mathbf{a}, l_2 \rangle \And \ldots \And \mathbf{A}@\langle \mathbf{a}, l_n \rangle$ and $\mathbf{A}@l_1 \And \mathbf{A}@l_2 \And \ldots \And \mathbf{A}@l_n$ will be abbreviated as $\mathbf{A}@\langle \mathbf{a}, \{l_1, l_2, \ldots, l_n\} \rangle$ and $\mathbf{A}@\{l_1, l_2, \ldots, l_n\}$, respectively, of which the latter means that all personalities listed in $\{l_1, l_2, \ldots, l_n\}$ believe $\mathbf{A}$.

$\lambda, \lambda_1, \lambda_2, \ldots$ are used as metasymbols of tense, label, and pair or set of them following @ in a formula, e.g., $\langle \mathbf{a}, l \rangle, \{l_1, l_2, \ldots, l_n\}$, etc.

2.3 Proof System

Axioms. The following axioms are added to those of $PA(\infty)$ as the logical, or *proper* axioms, where **false** is an abbreviation of $0 = 1$.

1. The equality substitution for @: $x = y \supset \mathbf{A}@x \supset \mathbf{A}@y$.
2. Elimination of tense 0: $\mathbf{A}@0 \equiv \mathbf{A}$.
3. Inductive valuation:
   (a) $\mathbf{false}@x \equiv x = \infty$,
   (b) $(x \le y)@\lambda \equiv \mu z (z = x@\lambda) \le \mu z (z = y@\lambda)$,
   (c) $\mathbf{A}@x@y \equiv \mathbf{A}@y; x$,
   (d) $\mathbf{A}@x@l \equiv \mathbf{A}@\langle x, l \rangle$,
   (e) $x < \infty \supset ((\neg \mathbf{A})@x \equiv \neg(\mathbf{A}@x))$,
   (f) $\neg(\mathbf{A}@l) \equiv (\neg \mathbf{A})@l$,
   (g) $(\mathbf{A} \And \mathbf{B})@\lambda \equiv \mathbf{A}@\lambda \And \mathbf{B}@\lambda$,
   (h) $(\forall y \mathbf{A})@\lambda \equiv \forall y (\mathbf{A}@\lambda)$.

Remark 1. In studying artificial intelligence it seems natural to introduce some structure into Σ making it a space, including the use of N, rather than an alphabet, although there are cases when they can be simply treated as abbreviations. E.g., a formula of Σ^2-labeled @-calculus, even dispensing with the tense, like $\mathbf{A}@\langle l_1, l_2 \rangle$ can mean $\mathbf{A}@l_1@l_2$ by a straightforward generalization of axiom 3(d), $\langle l_1, l_2 \rangle$ representing l_1's belief as believed by l_2; and $\mathbf{A}@\langle l, l \rangle \equiv \mathbf{A}@l$ makes the introspective (reflection) axiom. It can be more faithful in a certain respect to apply this generalization to the case treated in the present paper too.

Inference Rules. The rules of NK [9] are used with the only one restriction as follows.

  1. *Restriction of ∀-E rule.* In ∀-E(elimination) rule:

$$\frac{\forall x(\mathbf{A}[x])}{\mathbf{A}[\mathbf{a}]}$$

only a pseudo-arithmetic expression $\mathbf{a}$ can be substituted for x if x occurs in a subformula of the form $\mathbf{B}[x]@\langle \mathbf{b}, l \rangle$ or $\mathbf{B}[x]@\mathbf{b}$ of the upper formula, while the occurrence of x in $\mathbf{b}$ does not matter (see [17] for the detail).

Moreover, the following two rules are added.

  2. @-I(introduction rule):

$$\frac{\mathbf{A}}{\mathbf{A}@x}$$

where every assumption of $\mathbf{A}$ does not have any special constants.

  3. @-E(elimination rule):

$$\frac{\mathbf{A}@\mathbf{a}}{\mathbf{A}}$$

where no special constant occurs in $\mathbf{A}$.

Remark 2. It is likely that the metatheory, including the proof of the soundness, of the original @-calculus in [17] can be generalized for the labeled @-calculus in a straightforward manner. It must be noted that the soundness of *TA* is shown in [15] and [21], latter relying upon the elimination of ∞ (see also Conclusion).

2.4   Axiom Tableaux

Using *axiom tableaux* [17], actions or changes of states of programs for verification can be readily represented by *program axioms*. First, some primitive notions for program axioms and then axiom tableaux are introduced.

Spurs. *Spurs* [13] are generalizations of schedulers, the 'next' operator in temporal logic, i.e. "$\bigcirc$", etc. α, β, γ, ... , κ are used for the metasymbols of spurs. A spur α is defined as a term by

$$\alpha = \mu y(0 < y \& 0 < J@y), \quad \text{for an arbitrarily chosen and fixed special constant } J.$$

Each component process, or even certain external object, of a multi-CPU parallel (or interleaving) program system is assigned a distinct spur usually, and the whole program is described in accordance with the causality, i.e. the relationships between the spurs as 'motives' and the changes of values.

Program Labels. a, a_1, a_2, ... are *program labels*, expressed by mutually exclusive special, or tense-sensitive, boolean constants defined as:

$$a \equiv L = 0, \quad a_i \equiv L = i \quad (i = 1,\ 2,\ \ldots).$$

where L is an arbitrary special constant.

Conservation Axioms. We adopt the following two axiom schemata as follows [17].

$(CA1)$ $\qquad\qquad J = z \ \supset \ x < \alpha \ \& \ x < \beta \ \& \ \ldots \ \& \ x < \kappa \supset \ J = z@x,$

for each special constant J and all spurs α, β, ..., which means that the value of J does not change until (unless) any spur rises, i.e. the next step of any process rises, where each spur α, β, ... indicates the tense when the next step of the process corresponding to the spur rises.

$(CA2)$ $\quad J = z@a \ \supset \ \alpha < \beta \ \& \ \ldots \ \& \ \alpha < \kappa \ \supset \ a \leq x \leq \alpha \ \supset \ J = z@x,$

$$\text{each } a \text{ and } J$$

such that J does not occur in the 'act' part of corresponding program axiom, which means that J does not change within the block corresponding to a.

Remark 3. Theoretically these axioms belong to the so-called frame axioms [12]: the value of a variable, like J, in a program does not change unless there is an execution of statement or command, i.e. the cause or motive, to change it positively.

Special Implication Symbol. Most of program axioms are of the form $(\mathbf{A} \supset \mathbf{B})@\lambda$ or $\mathbf{A} \supset \mathbf{B}@\lambda$ (an abbreviation of $\mathbf{A} \supset (\mathbf{B}@\lambda)$, since the precedence of @ is greater than other logical symbols), the former intuitively read that if $\mathbf{A}$ holds at time designated by λ, then the action represented by $\mathbf{B}$ is carried out, while the latter that if $\mathbf{A}$ holds *now* then $\mathbf{B}$ is carried at λ. Though in the usual logical sense of the implication symbol, these axioms allow that $\mathbf{B}$ can be done even if $\neg\mathbf{A}$, it is prohibited by the conservation axioms. For avoiding such an ambiguity, the special implication symbol "$\Rightarrow$" will be used for the implication associated with conservation axioms [2].

Labels in Axiom Tableau. $(\mathbf{A} \Rightarrow \mathbf{B})@\lambda$ is represented by the tableau as

index	condition	action	tense	personality
i	$\mathbf{A}$	$\mathbf{B}$	a	$[*]\ l_1,\ l_2,\ \ldots\ ,\ l_n$

---

[2] In the intuitionistic logic [26], "$\mathbf{A} \Rightarrow \mathbf{B}$" is often used as "if $\mathbf{A}$ then $\mathbf{B}$", which is essentially the same as the notation of this paper.

and $\mathbf{A} \Rightarrow \mathbf{B}@\lambda$ is

index	condition	action	tense	personality
i	A, global	B	a	[*] $l_1, l_2, \ldots, l_n$

where λ is $\langle \mathbf{a}, \{l_1, l_2, \ldots, l_n\}\rangle$, including the case $\mathbf{a}$ is missing, while " $*$ " will be used to indicate the case λ is $\mathbf{a}$.

Remark 4. Those program axioms that reflect the target program but no physical pehnomena are sound, since there is a unform translation from programs into (labeled) @-calculus [15], [17].

3   Case Study: The Shigaraki Kougen Railway Accident

In the following two sections, an actual traffic accident "Shigaraki Kougen Railway accident" will be analyzed in labeled @-calculus.

3.1   Outline of the Accident

It occurred about at 10:35, 14th May 1991, between Onotani signal station and Shigaraki-no-Miya station on Shigaraki Kougen Line that was a single track, Shiga Prefecture, Japan. An up train of Shigaraki Kougen Railway (SKR) for Kibukawa Station collided with a down train of Japan Railway (JR) for Shigaraki station. The cause was as follows. When the up train was to depart from Shigaraki station at 10:14 as scheduled, the signal at that station was still red. After a fairly reasonable (logical indeed) consideration, the responsible person of SKR decided that the up train depart 11 minutes after the scheduled time, with confidence of safety. But the departure signal for the down trains at Onotani signal station was still green. Therefore, the down train that departed from Kibukawa station at 10:16 on schedule did not wait at the signal station and entered the interval between the signal station and Shigaraki station. Thus, two trains collided.

3.2   Formalization

The formalization of this accident appeared below is along the analysis report [5] by the counselors of the victims, some details excluded for avoiding unnecessary complication.

Coordinates. Let us consider the one-dimensional coordinate axes along the line as Figure 1, in which the origin is at Kibukawa station, and the coordinate of Onotani signal station is d and that of Shigaraki station is l.

Axiom Tableau. Table 2 shows the axiom tableau representing the train control system, where S and J are labels indicating the personalities SKR and JR, respectively. The indices will be parenthesized in the text. (1) to (20) are the program axioms of the system, from which (21) and (22) are inferred. (5') means (5) did not occur, which did not occur actually. From (1) to (20) without (5), (23) and (24) are inferred, which occurred actually.

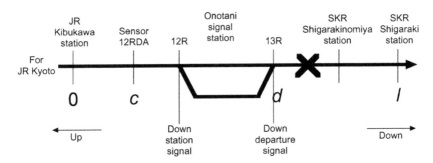

Fig. 1. SKR line

Each line in this tableau is explained below with its corresponding normal expression of the calculus, in which $@\tau_0$ is abbreviated uniformly from the end of each formula, where $\tau_0 < \infty$ is the start time of the train controlling system. $\hat{@}r$ is an abbreviation of $@\langle r, S\rangle$, while $@\langle r, \{S, J\}\rangle$ is not specifically mentioned for the explanation of the also correct $@r$ will suffice.

$$Clock = r@r \tag{1}$$

This means that the value of the program variable *Clock* is r at time $r[\mathrm{s}]$ after the system starts.

$$(A = l \ \& \ B \leq 0 \ \& \ \neg 13R \ \& \ \neg lock)@Clock = 0 \tag{2}$$

This is the initial condition at time when the system starts, i.e. $Clock = 0$. $13R$ and *lock* are program variables which take truth values. The former expresses the status of the signal $13R$ for down trains to be allowed to depart from Onotani signal station, while the latter that of the block signal to $13R$ operated when an up train departs from Shigaraki station. When the signal $13R$ is green (red), the value is true (false). The program variables A and B represent the positions of the up train of SKR for Kibukawa and the down train of JR for Shigaraki, respectively. This axiom shows that the up train is at the Shigaraki station and the down train is not between Shigaraki and Kibukawa, and also that $13R$ is red and the block signal is open at the time when the system starts.

$$r^+ \equiv r \leq Clock \tag{3}$$

This defines r^+. It is a formula meaning that the value of *Clock* is greater than or equal to r at every observation time. The expressions like "10:16", "10:25", etc are abbreviations of r^+ when r represents the difference between the actual value of τ_0 and that of 10:16, 10:25, etc.

$$\alpha = 10 : 25 + 1\hat{@}A = l \tag{4}$$

This means that the up train departs just after $(+1[\mathrm{s}])$ 10:25 when it is at Shigaraki station, where α is a spur to control the up train. It is a fact but only SKR staff knows it.

$$\gamma = \alpha\hat{@}A = l \tag{5}$$

Table 2. Axiom tableau representing the railway system

index	condition	action	tense	personality
1	$Clock = r$, **global**		r	* $\quad S, J$
2	$A = l$, $B \leq 0$, $\neg 13R$, $\neg lock$, **global**		$Clock = 0$	* $\quad S, J$
3	$r^+ \equiv r \leq Clock$, **def**			* $\quad S, J$
4		$\alpha = 10 : 25 + 1$	$A = l$	* $\quad S$
5		$\gamma = \alpha$	$A = l$	S
5'	-	-	-	-
6		$\gamma = \alpha$	$A = d + u$	S
7		$\kappa = \alpha$	$A = d$	S
8	$\neg 13R$	$\gamma = lock$		* $\quad S$
9	$lock$	$\gamma = \neg lock$		* $\quad S$
10	$0 < i < imax$	$\alpha^{i+1} = \alpha^i + 1$		* $\quad S, J$
11	$0 < i \leq imax$	$A = l - i \cdot u$	α^i	* $\quad S, J$
12		$\beta = 13R + 1$	$B = d$	* $\quad S, J$
13	$0 < j < jmax$, $j \neq jmid$	$\beta^{j+1} = \beta^j + 1$		* $\quad S, J$
14	$0 < j \leq jmid$	$\beta = j \cdot v$	β^j	* $\quad S, J$
15	$jmid < j \leq jmax$	$B = d + j \cdot w$	β^j	* $\quad S, J$
16		$\beta = \kappa$	$B = c$	* $\quad S, J$
17		$\kappa = 12R$		* $\quad S, J$
18	$\neg lock$	$\kappa = 13R$		* $\quad S, J$
19	$10 : 16 \leq (B = 0)$, **global**			* $\quad S, J$
20	$v \leq c/(9 \cdot 60)$, **global**			* $\quad S, J$
21	$d < A < l \Rightarrow B \leq d$		$x + 10 : 25 + 1$	S
22	$\neg Crash$		$x + 10 : 25 + 1$	S
23	$d < A < l$ & $d < B < l$		$\exists x.x + 10 : 25 + 1$	*
24	$Crash$		$\exists x.x + 10 : 25 + 1$	*

This is also a knowledge only of SKR staff. When the up train departs from Shigaraki station, the spur γ to control the block signal rises. In reality, this axiom did not work properly; γ did not rise, thus the signal $13R$ did not turn red, i.e. $\gamma = \infty$, therefore both up and down trains were in the section between Shigaraki and Onotani that caused the accident.

$$\gamma = \alpha \hat{@} A = d + u \tag{6}$$

The spur γ to control $13R$ rises just before (-1[s]) the up train reaches Onotani signal station, where u[m/s] is the speed of the up train. In the axiom, u means not a speed but a distance u[m] since every action rises every 1[s] from axioms (10) and (13) below. In reality, the spur did not rise since the up train did not reach Onotani by the accident.

$$\kappa = \alpha \hat{@} A = d \tag{7}$$

The spur κ for controlling $12R$ rises when the up train reaches Onotani signal station. In reality, this spur also did not rise by the accident.

$$(\neg 13R \Rightarrow \gamma = lock)\hat{@}0 \tag{8}$$

$13R$ becomes locked if it is red when γ rises by (5).

$$(lock \Rightarrow \gamma = \neg lock)\hat{@}0 \tag{9}$$

$13R$ becomes open if it is locked when γ rises by (6).

$$0 < i < imax \Rightarrow \alpha_{(i+1)-th} - \alpha_{i-th} = 1 \tag{10}$$

$$(0 < i \leq imax \Rightarrow A = l - i \cdot u)@\alpha_{i-th} \tag{11}$$

Let u be the speed of the up train and $imax$ be $(l - d)/u$. α_{i-th} (and also α^i) means the time of i-th rise of α from the observation time. Hence, $\alpha_{(i+1)-th} - \alpha_{i-th}$ in (10) is the interval between the rises of α. (10) shows that the period of the rise of α is 1[s] when the up train is between Shigaraki and Onotani, while (11) indicates the position of the train.

$$\beta = 13R + 1@B = d \tag{12}$$

This means the down train departs just after ($+1$ [s]) the time when $13R$ turns green, where β is a spur for controlling the down train.

$$0 < j < jmax \ \& \ j \neq jmid \Rightarrow \beta_{(j+1)-th} - \beta_{j-th} = 1 \tag{13}$$

$$(0 < j < jmid \Rightarrow B = j \cdot v)@\beta_{j-th} \tag{14}$$

$$(jmid < j < jmax \Rightarrow B = d + j \cdot w)@\beta_{j-th} \tag{15}$$

Let v and w be the speeds of the down train between Kibukawa and Onotani, and between Onotani and Shigaraki, respectively, and $jmid$ and $jmax$ be d/v and $d/v + (l - d)/w$, respectively. Similarly to (10) and (11), (13) shows that the period of the rise of β is 1[s] except the case that the train is at Onotani, which is shown in (12), and (14) and (15) show the position of the train.

$$\kappa = \beta@B = c \tag{16}$$

When the down train reaches c, the spur κ rises to control $12R$ and $13R$, where c is the position of the sensor $12RDA$ to control the signals between Kibukawa and Onotani.

$$\kappa = 12R \tag{17}$$

$$\neg lock \Rightarrow \kappa = 13R \tag{18}$$

When κ rises, $12R$ turns green, and simultaneously, $13R$ turns green if it is open.

$$10 : 16 \leq (B = 0) \tag{19}$$

The down train did not reach Kibukawa station till 10:16.

$$v \leq c/(9 \cdot 60) \tag{20}$$

This indicates the upper limit of the speed of the down train. From this and (19), the fact that the train did not reach c till 10:25 is obtained.

4   Analysis of the Accident

4.1   Inference by SKR

The responsible person of SKR inferred the conclusion that no collision could happen even if the up train departed at 10:25 by his own knowledge. Namely, he testified that he deduced as follows:

Suppose that the up train will depart at 10:25 as (4).

$13R$ will be locked from (5) and (8).

Even if down train will reach c after the above two things as (16), $13R$ will not turn green.

The down train will not reach c till 10:25 from (19) and (20).

Thus, $13R$ will not turn red.

On the other hand, by (12), the down train will not beyond Onotani.

Hence the two train will not be in the section between Onotani and Shigaraki at the same time:

$$(d < A < l \Rightarrow B \leq d)\hat{@}x + 10 : 25 + 1, \tag{21}$$

which is a kind of the *mutual exclusion*, where x is a logical variable and the tense $x + 10 : 25 + 1$ indicates the proposition holds any time after 10:25.

Therefore, any crash will not happen:

$$\neg Crash\hat{@}x + 10 : 25 + 1, \tag{22}$$

where $Crash \equiv d < A < l \ \& \ d < B < l \ \& \ |A - B| < \delta$.

Remark 5. The assertion (22) can be also expressed as

$$0; \ Crash = \infty \hat{@}10 : 25 + 1, \tag{22'}$$

i.e. the tense when $Crash$ occurs is infinity. It must be noted that it is possible to introduce an abbreviation to omit "0; " from the term 0; $Crash$, so that any formula like $Crash$ can be regarded as a term to mean the tense when it occurs, so that (22') can be written as

$$Crash = \infty \hat{@}10 : 25 + 1. \tag{22''}$$

4.2   The Actual Action

As inferred in the previous subsection, the responsible person of SKR decided that the up train depart at 10:25 as (4).

But against (5), the block signal did not rise. Hence $13R$ was not locked (against (8)), and $13R$ turned green by (18).

Thus the up train entered the interval between Onotani and Shigaraki by (12). Therefore, there exists a tense x that (21) and (22) do not hold. Namely,

$$\exists x((d < A < l \ \& \ d < B < l)@x + 10 : 25 + 1) \tag{23}$$

and

$$\exists x(Crash@x + 10 : 25 + 1) \tag{24}$$

hold, i.e., the crash occurred.

These inferences can be carried out in labeled @-calculus precisely, not relying very much on mathematical induction on time values or positions, while the formal derivations are omitted from this paper for the lack of space.

5   Conclusion and Discussions

This paper introduced a formalization of verification and analysis of the complicated control system involving human factor, especially misunderstanding and inappropriate decision, with the demonstration of the analysis of the actually occurred train accident. We consider this formal system is one of the earliest formalism for "time-concerned" or time-dependent knowledge with real-timing controlled hybrid systems. Since @-calculus, the basis of labeled @-calculus, is convenient to analyze such complicated systems as shown in [17], so is labeled @-calculus as shown in sections 3 and 4. The former calculus has been successfully applied to deal with a computerized ensemble system for musical performance, which is another sophisticated example of time-concerned system apparently, bearing much resemblance to cooperative autonomous vehicles in logical and mathematical representations [11], [18].

As mentioned in Introduction, there are many reasoning systems for knowledge and belief with time. To compare them with the formal system introduced in this paper, most of them are based on propositional logic, and the rest on predicate logic without any reference to specific and rigid mathematical foundation, whereas the latter is based on the concrete and very basic mathematical theory PA, which develops precise mathematics. And moreover, it is very likely that the formal system is a conservative extension of PA, since it has been proved that ∞ can be eliminated from the present formalism [21]. Hence, verification and analysis of complicated control systems with time can be carried out in a consistent formal theory. The authors think that automated verification with human assistance of (labeled) @-calculus can be devised relatively easily, since it is based on NK with PA, for which automatic verification methods have been studied well.

The way that natural numbers are used so as to represent the time-dependent position of trains as axioms in section 3 seems to be a promising solution for the difficulty of treating *continuous phenomena* and discrete changes like decisions, computer operations, etc together, while the *external variables* with very elementary differential equations were introduced in [17] and [23] to cope with this problem. It must be noted that mathematical induction on the positions or time values does not appear explicitly in the formal proofs.

This research was first reported in [19], [24], which had been resumed after the 2005 overturn accident in Amagasaki on Fukuchiyama line of West Japan

Railway Co. (JR West), killing 107 people. This accident as well as more than one handred elevator accidents, (many of them apparently caused by errors of software, CPUs or their conjoint actions with sensors or actuators), recently reported in Japan, however, seems to be much simpler than the Shigaraki one theoretically in so far as more or less social aspects like labor management, maintenance practice, etc are excluded from analyses, while some social matters did affect the latter.

References

1. Alechina, N., Logan, B. and Whitsey, M. : Modelling Communicating Agents in Timed Reasoning Logics, *9th European Conference on Logics in Artificial Intelligence, Lecture Notes in Artificial Intelligence* **3229** (2004), pp. 95–107.
2. Aviation Safety Network : Aircraft accident description McDonnell Douglas DC-10-40 JA8546 - off Shizuoka Prefecture,
   URL: http://aviation-safety.net/database/record.php?id=20010131-2.
3. Bennett, B., Dixon, C., Fisher M., Hustadt, U., Franconi, E., Horrocks, I. and de Rijke, M. : Combinations of Modal Logics, *AI Review*, **17** (2002), pp. 1–20,
4. Damm, W., Hungar, H. and Olderog, E.-R. : Verification of Cooperating Traffic Agents, *International Journal of Control*, **79** (2006), pp. 395–421.
5. Bereaved and Counselors of Shigaraki Accident (ed.): Shigaraki Kougen Railway Accident, *Gendaijinbunsha*, 2005 (in Japanese).
6. Dixon, C., Nalon, C. and Fisher, M. : Tableaux for Temporal Logics of Knowledge: Synchronous Systems of Perfect Recall or No Learning, *the Proceedings of TIME-ICTL 2003* (2003), pp. 62–71
7. Dixon, C., Gago, M.-C. F., Fisher,M. and van der Hoek, W. : Using Temporal Logics of Knowledge in the Formal Verification of Security Protocols, *Proceedings of TIME 2004*, (2004), pp. 148–151.
8. Fagin. R., Halpern, J.Y., Moses, Y. and Vardi, M. Y. : Reasoning About Knowledge, *The MIT Press*, 1995.
9. Gentzen, G. : Untersuchungen über das logische Schließen, *Mathematische Zeitschrift*, **39** (1935), 176–210, 405–431; Investigations into Logical Deduction, M. E. Szabo (ed.), *The Collected Papers of Gerhard Gentzen, Series of Studies in Logic and the Foundations of Mathematics, North-Holland Publ. Co., Amsterdam*, 1969, pp. 68–131.
10. Halpern, J. Y. and Vardi, M. Y. : The Complexity of Reasoning about Knowledge and Time. I. Lower Bounds, *Journal of Computer and System Sciences*, **38** (1989), pp. 195–237.
11. Hiraga, R. and Igarashi, S. : Psyche : Computer Music Project, University of Tsukuba, *International Computer Music Conference*, (1997), pp. 297–300.
12. Igarashi, S., London R. and Luckhum, D. : Automatic Program Verification I: Logical Basis and its Application, *Acta Informatica*, **4** (1975), pp. 145–182.
13. Igarashi, S. : The ν-Conversion and an Analytic Semantics, R. E. A. Mason (ed.), *Information Processing 83, Elsevier Sci. Publ. B. V.*, (1983), pp. 769–774.
14. Igarashi, S., Mizutani, T., Shirogane, T. and Shio M. : Formal Analysis for Continuous Systems Controlled by Programs, *Second Asian Computing Science Conference, ASIAN'96, Lecture Notes in Computer Science* **1179** (1996), pp. 347–348.
15. Igarashi, S., Shirogane, T., Shio, M. and Mizutani, T. : Tense Arithmetic I: Formalization of Properties of Programs in Rational Arithmetics, *Tensor, N. S.*, **59** (1998), pp. 132–152.

16. Igarashi, S. : *Science of Music Expression*, YAMAHA Music Media, 2000 (in Japanese).

17. Igarashi, S., Mizutani, T., Ikeda, Y. and Shio, M. : Tense Arithmetic II: @-calculus as an Adaptation for Formal Number Theory, *Tensor, N. S.*, **64** (2003), pp. 12–33.

18. Igarashi, S., Shio, M., Mizutani, T. and Ikeda, Y. : Specification and Verification of Cooperative Real-Timing Processes in @-Calculus, to appear.

19. Ikeda, Y. : Mathematics of Intellect and Sensibility: Logical System for Specifications, *2005 China-Japan Symposium on Advanced Science and Technology, Xiamen Univ.*, 2005.

20. McCarthy, J., Sato, M. and Igarashi, S. : On the Model of Knowledge, *Stanford Artificial Intelligence Lab. Memo*, **AIM-312** (1977), pp. 1–11; URL: `http://www-formal.stanford.edu/jmc/model/`.

21. Machi, T., Ikeda, Y., Tomita, K., Hosono, C. and Igarashi, S. : Tense Calculus and the Refinement without ∞, *Tensor, N. S.*, **67** (2006), pp. 79–93.

22. Mizutani, T., Igarashi, S. Tomita, K. and Shio, M. : Representation of Discretely Controlled Continuous Systems in Software-Oriented Formal Analysis, *Third Asian Computing Science Conference ASIAN'97, Lecture Notes in Computer Science*, **1345**, 1997, pp. 110–120.

23. Mizutani, T., Igarashi, S. and Shio, M. : Representation of a Discretely Controlled Continuous System in Tense Arithmetic, *Electr. Notes in Theor. Comp. Sci.*, **42** (2001), URL: `http://www.elsevier.com/locate/entcs/volume42.html`.

24. Mizutani, T. : Mathematics for Intellect and Sensibility: Case Study: Formal Representation of Shigaraki Kougen Railway Accident, *2005 China-Japan Symposium on Advanced Science and Technology, Xiamen Univ.*, 2005.

25. Shoenfield, J. R. : Mathematical Logic, *Addison-Wesley Publishing Company, Massachusetts*, 1967.

26. van Dalen, D. : Intuitionistic logic, In D. M. Gabbay and F. Guenthner (Eds.), *Handbook of Philosophical Logic, 2nd Edition*, **5** (2002), pp. 1–114.

Enhanced Theorem Reuse
by Partial Theory Inclusions

Immanuel Normann

Computer Science, Internaitonal University Bremen
i.normann@iu-bremen.de

Abstract. Finding new theorems is essential for the general progress in mathematics. Apart from creating and proving completely new theorems progress can also be achieved by theorem reuse; i.e. by translating theorems from related theories into the target theory and proving its translated premises under which the theorem was proven. This approach is pursued by the "little theories" and the development graph paradigms. This work suggests an improvement in this direction in two aspects: partial theory inclusions enhances theorem reuse and formula matching to support automated detection of theory inclusions. By representing theory axioms as facts and partial theories (i.e. theorems and their minimal set of premises) as horn clauses, reusable theorems correspond to derived facts such that model generator like KRHyper can be used for this task.

1 Introduction

Finding new theorems is essential for the general progress in mathematics. For this we postulate new assertions within a certain theory T and try to prove them based on the axioms of T and the already proven theorems therein. Sometimes, however, we could alleviate this burden if the proofs are already done in some other theory S: For instance Jane, an undergraduate math student, has to prove for an arithmetic course the following assertion: (1) $\forall m, n. \gcd(m, n) = m \Rightarrow \mathrm{lcm}(m, n) = n$. She may attack this problem using only her knowledge from arithmetics. Or she is smart and uses her knowledge acquired in a course about lattices. From there she remembers the lemma (2) $\forall a, b. a \sqcup b = a \Rightarrow a \sqcap b = b$. This can easily be mapped on the assertion (1) with the signature morphism $\sigma := [\sqcup \to \gcd, \sqcap \to \mathrm{lcm}]$. Of course mapping only the theorem does not turn the original assertion into a theorem. But as soon as we know that all the axioms from which (2) can be derived also hold in our target theory we are done. To finish the example, Jane happens to know that "gcd" and "lcm" are commutative and associative functions, and finally she proves the absorption rules between these two functions. As these are exactly the lattice axioms translated into arithmetics with σ she has implicitly proven the initial assertion (1); i.e. she *reused* the theorem from lattice theory in arithmetics.

In mathematical parlance we would say Jane proved that the algebraic structure $(\mathbb{N}, \gcd, \mathrm{lcm})$ is a lattice. Such phrases are quite common in mathematics:

J. Calmet, T. Ida, and D. Wang (Eds.): AISC 2006, LNAI 4120, pp. 40–52, 2006.

We say for instance $(\mathbb{Z}, +, 0)$ is a group as well as $(\mathbb{Q}^+, *, 1)$ is where we actually mean that the axioms from group theory can be proven in the theory of integers and positive rational numbers respectively after translating the symbols appropriately.

The "little theory" paradigm [4] is designed to make use of such theory embedding. A further elaborated variant called "development graph" [1] pursues this goal by means of "theory inclusions" [1]. The basic idea behind that is firstly to translate a complete source theory into the target theory and secondly to prove the translated axioms of the source theory inside the target theory. Whenever this is possible all the theorems from the source theory become automatically theorems in the target theory. IMPS, MAYA, and HETS are systems implementing the little theories paradigm and the development graph respectively. For brevity we call them just dg-systems – more about this topic is provided in section 2.

The work reported here presents an improvement of theorem reuse in dg-systems in two aspects: 1) the theorem reuse factor is enhanced by embedding into the target theory only the minimal premises of a theorem to be reused instead of the whole set of axioms of the source theory. We call these embeddings partial theory inclusions. 2) Consistent many to many formulae matching is used to detect appropriate signature morphisms for such embeddings. Finally it is shown how elaborated model generator like KRHYPER [11] can be effectively used to generate the transitive closure of partial theory inclusions.

In current dg-systems it is left to the user to choose a source theory $\mathcal{S}$ for a given target theory $\mathcal{T}$ and to find (if possible at all) an appropriate signature morphism σ and finally prove that $\sigma\mathcal{S}$ is included in $\mathcal{T}$. However, finding all signature morphisms σ that map all axioms from $\mathcal{S}$ into valid statements in $\mathcal{T}$ is a many to many formulae matching task that can be automated. Thus some theory inclusions can be automatically found without generating any proof obligations. In fact we will define constructively (section 3) the set of all signature morphisms σ with $\Phi \supseteq \sigma\Psi$ for any sets of formulae Φ and Ψ. This can already be used to build the transitive closure of a theory based on total[2] theory inclusion. Even more theorems can be reused if we take partial theory inclusions into account: It is always sufficient to prove only the assumptions of the theorem inside the target theory to reuse this theorem. A theorem φ together with its minimal set of assumptions Γ actually used in a proof can be considered as a partial theory of some other theory. We will call $\Gamma \vdash \varphi$ a *sequent*. Every theorem caught by total theory inclusion is also caught by partial theory inclusion whereas the opposite does not hold – a formal argument will be given in section 4 and an illustrating example in section 6.

We compute the transitive closure of a theory by forward chaining: All theorem instances are added to the theory whose assumptions in a sequent match

---

[1] The term "theory inclusion" from in the development graph is called "theory interpretation" in in the little theory paradigm.

[2] By "total theory inclusion" we mean the (global or local) theory inclusions from the development graph and we use the attribute "total" just to distinguish them from our "partial theory inclusion".

the axioms of the target theory or theorems added before in the same manner. To enhance this procedure we do not operate on the formulae themselves, but on their abstracted representatives (section 5). On the abstract level the theory closure task corresponds to model generation of all ground terms derivable from the initial input facts representing the axioms of the target theory. Hence systems specialized on model generation like SATCHMO [7] and KRHYPER [11] can be used to build the transitive closure of a theory. A mini session in KRHYPER is given in section 6 to demonstrate this possibility.

2   Theorem Reuse in the Development Graph

The basic goal in this paradigm is to find a source theory S and an appropriate translation σ such that every model of the translated source theory is also a model of the target theory T – formally $T \vDash \sigma S$ which we call *theory inclusion* where T is said to *include* S via the signature morphism σ. Obviously, if we have $T \vDash \sigma S$ then we can be sure that for any theorem φ in S its translation $\sigma\varphi$ must be a theorem in T too. In order to prove $T \vDash \sigma S$ it is sufficient to prove that all axioms from S translated with σ are theorems in T. Thus one proves $T \vDash \sigma(\mathrm{Ax}(S))$ once and gets all theorems from S for free in T; i.e. just by translating them with the guarantee that these translated formulae are really theorems in T.

Of course every target theory may serve as a source theory for some other theory and so we may find paths of theory inclusions. Furthermore target theories may include several source theories. Eventually leading to a directed graph of theory inclusions. The gain of theorems for a given target theory T is then the union of all theorems contained in those theories which are recursively included in T.

So far we have considered theory inclusions as edges and theories as nodes of a graph where theories contain axioms[3] and locally proven theorems. Adding to the graph theory imports as another sort of edges we get essentially a data structure known under the notion of development graph [1]. A *theory import* of a source theory S into a target theory T simply means that all axioms of S hold in T by definition. Thus nodes representing theories in the development graph can have local axioms and non-local axioms imported from other theories.

Systems implementing the development graph[4] are IMPS [3], MAYA [2], and HETS [10]. For IMPS its development graph is rather an add on feature, but its main focus is on proof assistance for local theorem proofs in a fixed logic. MAYA and HETS in contrast abstract from a fixed logic as they are theoretically based on entailments [8] and institutions [5]. The development graph is their central

---

[3] For brevity we subsume definitions and axioms under the term "axiom" for the rest of this paper.

[4] Actually they implement different variants of the development graph, e.g. MAYA differentiates in contrast to IMPS and HETS between theory imports that import only from direct processors and those which import recursively along a whole path of theory imports. A similarly global and local theory inclusions are distinguished.

data structure which is also equipped with a development graph transformation calculus. Roughly spoken these transformations are intended to reduce as much as possible proof obligations that are necessary to keep the whole graph valid after an external graph modification performed like postulating a new theory inclusion or adding, removing local theorems or axioms to a theory.

3   Mapping Sets of Formulae on Sets of Formulae

As already mentioned in the introduction if we have found a signature morphism σ such that $\mathcal{T} \supseteq \sigma(\text{Ax}(\mathcal{S}))$ then we know that the target theory $\mathcal{T}$ includes the translated source theory $\sigma\mathcal{S}$; i.e. all theorems of $\mathcal{S}$ (translated via σ) can be reused in $\mathcal{T}$. Hence for automating theorem reuse it is helpful to automatically construct the set $\{\sigma | \mathcal{T} \supseteq \sigma(\text{Ax}(\mathcal{S}))\}$ which is a missing feature in current dg-systems. For our partial theory inclusion we need a little more general set: $\{\sigma | \Phi \supseteq \sigma\Psi\}$ for given sets of formulae Φ and Ψ. In order to make the construction of this set explicit we give an alternative constructive definition for this set: Let μ be a matcher function that takes two formulae ψ and φ and returns a signature morphism σ such that $\sigma\psi = \varphi$ if possible and otherwise σ_0 which denotes a failure. We define for two signature morphisms σ and σ' the merge $\sigma \oplus \sigma'$ as the union of them if they are compatible[5] and otherwise σ_0. The merge of many signature morphisms can be expressed as $\bigoplus_{j=1}^{n} \sigma_j := \sigma_1 \oplus \ldots \oplus \sigma_n$. For two sets of formulae Ψ and Φ we define the set $\Sigma_{\Psi \to \Phi}$ of all signature morphisms whose elements map all formulae from Ψ on a subset of Φ (i.e. $\Phi \supseteq \sigma\Psi$ for all $\sigma \in \Sigma_{\Psi \to \Phi}$):

$$\sum_{\Psi \to \Phi} := \{\sigma | \exists m : \Psi \to \Phi . \sigma = \bigoplus_{\psi \in \Phi} \mu(\psi, m(\psi)) \neq \sigma_0\}$$

The set $\Sigma_{\Phi \to \Psi}$ and hence the effect of theorem reuse can be even enlarged if we extend the matcher function μ by requiring $\sigma\psi = \varphi$ *modulo normalization* as investigated in [9]. Another aspect to be mentioned here concerns the efficiency of building such a set. In the extreme case this set can contain as many signature morphisms as there are mappings $m : \Phi \to \Psi$. In practice, however, for most of these mappings $m : \Phi \to \Psi$ there is no consistent signature morphism with $\Phi \supseteq \sigma\Psi$. In principle, term indexing is the state of the art for matching many to many formulae as described in [6]. It is not the purpose of this paper to invent new term indexing methods, but rather to suggest an adjustment to the particular task: Term indexing methods are typically used to match quantifier free formulae where all variables from one formula are tried to match consistently against all variables of the other formula. Our task in contrast is to match quantified formulae against quantified formulae. All variables involved are bound variables and not subjected to matching whereas all symbols from the signature are subjected. Thus it makes sense in our scenario not to apply term indexing methods directly on the formulae from the theories, but on their abstraction

---

[5] σ and σ' is said to be compatible if $\sigma\varphi = \sigma'\varphi$ for every $\varphi \in \text{support}(\sigma) \cap \text{support}(\sigma')$.

where all signature symbols are considered as variables and their variables as constants. We will go more in detail in section 5.

4   Transitive Closure of a Theory Based on Partial Theories

In the last section we considered for theorem reuse only two sets of formulae one being the source of theorems and the other the consumer of theorems so to speak. In the development graph these sets are theories. In our approach we consider partial theories and consequently partial theory inclusions as opposed to total theory inclusions. Both kind of inclusions have transitivity in common, of course. To make the best theorem reuse from all source theories connected by theory inclusion to the target theory one builds the transitive closure of it. We want to express this more formally that one can easily see how this transitive closure can be constructed. Thereby it becomes even clearer why the yield of theorem reuse is higher when based on partial theory inclusion than on total theory inclusion.

As already introduced above, a *sequents* $P \vdash \varphi$ represents a proof where all premises from the set of formulae P were used, but no more, to derive the theorem φ. We define the reflexive transitive closure modulo signature morphism on a set of formulae Γ based on sequents:

$$\Gamma_0 := \Gamma$$
$$\Gamma_n := \Gamma_{n-1} \cup \{\sigma\varphi | \Gamma_{n-1} \supseteq \sigma P \wedge P \vdash \varphi\}$$
$$\Gamma^* := \bigcup_n \Gamma_n$$

From the last section we know how to construct $\Gamma_{n-1} \supseteq \sigma P$, thus the recursive construction of Γ_n is clear. And Γ^* is the fixed point which exists essentially due to the finite number of sequents.

It should be mentioned that we are not bound to a particular logic[6] in our definition of Γ^*. We rely only on entailment systems and consider the sequent relation as a subrelation of the entailment relation. From the entailment axioms it can be shown that in fact Γ^* is entailed in Γ; i.e. all statements in Γ^* are derivable from Γ in the entailment system. In particular for $\Gamma_{n-1} \supseteq \sigma P$ and $P \vdash \varphi$ we know that Γ_n entails $\sigma(P \cup \varphi)$ and since $P \cup \varphi$ can be viewed as a partial theory we call this entailment a partial theory inclusion.

Let us now assume that we build our transitive closure, now denoted by Γ'^*, only with total theory inclusions. This would imply that P in the definition of Γ_n would not be just the assumptions needed to derive φ, but all axioms from the theory where φ was derived; i.e. often more than needed for an actual proof. Γ_n can become only smaller or at best stay the same when P is increased in $\Gamma_{n-1} \supseteq \sigma P$. Therefore Γ'^* is always at best equal to Γ^*, but often smaller. Which means that partial theory inclusion yields a better theorem reuse.

---

[6] Moreover we are not interested in the details of the proofs.

5 Abstraction of Formulae and Sequents

For efficiency reasons as mentioned in section 3 we will now introduce a notion of formula abstraction and extend it to abstraction of a sequent. The basic idea is to separate structure and content of a formula where we mean by content the symbols of a signature occurring in a formula.

Abstraction of formulae. We consider only closed formulae (i.e. formulae with free variables universally quantified). We classify the symbols of our logic into constants (subsuming also function and relation symbols), variables (possibly higher-order), and logical constants.

Let c be a constant, v be a variable, x be a variable or a logical constant, $\circ$ be any symbol, Q be a quantifier, and φ_j be a formula. We define recursively a function $\mathfrak{f}$ that maps to each formula its structure:

$$\mathfrak{f}(c) = \Box, \qquad \mathfrak{f}(x) = x, \qquad \mathfrak{f}(Qv.\varphi) = Qv.\mathfrak{f}(\varphi)$$

$$\mathfrak{f}(\circ(\varphi_1, \ldots, \varphi_n)) = \mathfrak{f}(\circ)(\mathfrak{f}(\varphi_1), \ldots, \mathfrak{f}(\varphi_n))$$

Thus $\mathfrak{f}$ just removes all signatures symbols out of the formula and replaces it by a placeholder. Moreover we assume that our $\mathfrak{f}$ renames all bound variables in a standardized way[7] such that structural identity is relieved from α-equivalence. For readability we abandon this α-standardizations in our examples, but choose instead our bound variables conveniently which makes α-standardization unnecessary.

Our second function $\mathfrak{p}$ collects in a list all the signature symbols from a formula. Here is its recursive definition[8]:

$$\mathfrak{p}(c) = (c), \qquad \mathfrak{p}(x) = (), \qquad \mathfrak{p}(Qv.\varphi) = \mathfrak{p}(\varphi)$$

$$\mathfrak{p}(\circ(\varphi_1, \ldots, \varphi_n)) = \mathfrak{p}(\circ) :: \mathfrak{p}(\varphi_1) :: \ldots :: \mathfrak{p}(\varphi_n)$$

The following example demonstrates $\mathfrak{f}$ and $\mathfrak{p}$ on a formula φ. Note, in our definitions of $\mathfrak{f}$ and $\mathfrak{p}$ the formulae are supposed to be in prefix notation. For a better readability, however, we give the example in usual infix notation[9]:

$$\text{The formula}: \quad \varphi := \forall a, b, c.a * (b + c) \simeq (a * b) + (a * c)$$
$$\text{its structure}: \quad \mathfrak{f}(\varphi) = \forall a, b, c.a\Box(b\Box c) \simeq (a\Box b)\Box(a\Box c)$$
$$\text{its constants}: \quad \mathfrak{p}(\varphi) = (*, +, *, +, *)$$

Let $\mathcal{F}$ be the set of all formulae, $\mathcal{N}$ a set of identifiers, and ind : $\mathfrak{f}(\mathcal{F}) \leftrightarrow \mathcal{N}$ a bijective mapping. We define the formula abstraction λ on $\mathcal{F}$ thus

$$\lambda(\varphi) := (\mathrm{ind}(\mathfrak{f}(\varphi)), \mathfrak{p}(\varphi))$$

---

[7] This actually one step of normalization as described in [9].

[8] The symbol ”::” denotes the list append operator.

[9] Equality in the meta language is denoted by ”=” and by ”$\simeq$” in the object language.

Let's take for example φ from above and assume that $\mathrm{ind}(\mathfrak{f}(\varphi)) = \mathrm{dist}$ then we get from φ the abstraction $\lambda(\varphi) := (\mathrm{dist}, (*, +, *, +, *))$. Instead of this tuple notation we also use the meta term notation : "$\mathrm{dist}(*, +, *, +, *)$".

The inverse of formula abstraction is called formula instantiation. For example the instantiation of the meta term $\mathrm{dist}(\cap, \cup, \cap, \cup, \cap)$ would give the concrete formula $\forall a, b, c. a \cap (b \cup c) \simeq (a \cap b) \cup (a \cap c)$.

Let's return to the efficiency issue mentioned in section 3: The constant symbols from the signature are the only symbols subjected to the signature morphisms and these are exactly the only visible symbols in the abstract level. Since there is a bijection between the abstract and the concrete level we can apply the signature morphisms as well on abstract level as on the concrete level. However, the formulae on the abstract level are much more compact which makes matching faster.

Abstraction of sequents. Let $\mathfrak{s}$ be a function that takes a formula φ and collects in a set all signature symbols occurring in φ. The abstraction of formulae is then extended to abstraction of sequents as follows:

$$\lambda(\varphi_1, \ldots, \varphi_n \vdash \psi) = \forall s_1, \ldots, s_m. \lambda(\varphi_1) \wedge \ldots \wedge \lambda(\varphi_n) \Rightarrow \lambda(\psi)$$

where $\{s_1, \ldots, s_m\} = \bigcup_{k=1}^{n} \mathfrak{s}(\varphi_k) \cup \mathfrak{s}(\psi)$. Note, the logical symbols occurring on the right hand side of equality are part of the abstract sequent – they should not be mixed up with the logical symbols from the concrete level.

For example given the following concrete sequent:

$$\forall a. a + 0 \simeq a, \forall a. \exists b. a - b \simeq 0 \vdash \forall a. \exists b. a + (a - b) = a$$

and furthermore assume we have for the formulae involved the following abstractions:

Table 1. Example of formulae abstractions

φ	$\lambda(\varphi)$	$\mathfrak{s}(\varphi)$
$\forall a. a + 0 \simeq a$	$\mathrm{neut}(+, 0)$	$\{+, 0\}$
$\forall a. \exists b. a - b \simeq 0$	$\mathrm{inv}(-, 0)$	$\{-, 0\}$
$\forall a. \exists b. a + (a - b) = a$	$\mathrm{abs}(+, -)$	$\{+, -\}$

This would turn the concrete sequent from above into this abstracts sequent:

$$\forall +, -, 0. \mathrm{neut}(+, 0) \wedge \mathrm{inv}(-, 0) \Rightarrow \mathrm{abs}(+, -)$$

Abstraction of typed formulae. For typed formulae the abstraction process must be slightly extended. The basic intuition is that types can be considered as (untyped) formulae. Instead of developing here the whole mechanism formally we want to illustrate the idea by example. The first step is exactly as in the untyped situation:

The formula : $\varphi := \forall f : i \to i. \exists a : i, g : i \to i.a * g > f$

its structure : $\mathfrak{f}(\varphi) = \forall f : i \to i. \exists a : i, g : i \to i.(a \square g) \square f$

its constants : $\mathfrak{p}(\varphi) = (* : i \times i \to i, > : i \times i \to o)$

In the second step we treat the types as formulae:

$$\mathfrak{f}(* : i \times i \to i) \;=\; \mathfrak{f}(> : i \times i \to o) \;=\; \square : \square \times \square \to \square$$

$$\mathfrak{p}(* : i \times i \to i) = (*, i, i, i) \text{ and } \mathfrak{p}(> : i \times i \to o) = (>, i, i, o)$$

Let us assume the following names for the two structures:

$$\text{mult} := \text{ind}(\mathfrak{f}(\varphi)) \text{ and op} := \text{ind}(\square : \square \times \square \to \square)$$

Then we get finally the abstracted formula:

$$\text{mult}(\text{op}(*, i, i, i), \text{op}(>, i, i, o))$$

From this information we can reconstruct the original formula (modulo α-renaming of the bound variables).

6   An Illustrative Example

In this section we want to demonstrate on some example theories how a system would accomplish theorem reuse which implements the methods from above. Assume we have a collection consisting of the following theories $\mathcal{S}$ and $\mathcal{S}'$: [10]

Table 2. Two source theories $\mathcal{S}$ and $\mathcal{S}'$

	$\mathcal{S}$	$\mathcal{S}'$
axioms	$(s1)\ \forall a,b.a \sqcup b \simeq b \sqcup a$ $(s2)\ \forall a,b.a \sqcap b \simeq b \sqcap a$ $(s3)\ \forall a,b.a \sqcup (a \sqcap b) \simeq a$ $(s4)\ \forall a,b.a \sqcap (a \sqcup b) \simeq a$	$(s'1)\ \forall a.a\vert a \simeq a$ $(s'2)\ \forall a,b.a \leqslant b \Leftrightarrow a\vert b \simeq a$
theorems	$(s5)\ \forall a.a \sqcup a \simeq a$ $(s6)\ \forall a,b.a \sqcup b \simeq a \Rightarrow a \sqcap b \simeq b$	$(s'3)\forall a.a \leqslant a$

Now we create a new theory $\mathcal{T}$ containing only axioms. We want the system to reuse in our target theory $\mathcal{T}$ as many theorems from our source theories $\mathcal{S}$ and $\mathcal{S}'$ as possible in the way described above. At first we observe that there is neither for $\mathcal{S}$ nor for $\mathcal{S}'$ a signature morphism σ such that $\mathcal{T} \supseteq \sigma(\text{Ax}(\mathcal{S}))$ or $\mathcal{T} \supseteq \sigma(\text{Ax}(\mathcal{S}'))$. Hence the method to make theorem reuse via total theory inclusion fails completely.

Fortunately we can do better with partial theory inclusions under the assumption that the user or theorem prover records for each proven theorem in the theory its minimal set of premises; i.e. the sequent. In our example we have:

---

[10] For readability of formulae we use in this section again usual infix instead of prefix notation.

Table 3. A new target theory $\mathcal{T}$

	$\mathcal{T}$
axioms	(t1) $\forall a,b.a \cup (a \cap b) \simeq a$ (t2) $\forall a,b.a \cap (a \cup b) \simeq a$ (t3) $\forall a,b.a \subseteq b \Leftrightarrow a \cap b \simeq a$
theorems	?

- $s3, s4 \vdash s5$
- $s1, s2 \vdash s6$
- $s'1, s'2 \vdash s'3$

Formula abstraction. At first the database of all formula abstractions needs to be updated:

Table 4. Abstraction of all recorded formulae

φ	$f(\varphi)$	$\mathrm{ind}(f(\varphi))$
$s1, s2$	$\forall a,b.a\square b \simeq b\square a$	com
$s3, s4, t1, t3$	$\forall a,b.a\square(a\square b) \simeq a$	abs
$s5, s'1$	$\forall a.a\square a \simeq a$	idem
$s'2, t3$	$\forall a,b.a\square b \Leftrightarrow a\square b \simeq a$	ord
$s6$	$\forall a,b.a\square b \simeq a \Rightarrow a\square b \simeq b$	dual
$s'3$	$\forall a.a\square a$	refl

For every row of this table the entries in the first column lists all formulae which have the structure recorded in the second column and in the third column the identifiers[11] are recorded giving each formula structure a name.

Sequent abstraction. From the formula abstraction database and the concrete sequents listed above the system computes the abstract sequents:

Table 5. Abstraction of all recorder sequents

concrete sequents	abstract sequents
$s3, s4 \vdash s5$	$\forall R_1, R_2, R_3.\, \mathrm{abs}(R_1, R_2) \wedge \mathrm{abs}(R_2, R_1) \Rightarrow \mathrm{idem}(R_3)$
$s1, s2, s3 \vdash s6$	$\forall R_1, R_2.\, \mathrm{com}(R_1) \wedge \mathrm{com}(R_2) \wedge \mathrm{abs}(R_1, R_2) \Rightarrow \mathrm{dual}(R_1)$
$s'1, s'2 \vdash s'3$	$\forall R_1, R_2.\, \mathrm{idem}(R_1) \wedge \mathrm{ord}(R_1, R_2) \Rightarrow \mathrm{refl}(R_2)$

Abstraction of the target theory. From the formula abstraction database and the concrete axioms of the target theory $\mathcal{T}$ the system computes the abstracted version of $\mathcal{T}$:

---

[11] For readability we chose intuitive names where actually generic identifiers would be generated.

Table 6. Abstraction of the target theory

	$\mathcal{T}$
axioms	$(t1)$ abs$(\cup,\cap)$ $(t2)$ abs$(\cap,\cup)$ $(t3)$ ord$(\subseteq,\cap)$

The transitive closure of the abstract target theory. With the abstract target theory as facts and the abstract sequents as rules the system computes the transitive closure of the target theory as the Herbrand model for these facts and rules. Delegating this job to the model generator KRHYPER yields the following session:

Table 7. The transitive closure of the abstract target theory in a KRHYPER session

Input of facts and rules	model as output
KRH> abs(cap,cup).	abs(cap,cup)
KRH> abs(cup,cap).	abs(cup,cap)
KRH> ord(sub,cap).	ord(sub,cap)
KRH> idem(A):-abs(A,B),abs(B,A).	idem(cap)
KRH> dual(A):-abs(A,B),com(A),com(B).	idem(cup)
KRH> refl(A):-idem(B),ord(A,B).	refl(sub)

The transitive closure of the concrete target theory. Finally the system translates the result from the abstract level back to the concrete level:

Table 8. The final target theory $\mathcal{T}$ with theorems reused from the source theories $\mathcal{S}$ and $\mathcal{S}$'

	$\mathcal{T}$
axioms	$(t1)\ \forall a,b.a \cup (a \cap b) \simeq a$ $(t2)\ \forall a,b.a \cap (a \cup b) \simeq a$ $(t3)\ \forall a,b.a \subseteq b \Leftrightarrow a \cap b \simeq a$
theorems	$(t4)\ \forall a.a \cap a \simeq a$ $(t5)\ \forall a.a \cup a \simeq a$ $(t6)\ \forall a.a \subseteq a$

Thus the system has derived three theorems from the axioms of $\mathcal{T}$ by reusing two theorems from the source theories $\mathcal{S}$ and $\mathcal{S}'$. Note that the idempotence theorem was reused in two variants.

Cyclic theory inclusions The simple example from above may raise the question how this forward chaining copes with cyclic theory inclusions. Such cyclic inclusions are not at all pathological. A very popular example can be found in basic group theory: Having the left neutral as axiom we can prove the right neutral property and vice versa. Translating this to abstract sequents would yield cyclic rules. However, this kind of cycles would not cause KRHyper to run forever. Only rules of the kind `f(f(X)):-f(X)` generate infinite Herbrand models. But this kind of rules are never the result of sequent abstraction.

Remarks on the triviality of the given examples. A working mathematician may immediately object that theorems such as that from the example above are not at all impressing. In fact every theorem found by our method is presumably trivial from a mathematicians perspective. This is not surprising since consistent formula matching and applying Horn clauses never constitutes a sophisticated proof as mathematicians appreciate. Finding a sophisticated proof is not at all the primary goal.

The actual strength of the method presented here is the ability of scanning masses of formulae. Mathematicians are unsurpassable in their dedicated field, but machines are good in precision and mass processing - they can discover useful things which are simply overlooked by humans. In this sense the proposed method rather aims to be an assistant tool: It may find almost only theorems which are folklore to mathematician in a research area where this theorem comes from. Nevertheless this seemingly trivial theorem is exactly what another mathematician from a different research area with a different background is looking for.

7   Limitations and Future Work

First of all it must be conceded that important parts of the methods described here need to be implemented. The current implementation can load formulae in the CASL format[12], normalize and abstract them, and build up in the working memory a database of abstracted formulae as well abstracted sequents. All the implementations are well integrated into HETS which is written in Haskell. A communication with KRHyper is not yet implemented and a persistent database for these abstracted formulae and sequents does not exist either.

All this is planned in the near future as well as case studies on large real world math libraries. Unfortunately, the author hase not found so far a library which explicitly keeps track of the minimal premises for each of its theorem. May be the potential use of such information has been simply overseen by the creators of the libraries. Although many libraries store the proofs of their theorem it is a tedious and very library specific programming task to extract the sequent information out of the proof. This is not done so far.

Apart from these rather implementational issues there are some rather theoretical:

So far we have considered only *simple signature morphisms* meaning morphism which only rename symbols of a theories signature. In fact the HETS system which is the current test bed for this work does support only simple signature morphisms. A generalization to signature morphisms mapping symbols to terms is subject of future work.

---

[12] The load process is actually executed by HETS.

Concerning typed formulae the current implementation of formula abstraction is restricted to a *simple type system* with overloading. There are certainly many other non-simple type systems around worth to be supported. With the general approach handling types like untyped formulae it should not be a fundamental problem to support other non-simple type systems. This will be done if required e.g. in a case study.

After a new theorem is proven in a certain theory $\mathcal{T}$ of the development graph one has in a dg-system almost immediately access to this theorem from all other theories which are connected to $\mathcal{T}$ via theory inclusions or imports. The only operation executed is a fast translation between the theories. Our model generation is inferior in this respect: Proving a new theorem would mean upgrading the database by one abstract sequent. Calculating from the upgraded database the transitive closure of a theory would not require a new model generation which involves a lot of backtracking. But after all this additional time expense is rewarded with the higher theorem reuse factor.

8   Conclusion

We presented a method for theorem reuse based on partial theory inclusion which is related to total theory inclusion from development graph. We have shown that partial theory inclusion can increase the theorem reuse factor of that from total theory inclusion. This new method makes use of many to many formulae matching for the search of theory inclusions - a feature not yet supported by current system implementing the development graph. For efficiency all formulae and partial theories (=sequents) are abstracted before Matching is conducted. On this abstract level theory axioms correspond to Prolog facts and sequents to Horn clauses. The maximum of theorem reuse in some target theory $\mathcal{T}$ by this method is building a Herbrand model from the facts representing the axioms of $\mathcal{T}$ and the Horn clauses representing all partial theories of the whole theory collection. As a proof of concept this method was tested on a small collection of partial theories with the model generator KRHYPER.

References

1. Serge Autexier and Dieter Hutter. Maintenance of formal software developments by stratified verification. `http://www.dfki.de/vse/papers/ah02.ps.gz`, 2002.
2. Serge Autexier, Dieter Hutter, Till Mossakowski, and Axel Schairer. The development graph manager maya (system description). In Hélene Kirchner, editor, *Proceedings of 9th International Conference on Algebraic Methodology And Software Technology (AMAST'02)*. Springer Verlag, 2002.
3. W. M. Farmer, J. D. Guttman, and F. J. Thayer. The IMPS user's manual. Technical Report M-93B138, The MITRE Corporation, 1993.
4. William Farmer, Josuah Guttman, and Xavier Thayer. Little theories. In D. Kapur, editor, *Proceedings of the 11th Conference on Automated Deduction*, volume 607 of LNCS, pages 567–581, Saratoga Springs, NY, USA, 1992. Springer Verlag.

5. J. A. Goguen and R. M. Burstall. Introducing institutions. In E. Clarke and D. Kozen, editors, *Lecture Notes in Computer Science 164, Logics of Programs*. Springer Verlag, 1983.
6. Peter Graf and Christoph Meyer. Advanced indexing operations on substitution trees. In *Conference on Automated Deduction*, pages 553–567, 1996.
7. Rainer Manthey and François Bry. SATCHMO: A theorem prover implemented in Prolog. In Ewing L. Lusk and Ross A. Overbeek, editors, *Proceedings of the 9th Conference on Automated Deduction*, number 310 in LNCS, pages 415–434, Argonne, Illinois, USA, 1988. Springer Verlag.
8. J. Meseguer. General logics. In *Logic Colloquium 87*, pages 275–329. North Holland, 1989.
9. Immanuel Normann. Extended normalization for e-retrieval of formulae. to appear in the proceedings of Communicating Mathematics in the Digital Era, 2006.
10. Klaus Lttich Till Mossakowski, Christian Maeder and Stefan Wlfl. Hets. http://www.informatik.uni-bremen.de/agbkb/forschung/formal_methods/CoFI/hets/
11. C. Wernhard. System description: Krhyper, 2003.

Extension of First-Order Theories into Trees

Khalil Djelloul[1] and Thi-Bich-Hanh Dao[2]

[1] Faculty of Computer Science, University of Ulm, Germany
DFG research project "Glob-Con"
[2] Laboratoire d'Informatique Fondamentale d'Orléans.
Bat. 3IA, rue Léonard de Vinci. 45067 Orléans, France

Abstract. We present in this paper an automatic way to combine any first-order theory T with the theory of finite or infinite trees. First of all, we present a new class of theories that we call *zero-infinite-decomposable* and show that every decomposable theory T accepts a decision procedure in the form of six rewriting which for every first order proposition give either true or false in T. We present then the axiomatization T^* of the extension of T into trees and show that if T is flexible then its extension into trees T^* is zero-infinite-decomposable and thus complete. The flexible theories are theories having elegant properties which enable us to eliminate quantifiers in particular cases.

1 Introduction

The theory of finite or infinite trees plays a fundamental role in programming. Recall that Alain Colmerauer has described the execution of Prolog II, III and IV programs in terms of solving equations and disequations in this theory [6, 9, 2]. He has first introduced in Prolog II the unification of infinite terms together with a predicate of non-equality [8]. He has then integrated in Prolog III the domain of rational numbers together with the operations of addition and subtraction and a linear dense order relation without endpoints [5, 7]. He also gave a general algorithm to check the satisfiability of a system of equations, inequations and disequations on a combination of trees and rational numbers. Finally, in Prolog IV, the notions of list, interval and Boolean have been added [10, 2].

We present in this paper an idea of a general extension of the model of Prolog IV by allowing the user to incorporate universal and existential quantifiers to Prolog closes and to decide the validity or not validity of any first-order proposition (sentence) in a combination of trees and first-order theories. For that:

(1) we give an automatic way to generate the axiomatization of the combination of any first order theory T with the theory of finite or infinite trees,

(2) we present simple conditions on T and only on T so that the combination of T with the theory of finite or infinite trees is complete and accepts a decision algorithm in the form of six rewriting rules which for every proposition give either true or false.

One of major difficulties in this work resides in the fact that the two theories can possibly have non-disjoint signatures. Moreover, the theory of finite or infinite trees does not accept full elimination of quantifiers.

J. Calmet, T. Ida, and D. Wang (Eds.): AISC 2006, LNAI 4120, pp. 53–67, 2006.
© Springer-Verlag Berlin Heidelberg 2006

The emergence of general constraint-based paradigms, such as constraint logic programming [19], constrained resolution [3] and what is generally referred to as *theory reasoning* [1], rises the problem of combining decision procedure for solving general first order constraints. Initial combinations results were provided by R. Shostak in [27] and in [28]. Shostak's approach is limited in scope and not very modular. A rather general and completely modular combination method was proposed by G. Nelson and D. Oppen in [21] and then slightly revised in [22]. Given, for $i = 1, ..., n$ a procedure P_i that decides the satisfiability of quantifier-free formulas in the theory $T_1 \cup ... \cup T_n$. A declarative and non-deterministic view of the procedure was suggested by Oppen in [24]. In [30], C. Tinelli and H.Harandi followed up on this suggestion describing a non-deterministic version of the Nelson-Oppen approach combination procedure and providing a simpler correctness proof. A similar approach had also been followed by C. Ringeissen in [26] which describes the procedure as a set of a derivation rules applied non-deterministically.

All the works mentioned above share one major restriction on the constraint languages of the component reasoners: they must have disjoint signatures, i.e. no function and relation symbols in common. (The only exception is the equality symbol which is however regarded as a logical constant). This restriction has proven really hard to lift. A testament of this is that, more than two decades after Nelson and Oppen's original work, their combination results are still state of the art.

Results on non-disjoint signatures do exists, but they are quite limited. To start with, some results on the union of non-disjoint equational theories can be obtained as a by-product of the research on the combination of term rewriting systems. Modular properties of term rewriting systems have been extensively investigated (see the overviews in [23] and [18]). Using some of these properties it is possible to derive combination results for the word problem in the union of equational theories sharing constructors[1]. Outside the work on modular term rewriting, the first combination result for the word problem in the union of non-disjoint constraint theories were given in [16] as a consequence of some combination techniques based on an adequate notion of (shared) constructors. C. Ringeissen used similar ideas later in [25] to extend the Nelson-Oppen method to theories sharing constructors in a sense closed to that of [16].

Recently, C. Tinelli and C. Ringeissen have provided some sufficient conditions for the Nelson-Oppen combinability by using a concept of stable Σ-freeness [29], a natural extension of Nelson-Oppen's stable-infiniteness requirement for theories with non-disjoint signatures. As for us, we present a natural way to combine the theory of finite or infinite trees with any first order theory T which can possibly have a non-disjoint signature. A such theory is denoted by T^* and does not accept full elimination of quantifiers which makes the decision procedure

---

[1] The word problem in an equational theory T is the problem of determining whether a given equation $s = t$ is valid in T, or equivalently, whether a disequation $\neg(s = t)$ is (un)satisfiable in T. In a term rewriting system, a constructor is a function symbol that does not appear as the top symbol of a rewrite rule's left-hand-side.

not evident. To show the completeness of T^* we give simple conditions on T and only on T so that its combination with the theory of finite or infinite trees, i.e. T^*, is complete and accepts a decision procedure which using only six rewriting rules is able to decide the validity or not validity of any first order constraints in T^*.

This paper is organized in five sections followed by a conclusion. This introduction is the first section. In Section 2, we recall the basic definitions of signature, model, theory and vectorial quantifier. In Section 3, after having presented a new quantifier called *zero-infinite*, we present a new class of theories that we call *zero-infinite-decomposable*. The main idea behind this class of theories consists in decomposing each quantified conjunction of atomic formulas into three embedded sequences of quantifications having very particular properties, which can be expressed with the help of three special quantifiers denoted by $\exists$?, $\exists!$, $\exists_{o\,\infty}^{\Psi(u)}$ and called *at-most-one, exactly-one, zero-infinite*. We end this section by six rewriting rules which for every zero-infinite-decomposable theory T and for every proposition φ give either true or false in T. The correctness of our algorithm shows the completeness of the zero-infinite-decomposable theories. In Section 4, we give a general way to generate the axioms of T^* using those of T and show that if T is flexible then T^* is zero-infinite-decomposable and thus complete. The flexible theories are theories having elegant properties which enable us to eliminate quantifiers in particular cases. We end this section by some fundamental flexible theories.

The zero-infinite-decomposable theories, the decision procedure in zero-infinite-decomposable theories, the axiomatization of T^* and the flexible theories are our main contribution in this paper. A full version of this paper with detailed proofs can be found in [13] and in the Ph.D thesis of K. Djelloul [14] (chapters 3 and 4).

2   Preliminaries

Let V be an infinite set of *variables*. Let S be a set of symbols, called *a signature* and partitioned into two disjoint sub-sets: the set F of *function symbols* and the set R of *relation symbols*. To each function symbol and relation is linked a non-negative integer n called *its arity*. An n-ary symbol is a symbol of arity n. A 0-ary function symbol is called *a constant*.

An *S-formula* is an expression of the one of the eleven following forms:

$$s = t, \; rt_1 \ldots t_n, \; \text{true}, \; \text{false},$$
$$\neg\varphi, \; (\varphi \wedge \psi), \; (\varphi \vee \psi), \; (\varphi \rightarrow \psi), \; (\varphi \leftrightarrow \psi), \tag{1}$$
$$(\forall x\, \varphi), \; (\exists x\, \varphi),$$

with $x \in V$, r an n-ary relation symbol taken from F, φ and ψ shorter S-formulas, s, t and the t_i's S-terms, that are expressions of the one of the two following forms

$$x, \; ft_1 \ldots t_n,$$

with x taken from V, f an n-ary function symbol taken from F and the t_i shorter S-terms

The S-formulas of the first line of (1) are called *atomic*, and *flat* if they are of the one of the five following forms:

$$\text{true, false, } x_0 = fx_1...x_n, \; x_0 = x_1, \; rx_1...x_n,$$

with the x_i's possibly non-distinct variables taken from V, $f \in F$ and $r \in R$.

If φ is an S-formula then we denote by $var(\varphi)$ the set of the free variables of φ. An S-*proposition* is an S-formula without free variables. The set of the S-terms and the S-formulas represent *a first-order language with equality*.

An S-*structure* is a couple $M = (D, F)$, where D is a non-empty set of *individuals* of M and F a set of functions and relations in D. We call *instantiation* or *valuation* of an S-formula φ by individuals of M, the $(S \cup D)$-formula obtained from φ by replacing each free occurrence of a free variable x in φ by the same individual i of D and by considering each element of D as 0-ary function symbol.

An S-*theory* T is a set of S-propositions. We say that the S-structure M is a model of T if for each element φ of T, $M \models \varphi$. If φ is an S-formula, we write $T \models \varphi$ if for each S-model M of T, $M \models \varphi$. A theory T is called *complete* if for every proposition φ, one and only one of the following properties holds: $T \models \varphi$, $T \models \neg\varphi$.

Let M be a model. Let $\bar{x} = x_1 \dots x_n$ and $\bar{y} = y_1 \dots y_n$ be two words on **v** of the same length. Let φ and $\varphi(\bar{x})$ be M-formulas. We write

$$
\begin{aligned}
&\exists \bar{x}\, \varphi && \text{for } \exists x_1...\exists x_n\, \varphi, \\
&\forall \bar{x}\, \varphi && \text{for } \forall x_1...\forall x_n\, \varphi, \\
&\exists ?\bar{x}\, \varphi(\bar{x}) && \text{for } \forall \bar{x} \forall \bar{y}\, \varphi(\bar{x}) \wedge \varphi(\bar{y}) \rightarrow \bigwedge_{i \in \{1,...,n\}} x_i = y_i, \\
&\exists !\bar{x}\, \varphi && \text{for } (\exists \bar{x}\, \varphi) \wedge (\exists ?\bar{x}\, \varphi).
\end{aligned}
$$

The word $\bar{x}$, which can be the empty word ε, is called *vector of variables*. Note that the formulas $\exists ?\varepsilon\varphi$ and $\exists !\varepsilon\varphi$ are respectively equivalent to *true* and to φ in any model M.

3   Zero-Infinite-Decomposable Theories

In this section, let us fix a signature $S^* = F^* \cup R^*$. Thus, we can allow ourselves to remove the prefix S^* from the following words: formulas, equations, theories and models. We will also use the abbreviation *wnfv* for *"without new free variables"*. We say that an S-formula φ is equivalent to a wnfv S-formula ψ in T if $T \models \varphi \leftrightarrow \psi$ and ψ does not contain other free variables than those of φ.

3.1   Zero-Infinite Quantifier [15]

Let M be a model and T a theory. Let $\Psi(u)$ be a set of formulas having at most one free variable u. Let φ and φ_j be M-formulas. We write

$$M \models \exists_o^{\Psi(u)} {}_\infty x\, \varphi(x), \tag{2}$$

if for each instantiation $\exists x\, \varphi'(x)$ of $\exists x\, \varphi(x)$ by individuals of M one of the following properties holds:

- the set of the individuals i of M such that $M \models \varphi'(i)$, is infinite,
- for every finite sub-set $\{\psi_1(u), .., \psi_n(u)\}$ of elements of $\Psi(u)$, the set of the individuals i of M such that $M \models \varphi'(i) \wedge \bigwedge_{j \in \{1,...,n\}} \neg \psi_j(i)$ is infinite.

We write $T \models \exists_{o\,\infty}^{\Psi(u)} x\, \varphi(x)$, if for every model M of T we have $M \models \exists_{o\,\infty}^{\Psi(u)} x\, \varphi(x)$.

This infinite quantifier holds only for infinite models, i.e. models whose set of elements are infinite. Note that if $\Psi(u) = \{false\}$ then (2) simply means that if $M \models \exists x\, \varphi(x)$ then M contains an infinity of individuals i such that $M \models \varphi(i)$. The intuitions behind this definition come from an aim to eliminate a conjunction of the form $\bigwedge_{i \in I} \neg \psi_i(x)$ in complex formulas of the form $\exists \bar{x}\, \varphi(x) \wedge \bigwedge_{i \in I} \neg \psi_i(x)$ where I is a finite (possibly empty) set and the $\psi_i(x)$ are formulas which do not accept full elimination of quantifiers.

3.2   Zero-Infinite-Decomposable Theory [15]

A theory T is called *zero-infinite-decomposable* if there exists a set $\Psi(u)$ of formulas, having at least one free variable u, a set A of formulas closed under conjunction, a set A' of formulas of the form $\exists \bar{x}\alpha$ with $\alpha \in A$, and a sub-set A'' of A such that:

1. every formula of the form $\exists \bar{x}\, \alpha \wedge \psi$, with $\alpha \in A$ and ψ a formula, is equivalent in T to a wnfv formula of the form:

$$\exists \bar{x}'\, \alpha' \wedge (\exists \bar{x}''\, \alpha'' \wedge (\exists \bar{x}'''\, \alpha''' \wedge \psi)),$$

   with $\exists \bar{x}'\, \alpha' \in A'$, $\alpha'' \in A''$, $\alpha''' \in A$ and $T \models \forall \bar{x}''\alpha'' \rightarrow \exists! \bar{x}'''\alpha'''$,
2. if $\exists \bar{x}'\alpha' \in A'$ then $T \models \exists?\bar{x}'\,\alpha'$ and for every free variable y in $\exists \bar{x}'\alpha'$, one at least of the following properties holds:
   - $T \models \exists?y\bar{x}'\,\alpha'$,
   - there exists $\psi(u) \in \Psi(u)$ such that $T \models \forall y\,(\exists \bar{x}'\,\alpha') \rightarrow \psi(y)$,
3. if $\alpha'' \in A''$ then
   - the formula $\neg \alpha''$ is equivalent in T to a wnfv formula of the form $\bigvee_{i \in I} \alpha_i$ with $\alpha_i \in A$,
   - for every x'', the formula $\exists x''\alpha''$ is equivalent in T to a wnfv formula which belongs to A'',
   - for every variable x'', $T \models \exists_{o\,\infty}^{\Psi(u)} x''\, \alpha''$,
4. every conjunction of flat formulas is equivalent in T to a wnfv disjunction of elements of A,
5. if the formula $\exists \bar{x}'\alpha' \wedge \alpha''$ with $\exists \bar{x}'\alpha' \in A'$ and $\alpha'' \in A''$ has no free variables then $\bar{x}$ is the empty vector, α' is the formula *true* and α'' is either the formula *true* or *false*.

3.3   A Decision Procedure for Zero-Infinite-Decompoable Theories [14]

Let T be a zero-infinite-decomposable theory. The sets $\Psi(u)$, A, A' and A'' are known and fixed.

Definition 3.3.1. *A normalized formula φ of depth $d \geq 1$ is a formula of the form $\neg(\exists \bar{x}\, \alpha \wedge \bigwedge_{i \in I} \varphi_i)$, where I is a finite possibly empty set, $\alpha \in A$, the φ_i are normalized formulas of depth d_i with $d = 1 + \max\{0, d_1, ..., d_n\}$, and all the quantified variables have distinct names and different form those of the free variables.*

Property 3.3.2. *Every formula is equivalent in T to a normalized formula.*

Definition 3.3.3. *A final formula is a normalized formula of the form*

$$\neg(\exists \bar{x}'\, \alpha' \wedge \alpha'' \wedge \bigwedge_{i \in I} \neg(\exists \bar{y}_i'\, \beta_i')), \tag{3}$$

with I a finite possibly empty set, $\exists \bar{x}' \alpha' \in A'$, $\alpha'' \in A''$, $\exists \bar{y}_i' \beta_i' \in A'$, α'' is different from the formula false, all the β_i''s are different from the formulas true and false.

Property 3.3.4. *Let φ be a conjunction of final formulas without free variables. The conjunction φ is either the formula true or the formula $\neg$true.*

Property 3.3.5. *Every normalized formula is equivalent in T to a conjunction of final formulas.*

Proof. We give bellow six rewriting rules which transform a normalized formula of any depth d into a conjunction of final formulas equivalent in T. To apply the rule $p_1 \Longrightarrow p_2$ on a normalized formula p means to replace in p, the subformula p_1 by the formula p_2, by considering the connector $\wedge$ associative and commutative.

(1) $\neg \begin{bmatrix} \exists \bar{x}\, \alpha \wedge \varphi \wedge \\ \neg(\exists \bar{y}\, \text{true}) \end{bmatrix} \quad \Longrightarrow \quad \text{true}$

(2) $\neg \begin{bmatrix} \exists \bar{x}\, \alpha \wedge \text{false} \wedge \varphi \end{bmatrix} \quad \Longrightarrow \quad \text{true}$

(3) $\neg \begin{bmatrix} \exists \bar{x}\, \alpha \wedge \\ \bigwedge_{i \in I} \neg(\exists \bar{y}_i\, \beta_i) \end{bmatrix} \quad \Longrightarrow \neg \begin{bmatrix} \exists \bar{x}' \bar{x}''\, \alpha' \wedge \alpha'' \wedge \\ \bigwedge_{i \in I} \neg(\exists \bar{x}'''\, \bar{y}_i\, \alpha''' \wedge \beta_i) \end{bmatrix}$

(4) $\neg \begin{bmatrix} \exists \bar{x}\, \alpha \wedge \varphi \wedge \\ \neg(\exists \bar{y}'\, \beta' \wedge \beta'') \end{bmatrix} \quad \Longrightarrow \quad \begin{bmatrix} \neg(\exists \bar{x}\, \alpha \wedge \varphi \wedge \neg(\exists \bar{y}'\, \beta')) \wedge \\ \bigwedge_{i \in I} \neg(\exists \bar{x}\bar{y}'\, \alpha \wedge \beta' \wedge \beta_i'' \wedge \varphi) \end{bmatrix}$

(5) $\neg \begin{bmatrix} \exists \bar{x}\, \alpha \wedge \\ \bigwedge_{i \in I} \neg(\exists \bar{y}_i'\, \beta_i') \end{bmatrix} \quad \Longrightarrow \neg \begin{bmatrix} \exists \bar{x}'\, \alpha' \wedge \alpha_*'' \\ \bigwedge_{i \in I'} \neg(\exists \bar{y}_i'\, \beta_i') \end{bmatrix}$

(6) $\neg \begin{bmatrix} \exists \bar{x}\, \alpha \wedge \varphi \wedge \\ \neg \begin{bmatrix} \exists \bar{y}'\, \beta' \wedge \beta'' \wedge \\ \bigwedge_{i \in I} \neg(\exists \bar{z}_i'\, \delta_i') \end{bmatrix} \end{bmatrix} \quad \Longrightarrow \begin{bmatrix} \neg(\exists \bar{x}\, \alpha \wedge \varphi \wedge \neg(\exists \bar{y}'\, \beta' \wedge \beta'')) \wedge \\ \bigwedge_{i \in I} \neg(\exists \bar{x}\bar{y}'\bar{z}_i\, \alpha \wedge \beta' \wedge \beta'' \wedge \delta_i' \wedge \varphi) \end{bmatrix}$

with α an element of A, φ a conjunction of normalized formulas and I a finite possibly empty set. In the rule (3), the formula $\exists \bar{x}\, \alpha$ is equivalent in T to a decomposed formula of the form $\exists \bar{x}' \, \alpha' \wedge (\exists \bar{x}'' \, \alpha'' \wedge (\exists \bar{x}''' \, \alpha'''))$ with $\exists \bar{x}' \, \alpha' \in A'$, $\alpha'' \in A''$, $\alpha''' \in A$, $T \models \forall \bar{x}'' \alpha'' \rightarrow \exists! \bar{x}''' \alpha'''$ and $\exists \bar{x}''' \, \alpha'''$ different from $\exists \varepsilon \, true$. All the β_i's belong to A. In the rule (4), the formula $\exists \bar{x} \, \alpha$ is equivalent in T to a decomposed formula of the form $\exists \bar{x}' \, \alpha' \wedge (\exists \bar{x}'' \, \alpha'' \wedge (\exists \varepsilon \, true))$ with $\exists \bar{x}' \, \alpha' \in A'$ and $\alpha'' \in A''$. The formula $\exists \bar{y}' \, \beta'$ belongs to A'. The formula β'' belongs to A'' and is different from the formula $true$. Moreover, $T \models (\neg \beta'') \leftrightarrow \bigvee_{i \in I} \beta_i''$ with $\beta_i'' \in A$. In the rule (5), the formula $\exists \bar{x} \, \alpha$ is not of the form $\exists \bar{x} \, \alpha_1 \wedge \alpha_2$ with $\exists \bar{x} \, \alpha_1 \in A'$ and $\alpha_2 \in A''$, and is equivalent in T to a decomposed formula of the form $\exists \bar{x}' \, \alpha' \wedge (\exists \bar{x}'' \, \alpha'' \wedge (\exists \varepsilon \, true))$ with $\exists \bar{x}' \, \alpha' \in A'$ and $\alpha'' \in A''$. Each formula $\exists \bar{y}_i' \, \beta_i'$ belongs to A'. The set I' is the set of the $i \in I$ such that $\exists \bar{y}_i' \beta_i'$ does not contain free occurrences of a variable of $\bar{x}''$. Moreover, $T \models (\exists \bar{x}'' \alpha'') \leftrightarrow \alpha_*''$ with $\alpha_*'' \in A''$. In the rule (6), $I \neq \emptyset$, $\exists \bar{y}' \, \beta' \in A'$, $\exists \bar{z}_i' \, \delta_i' \in A'$ and $\beta'' \in A''$.

Let ψ be a formula without free variables, the *decision* of ψ proceeds as follows:

1. Transform the formula ψ into a normalized formula φ which is equivalent to ψ in T.
2. While it is possible, apply the rewriting rules on φ. At the end, we obtain a conjunction ϕ of final formulas.

According to Property 3.3.5, the application of the rules on a formula ψ without free variables produces a wnfv conjunction ϕ of final formulas, i.e. a conjunction ϕ of final formulas without free variables. According to Property 3.3.4, ϕ is either the formula $true$, or the formula $\neg true$, i.e. the formula $false$.

Corollary 3.3.6. *If T is zero-infinite-decomposable then T is complete and accepts a decision procedure in the form of six rewriting rules which for every proposition give either true or false in T.*

4 Extension of First-Order Theories into Trees

4.1 The Structure of Finite or Infinite Trees

Trees are well known objects in the computer science world. Here are some of them:

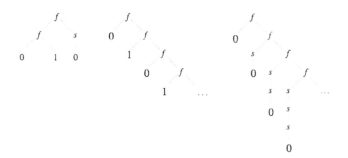

Their nodes are labeled by the symbols 0,1,s,f, of respective arities 0,0,1,2, taken from a set F of functional symbols which we assume to be infinite. While the first tree is a *finite tree* (it has a finite set of nodes), the two others are *infinite trees* and have an infinite set of nodes. We denote by A the set of all trees[2] constructed on F.

We introduce in A a set of construction operations[3], one for each element $f \in F$, which is the mapping $(a_1, ..., a_n) \rightarrow b$, where n is the arity of f and b the tree whose initial node is labeled by f and the sequence of suns is $(a_1, ..., a_n)$, and which be schematized as:

We thus obtain *the structure of finite or infinite trees* constructed on F, which we denote by (A, F).

4.2   Theory of Finite or Infinite Trees

Let S be a signature containing only an infinite set of function symbols F. Michael Maher has introduced the S-theory of finite or infinite trees [20]. The axiomatization of this S-theory is the set of the S-propositions of the one of the following forms:

$$1 \quad \forall \bar{x} \forall \bar{y}\ f\bar{x} = f\bar{y} \rightarrow \bigwedge_i x_i = y_i,$$
$$2 \quad \forall \bar{x} \forall \bar{y}\ \neg f\bar{x} = g\bar{y},$$
$$3 \quad \forall \bar{x} \exists! \bar{z}\ \bigwedge_i z_i = f_i(\bar{z}, \bar{x}),$$

with f and g two distinct function symbols taken from F, $\bar{x}$ a vector of variables x_i, $\bar{y}$ a vector of variables y_i, $\bar{z}$ a vector of distinct variables z_i and $f_i(\bar{x}, \bar{z})$ an S-term which begins with an element of F followed by variables taken from $\bar{x}\bar{z}$.

The first axiom is called *axiom of explosion*, the second one is called *axiom of conflict of symbols* and the last one is called *axiom of unique solution*.

We show that this theory has as model the structure of finite or infinite trees [12]. For example, using axiom 3, we have $T \models \exists! xy\ x = f1y \wedge y = f0x$. The individuals x and y represents the two following trees in the structure of finite or infinite trees:

---

[2] More precisely, we define first a node to be a word constructed on the set of strictly positive integers. A *tree* a on F, is then a mapping of type $a : E \rightarrow F$, where E is a non-empty set of nodes, each one $i_1 \ldots i_k$ (with $k \geq 0$) satisfies two conditions: (1) if $k > 0$ then $i_1 \ldots i_{k-1} \in E$ and (2) if the arity of $a(i_1 \ldots i_k)$ is n, then the set of the nodes E of the form $i_1 \ldots i_k i_{k+1}$ is obtained by giving to i_{k+1} the values $1, ..., n$.

[3] In fact, the construction operation linked to the n-ary symbol f of F is the mapping $(a_1, ..., a_n) \rightarrow b$, where the a_i's are any trees and b is the tree defined as follows from the a_i's and their set of nodes E_i's: the set E of nodes of a is $\{\varepsilon\} \cup \{ix | x \in E_i$ and $i \in \{1, ..., n\}$ and, for each $x \in E$, if $x = \varepsilon$, then $a(x) = f$ and if x is of the form iy, with i being an integer, $a(x) = a_i(y)$.

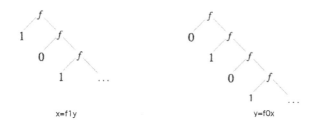

$$x=f1y \qquad\qquad y=f0x$$

4.3 Axiomatization of the Theory $T + Tree$ or T^*

Let us fix now a signature S containing a set F of function symbols and a set R of relation symbols, as well as a signature S^* containing:

- an infinite set $F^* = F \cup F_A$ where F_A is an infinite set of function symbols disjoint from F.
- a set $R^* = R \cup \{p\}$ of relation symbols, containing R, and an 1-ary relation symbol p.

Let T be an S-theory. The extension of the S-theory T into trees is the S^*-theory denoted by T^* and whose set of axioms is the infinite set of the following S^*-propositions, with $\bar{x}$ a vector of variables x_i and $\bar{y}$ a vector of variables y_i:

1. Explosion: for all $f \in F^*$:

$$\forall\bar{x}\forall\bar{y} \; \neg pf\bar{x} \wedge \neg pf\bar{y} \wedge f\bar{x} = f\bar{y} \rightarrow \bigwedge_i x_i = y_i$$

2. Conflict of symbols: Let f and g be two distinct function symbols taken from F^*:

$$\forall\bar{x}\forall\bar{y} \; f\bar{x} = g\bar{y} \rightarrow pf\bar{x} \wedge pg\bar{y}$$

3. Unique solution

$$\forall\bar{x}\forall\bar{y} \; \left(\bigwedge_i px_i\right) \wedge \left(\bigwedge_j \neg py_j\right) \rightarrow \exists!\bar{z} \bigwedge_k (\neg pz_i \wedge z_k = t_k(\bar{x}, \bar{y}, \bar{z}))$$

where $\bar{z}$ is a vector of distinct variables z_i, $t_k(\bar{x}, \bar{y}, \bar{z})$ an S^*-term which begins by a function symbol $f_k \in F^*$ followed by variables taken from $\bar{x}, \bar{y}, \bar{z}$, moreover, if $f_k \in F$, then the S^*-term $t_k(\bar{x}, \bar{y}, \bar{z})$ contains at least a variable from $\bar{y}$ or $\bar{z}$

4. Relations of R: for all $r \in R$,

$$\forall\bar{x} \; r\bar{x} \rightarrow \bigwedge_i px_i$$

5. Operations of F: for all $f \in F$,

$$\forall\bar{x} \; pf\bar{x} \leftrightarrow \bigwedge_i px_i$$

(if f is 0-ary then this axiom is written pf)

6. Elements not in T: for all $f \in F^* - F$,

$$\forall \bar{x} \; \neg p f \bar{x}$$

7. Existence of an element satisfying p (only if F does not contain 0-ary function symbols):

$$\exists x \; px,$$

8. Extension into trees of the axioms of T: all axioms obtained by the following transformations of each axiom φ of T : While it is possible replace all sub-formula of φ which is of the form $\exists \bar{x} \; \psi$, but not of the form $\exists \bar{x} \, (\bigwedge p x_i) \wedge \psi'$, by $\exists \bar{x} \, (\bigwedge p x_i) \wedge \psi$ and every sub-formula of φ which is of the form $\forall \bar{x} \; \psi$, but not of the form $\forall \bar{x} \, (\bigwedge p x_i) \rightarrow \psi'$, by $\forall \bar{x} \, (\bigwedge p x_i) \rightarrow \psi$.

4.4   Example: Extension of Linear Dense Order Relations Without Endpoints into Trees

Let F be an empty set of function symbols and let R be a set of relation symbols containing only the relation symbol $<$ of arity 2. If t_1 and t_2 are terms, then we write $t_1 < t_2$ for $< (t_1, t_2)$. Let T_{ord} be the theory of linear dense order relation without endpoints, whose signature is $S = F \cup R$ and whose axioms are the following propositions:

1 $\forall x \; \neg x < x,$
2 $\forall x \forall y \forall z \, (x < y \wedge y < z) \rightarrow x < z,$
3 $\forall x \forall y \; x < y \vee x = y \vee y < x,$
4 $\forall x \forall y \; x < y \rightarrow (\exists z \; x < z \wedge z < y),$
5 $\forall x \, \exists y \; x < y,$
6 $\forall x \, \exists y \; y < x.$

Let now F^* be an infinite set of function symbols and $R^* = \{<, p\}$ a set of relation symbols containing the 2-ary relation symbol $<$ and the 1-ary relation symbol p. Let S^* be the signature $F^* \cup R^*$. According to the transformations of axioms given in Section 4.3, the axiomatization of the extension of the theory T_{ord} into trees is the S^*-theory T_{ord}^* whose axioms are the following propositions:

1 $\forall \bar{x} \forall \bar{y} \; \neg p f \bar{x} \wedge \neg p f \bar{y} \wedge f \bar{x} = f \bar{y} \rightarrow \bigwedge_i x_i = y_i$
2 $\forall \bar{x} \forall \bar{y} \; f \bar{x} = g \bar{y} \rightarrow p f \bar{x} \wedge p g \bar{y}$
3 $\forall \bar{x} \forall \bar{y} \, (\bigwedge_i p x_i) \wedge (\bigwedge_j \neg p y_j) \rightarrow \exists ! \bar{z} \bigwedge_k (\neg p z_i \wedge z_k = f_k(\bar{x}, \bar{y}, \bar{z}))$
4 $\forall x \forall y \; x < y \rightarrow (px \wedge py),$
5 $\forall \bar{x} \; \neg p f \bar{x},$
6 $\exists x \; px,$
7 $\forall x \; px \rightarrow \neg x < x,$
8 $\forall x \forall y \forall z \; px \wedge py \wedge pz \rightarrow ((x < y \wedge y < z) \rightarrow x < z),$
9 $\forall x \forall y \, (px \wedge py) \rightarrow (x < y \vee x = y \vee y < x),$
10 $\forall x \forall y \, (px \wedge py) \rightarrow (x < y \rightarrow (\exists z \; pz \wedge x < z \wedge z < y)),$
11 $\forall x \; px \rightarrow (\exists y \; py \wedge x < y),$
12 $\forall x \; px \rightarrow (\exists y \; py \wedge y < x),$

where f and g are distinct function symbols taken from F*, x, y, z variables, $\bar{x}$ a vector of variables x_i, $\bar{y}$ a vector of variables y_i, $\bar{z}$ a vector of distinct variables z_i and $f_k(\bar{x}, \bar{y}, \bar{z})$ a term which begins by an element f_k of F* followed by variables taken from $\bar{x}\bar{y}\bar{z}$.

5   Completeness of T^*

We have given a general axiomatization of T^* using the axioms of T, what about the completeness of T^*? Are all the extensions into trees complete theories? While in [15] we have shown the completeness of a combination of trees and rational numbers, in this paper the challenge is to use general properties that hold not only for rational numbers but for a large set of different theories T_i and that make T_i^* zero-infinite-decomposable and thus complete.

Let $S = F \cup R$ be a signature and T an S-theory. Let $S^* = F^* \cup R^*$ be another signature with F^* an infinite set of function symbols containing F and $R^* = R \cup \{p\}$. Let T^* be the S^*-theory of the extension of T into trees. Suppose that the variables of V are ordered by a linear dense order relation without endpoints denoted $\succ$.

5.1   Flexible Theory

Definition 5.1.1. *We call* leader *of an S-equation α the greatest variable x in α, according to the order $\succ$, such that $T \models \exists! x \alpha$.*

Definition 5.1.2. *A conjunction of S-atomic formulas α is called* formatted *in T if*

- *α does not contain sub-formulas of the form $f_1 = f_2$ or $r f_1 ... f_n$ or $y = x$, where all the f_i's are 0-ary function symbols taken from F, $r \in R$ and $x \succ y$,*
- *each S-equation of α has a distinct leader which has no occurrences in other S-equations or S-relations of α,*
- *if α' is the conjunction of all the S-equations of α then for all $x \in var(\alpha')$ we have $T \models \exists? x\, \alpha'$.*

Definition 5.1.3. *The theory T is called* flexible *if for each conjunction α of S-equations and for each conjunction β of S-relations:*

1. *$\alpha \wedge \beta$ is equivalent in T to a formatted conjunction of atomic formulas wnfv,*
2. *the S-formula $\neg\beta$ is equivalent in T to a disjunction wnfv of S-equations and S-relations,*
3. *for all $x \in V$*
   - *the S-formula $\exists x\, \beta$ is equivalent in T to false, or to a wnfv conjunction of S-relations,*
   - *for all $x \in V$, we have $T \models \exists_0^{\{faux\}} {}_\infty x\, \beta$.*

Let us now present our main result

Theorem 5.1.4 *If T is flexible then T^* is zero-infinite-decomposable.*

5.2   Some Fundamental Flexible Theories

We present in this section the axiomatization of some fundamental flexible theories. Full proofs can be found in [14].

Infinite Clark equational theory: Let Cl be a theory together with an empty set of function and relation symbols and whose axioms is the infinite set of propositions of the following form:

$$(1_n) \quad \forall x_1...\forall x_n \exists y \, \neg(x_1 = y) \wedge ... \wedge \neg(x_n = y), \tag{4}$$

where all the variables $x_1...x_n$ are distinct and $(n \neq 0)$. The form (4) is called *diagram of axiom* and for each value of n there exists *an axiom* of Cl. This theory Cl has been introduced by Clark [4] and has an infinite set of models each one containing an infinite set of distinct individuals.

Additive rational or real numbers theory with addition and subtraction: Let $F = \{+, -, 0, 1\}$ be a set of function symbols of respective arities $2, 1, 0, 0$. Let $R = \emptyset$ be an empty set of relation symbols. The theory Ra of additive rational or real numbers together with addition and subtraction consists in the infinite set of propositions of the following form:

1   $\forall x \forall y \, x + y = y + x$,
2   $\forall x \forall y \forall z \, x + (y + z) = (x + y) + z$,
3   $\forall x \, x + 0 = x$,
4   $\forall x \, x + (-x) = 0$,
5_n $\forall x \, n.x = 0 \rightarrow x = 0$,
6_n $\forall x \, \exists! y \, n.y = x$,
7   $\forall x \forall y \forall z \, (x = y) \leftrightarrow (x + z = y + z)$,
8   $\neg(0 = 1)$.

with n an non-null integer. This theory has two usual models: rational numbers Q with addition and subtraction in Q and real numbers R with addition and subtraction in R.

Linear dense order theory without endpoints: Let F be an empty set of function symbols and R a set of relation symbols containing only the binary relation symbol $<$. The theory T_{ord} be the theory of the linear dense order without endpoints consists in the set of propositions of the following form:

1 $\forall x \, \neg x < x$,
2 $\forall x \forall y \forall z \, (x < y \wedge y < z) \rightarrow x < z$,
3 $\forall x \forall y \, x < y \vee x = y \vee y < x$,
4 $\forall x \forall y \, x < y \rightarrow (\exists z \, x < z \wedge z < y)$,
5 $\forall x \, \exists y \, x < y$,
6 $\forall x \, \exists y \, y < x$.

Ordered additive rational or real numbers theory with addition and subtraction: Let $F = \{+, -, 0, 1\}$ be a set of function symbols of respective arities $2, 1, 0, 0$. Let $R = \{<\}$ be a set of relation symbols containing only the binary relation symbol $<$. The theory T_{ad} of ordered additive rational or real numbers theory with addition and subtraction consists in the infinite set of propositions of the following form:

1   $\forall x \forall y \, x + y = y + x,$
2   $\forall x \forall y \forall z \, x + (y + z) = (x + y) + z,$
3   $\forall x \, x + 0 = x,$
4   $\forall x \, x + (-x) = 0,$
5_n $\forall x \, n.x = 0 \rightarrow x = 0, \quad (n \neq 0)$
6_n $\forall x \, \exists! y \, n.y = x, \quad (n \neq 0)$
7   $\forall x \, \neg x < x,$
8   $\forall x \forall y \forall z \, (x < y \land y < z) \rightarrow x < z,$
9   $\forall x \forall y \, (x < y \lor x = y \lor y < x),$
10   $\forall x \forall y \, x < y \rightarrow (\exists z \, x < z \land z < y),$
11   $\forall x \, \exists y \, x < y,$
12   $\forall x \, \exists y \, y < x,$
13   $\forall x \, \forall y \, \forall z \, x < y \rightarrow (x + z < y + z),$
14   $0 < 1.$

with n a non-null integer.

6   Conclusion

We have defined in this paper a general idea for the extension of the models of Prolog by giving an automatic way to combine any first order theory T with the theory of finite or infinite trees. To show the completeness of T^* we have introduced the flexible theories and have shown that if T is flexible then T^* zero-infinite-decomposable. The zero-infinite-decomposable theories are first order theories having elegant properties which enable us to decide the validity of any proposition using only six rewriting rules. The main idea behind this rules consists in a local decomposition of quantified conjunctions of hybrid atomic formulas, a partial elimination of quantifiers using the properties of the vectorial quantifiers, and a special distribution to decrease the depth of the formulas.

There exists many practical applications of the extensions into trees of first order theories. First-order constraints on trees can be expressed in a simpler way when they are in the extension into trees of another structure. For example, the constraints representing the moves in two players games introduced by Alain Colmerauer and Thi-Bich-Hanh Dao [11, 12] can be represented by a simpler constraint in the extension into trees of the integers together with the operations of addition and subtraction and a linear dense order relation.

On the other hand, our decision algorithm can decide the validity or not validity of big and complex propositions and can also be applied on formulas having free variables and produces in this case a Boolean combination of basic formulas which does not accept full elimination of quantifiers. Unfortunately, this algorithm is not able to detect formulas having free variables and being always equivalent to *false* or *true* in T^*. It does not warrant that a final formula having at least one free variable is neither true nor false in T^* and can not present the solutions of the free variables in a clear and explicit way. This is why our algorithm is called *decision procedure* and not a general algorithm for solving first order constraints. It would be interesting to transform our decision procedure into a general algorithm for solving any first order constraint in T^* and which presents the solutions of the free variables in a clear and explicit way, as it has been done in [11, 12] for the theory of finite trees and finite or infinite trees. This kind of algorithm needs another work completely different from this one, by introducing syntactic and semantic definitions much more complex than the definition of flexible theories given in this paper. The implementation of a such algorithm will enable us to extend the Prolog language by allowing the user to solve any complex first order constraint, with or without free variables, in many combinations of theories around trees.

Currently, we are trying to proof that every extension of a complete theory into trees is complete and may be zero-infinite-decomposable. For that, we expect to add new vectorial quantifiers in the decomposition such as $\exists^n$ which means *there exists n* and $\exists_{n,\infty}^{\Psi(u)}$ which means *there exists n or infinite*, in order to increase the size of the set of the zero-infinite-decomposable theories and may be get a much more simple definition than the one defined in this paper. We plan also with Thom Fruehwirth [17] to add to CHR a general mechanism to treat our normalized formulas. This will enable us to implement quickly and easily our algorithms and get a general idea on the expressiveness of first order constraints in combinations of trees and first order theories.

Acknowledgements. We thank Alain Colmerauer for our many discussions and his help in this work. We thank him too for the quality of her remarks and advice on how to improve the organisation of this paper. We dedicate to him this paper with our best wishes for a speedy recovery.

References

1. Baumgartner, T., Furbach, U. and Petermann, U. A unified approach to theory reasoning. Research report. pages 15-92. Univesitat Koblenz, Germany. 1992.
2. Benhamou, F., Colmerauer, A. and Van caneghem, M. the manuel of Prolog IV, PrologIA, Marseille, France, 1996.
3. Burckert, H.J. A resolution principle for constraint logics. Artificial intelligence. 66, pages 235-271. 1994.
4. Clark, K.L. Negation as failure. Logic and Data bases. Ed Gallaire, H. and Minker, J. Plenum Pub. 1978.
5. Colmerauer, A. Total precedence relations. JACM. Vol 17, pages 14-30. 1970
6. Colmerauer, A. Equations and inequations on finite and infinite trees. Proc. of the Int. Conf. on the Fifth Generation of Computer Systems, pages 85-99. Tokyo, 1984.
7. Colmerauer, A. Prolog in 10 figures. Communications of the ACM. Vol 28(12), pages 1296-1310. 1985.

8. Colmerauer, A. A view of the origins and development of Prolog. Communications of the ACM. Vol 31, pages 26-36. 1988.
9. Colmerauer, A. An introduction to Prolog III. *Communication of the ACM*, 33(7):68-90,1990.
10. Colmerauer, A. Specification of Prolog IV. LIM Technical Report. Marseille university, France. 1996.
11. Colmerauer A. and Dao, TBH., Expressiveness of full first-order constraints in the structure of finite or infinite trees, Constraints, Vol. 8, No. 3, 2003, pages 283-302.
12. Dao, TBH. Résolution de contraintes du premier ordre dans la théorie des arbres finis or infinis. Thèse d'informatique, Université de la Méditerranée, décembre 2000.
13. Djelloul, K. and Dao, TBH. Extension of first order theories into trees. Rapport de recherche LIFO RR-2006-07, http://www.univ-orleans.fr/lifo/rapports.php.
14. Djelloul, K. Complete theories around trees. Ph.D. thesis. Université de la Méditerranée, June 2006. (http://khalil.djelloul.free.fr)
15. Djelloul, K. About the combination of trees and rational numbers in a complete first-order theory. Proceeding of the 5th Int. Conf. on Frontiers of Combining Systems FroCoS 2005, Vienna Austria. LNAI, vol 3717, pages. 106–122.
16. Domenjoud, E., Klay, F. and Ringeissen, C. Combination techniques for non-disjoint equational theories. In proceedings of the 12th international conference on automated deduction. LNAI 814: 267-281. 1994.
17. Fruehwirth, T. and Abdelnnadher, S. Essentials of constraints programming. Springer Cognitive technologies. 2002.
18. Gramlich, B. On termination and confluence properties of disjoint and constructor sharing conditional rewrite systems. Theoretical computer science 165(1): 97-131. 1996.
19. Jaffar, J. and Maher, M. Constraint logic programing: a survey. Journal of logic programming. 19/20, pages 503-581. 1994.
20. Maher, M. Complete axiomatization of the structure of finite, rational and infinite trees. *Technical report, IBM*, 1988.
21. Nelson, G. and Oppen, D. Simplification by co-operating decision procedures. ACM Transaction on programming languages 1(2):245-251. 1979.
22. Nelson, G. Combining satisfiability procedures by equality sharing. Automated theorem proving: After 25 years vol 29: 201-211. 1984.
23. Ohelbush, E. Modular properties of composable term rewriting systems. Journal of symbolic computation, 20(1): 1-41. 1995.
24. Oppen, D. Complexity, convexity and combinations of theories. Theoretical computer science. Vol 12. 1980.
25. Ringeissen, C. co-operation of decision procedure for the satisfiability problem. In proceedings of the 1st international workshop on frontiers of combining systems. Kluwer pages 121-140. 1996.
26. Ringeissen, C. Combinaison de resolution de contraintes. These de doctorat, universite de Nancy. 1993.
27. Shostak, R. A practical decision procedure for arithmetic with function symbols. Journal of the ACM 26(2): 351-360. 1979.
28. Shostak, R. Deciding combinations of theories. Journal of the ACM 31: 1-12. 1984.
29. Tinelli, C. and Ringeissen, C. Unions of non-disjoint theories and combinations of satisfiability procedures. Theor. Comput. Sci. 290 (1): 291-353. 2003.
30. Tinelli, C. and Harandi, M. A new correctness proof for the Nelson-Oppen combination procedure. In proceedings of the 1st international workshop on frontiers of combining systems. Kluwer pages 103-120. 1996.

The Confluence Problem for Flat TRSs

Ichiro Mitsuhashi[1], Michio Oyamaguch[1], and Florent Jacquemard[2]

[1] Mie University
{ichiro, mo}@cs.info.mie-u.ac.jp
[2] INRIA-Futurs and LSV, UMR CNRS ENS Cachan
florent.jacquemard@lsv.ens-cachan.fr

Abstract. We prove that the properties of reachability, joinability and confluence are undecidable for flat TRSs. Here, a TRS is flat if the heights of the left and right-hand sides of each rewrite rule are at most one.

Keywords: Term rewriting system, Decision problem, Confluence, Flat.

1 Introduction

A term rewriting system (TRS) is a set of directed equations called rewrite rules. It defines a binary relation on terms by replacement of a subterm matching a left member of a rewrite rule by the corresponding right member. A TRS is called *confluent* (or Church-Rosser) if any two terms obtained, by the rewriting relation, from the same term are joinable. The confluence is a crucial property for the application of rewriting as a model for computation as it ensures the uniqueness of normal forms [1], and it has received much attention so far.

Confluence is undecidable in general, and even for restricted classes of TRS like monadic or semi-constructor TRSs [11]. On the other hand, decidability results have been established for several classes of TRSs, like e.g. ground (rewrite rules having no variables) TRSs [13, 4, 2], flat (left and right members of rewrite rules having height at most one) and rule-linear (a variable cannot occur more than once in a rewrite rule) TRSs [17], and more recently for flat and right-linear (a variable cannot occur more than once in a right member of rewrite rule) TRSs [6].

In this paper, we demonstrate that the above linearity restriction is necessary for decidability, showing that confluence is undecidable for flat TRSs, even with only one non-right-linear flat rewrite rule. A previous proof of this result has been published in [8]. However, we have found some technical flaws in this proof. This paper presents a correct and detailed undecidability proof, which is also significantly simpler than the one of [8].

The related properties of reachability (whether a given term can be reached from another given term by rewriting) and joinability (whether two given terms can be rewritten to the same term) are decidable for right-ground (right members of rewrite rules have no variable) TRSs [14], for right-linear monadic TRSs [15, 12], and for right-linear and finite-path-overlapping TRSs [16]. The latter two classes properly include the class of flat and right-linear TRSs. We show in this paper that

J. Calmet, T. Ida, and D. Wang (Eds.): AISC 2006, LNAI 4120, pp. 68–81, 2006.

reachability and joinability are undecidable if we drop the right-linearity condition, i.e. it is undecidable for general flat TRSs.

The paper is organized as follows: after giving the definitions and notations in Section 2, we show in Section 3 that reachability is undecidable for flat TRSs by reduction of the Post's correspondence problem. It follows as a corollary that joinability is also undecidable for the same class. Then, in Section 4, we show that confluence is undecidable for flat TRSs, by a reduction of reachability.

2   Preliminaries

We assume that the reader is familiar with the standard definitions of rewrite systems [5, 1] and we just recall here the main notations used in this paper.

Let ε be the empty string. Let X be a set of variables. Let F be a finite set of operation symbols graded by an arity function $\mathsf{ar}\colon F \to \mathbb{N}(= \{0, 1, 2, \cdots\})$, $F_n = \{f \in F \mid \mathsf{ar}(f) = n\}$. Let T be the set of terms built from X and F. A *substitution* is a finite mapping from X to T. As usual, we identify substitutions with their morphism extension to terms, and we use a postfix notation for the application of substitutions. We use x as a variable, f, h as function symbols, r, s, t as terms, θ as a substitution. A term s is ground if s has no variable. The *height* of a term is defined as follows: $\mathsf{height}(a) = 0$ if a is a variable or a constant and $\mathsf{height}(f(t_1, \ldots, t_n)) = 1 + \max\{\mathsf{height}(t_1), \ldots, \mathsf{height}(t_n)\}$ if $n > 0$.

A position in a term a sequence of positive integers, and positions are partially ordered by the prefix ordering $\geq$. Let $s|_p$ be the subterm of s at position p. Let $s \geq_{\mathsf{sub}} t$ if t is a subterm of s. For a position p and a term t, we use $s[t]_p$ to denote the term obtained from s by replacing the subterm $s|_p$ by t.

A *rewrite rule* $\alpha \to \beta$ is a directed equation over terms. A *TRS R* is a finite set of rewrite rules. A term s reduces to t at position p by the TRS R, denoted $s \xrightarrow[R]{p} t$ (p and R may be omitted), if $s|_p = \alpha\theta$ and $t = s[\beta\theta]_p$ for some rewrite rule $\alpha \to \beta$ and substitution θ. Let $\xrightarrow{=}$ be $\to \cup =$, $\leftarrow$ be the inverse of $\to$ and $\xrightarrow{*}$ be the reflexive and transitive closure of $\to$. The terms s and t are *joinable* if $s \xrightarrow{*} \cdot \xleftarrow{*} t$, which is denoted $s \downarrow t$. The term t is *reachable* from s if $s \xrightarrow{*} t$. The term r is *confluent* on the TRS R if for every peak $s \xleftarrow[R]{*} r \xrightarrow[R]{*} t$, we have $s \downarrow t$. The TRS R is *confluent* if every term is confluent on R. Let $\gamma\colon s_1 \xrightarrow{p_1} s_2 \cdots \xrightarrow{p_{n-1}} s_n$ be a *rewrite sequence*. This sequence is abbreviated by $\gamma\colon s_1 \xrightarrow{*} s_n$ and γ is called p-invariant if $p_i > p$ for every redex position p_i of γ; this is denoted by $\gamma\colon s_1 \xrightarrow{>p\ *} s_n$.

Definition 1. A rule $\alpha \to \beta$ is *flat* if $\mathsf{height}(\alpha) \leq 1$ and $\mathsf{height}(\beta) \leq 1$. A rule $\alpha \to \beta$ is *monadic* if $\mathsf{height}(\beta) \leq 1$. A term s is *shallow* if s is a variable or $s = f(s_1, \cdots, s_n)$ for some function symbol f and terms $s_1, \cdots, s_n$ such that every $s_i (1 \leq i \leq n)$ is either a variable or ground. A rule $\alpha \to \beta$ is *shallow* if both α and β are shallow. For $\mathcal{C} \in \{\text{flat, monadic, shallow}\}$, R is $\mathcal{C}$ if every rule in R is $\mathcal{C}$.

We are interested in the following decision problems:

Reachability. Given a TRS R and two terms s, t, does there exist a rewrite sequence $s \xrightarrow[R]{*} t$?

Joinability. Given a TRS R and two terms s, t, are s and t joinable, i.e., $s \downarrow_R t$?

Confluence. Given a TRS R, is every term confluent on R?

Definition 2. A *finite automaton* is a 5-tuple $(Q, \Sigma, \delta, F_Q, q_0)$ where Q is a finite set of states, Σ is a finite set of input symbols, $\delta : Q \times \Sigma \to Q$ is a function, $F_Q \subseteq Q$ is a finite set of final states, and $q_0 \in Q$ is the initial state.

3   Reachability and Joinability for Flat TRSs

In [8], it has been reported that reachability and joinability are also undecidable for flat TRSs. But, the undecidability proof of reachability contains a flaw. We propose a repaired proof of undecidability for reachability which is simpler than the former one. The proof is a reduction of the Post's Correspondence Problem (PCP) into the reachability of a constant 1 from a constant 0 using a a flat TRS R_1. This TRS, constructed from the given instance of PCP, is such that every rewrite sequence $0 \xrightarrow[R_1]{*} 1$ contains a representation of a solution of the PCP as a term t. This property is ensured, informally, by running separately several sub-TRS of R_1 on several copies of t, where all copies have different colors and are under a function symbol of arity 6 or 7. Moreover, some equality tests are performed during the rewrite sequence using R_1, by means of a flat rewrite rules in R_1 containing some non-linear variables.

Let $P = \{\langle u_i, v_i \rangle \in \Sigma^+ \times \Sigma^+ \mid 1 \leq i \leq k\}$ be an instance of PCP. The goal of the problem is to find a sequence of indices $i_1, \ldots, i_n$, possibly with repetitions, such that the concatenations $u_{i_1} \ldots, u_{i_n}$ and $v_{i_1} \ldots, v_{i_n}$ are equal. Note that the alphabet Σ is fixed. Let $l_P = \max_{1 \leq k \leq k}(|u_i|, |v_i|)$. Let $\_$ be a new symbol and $\Delta = \{1, \cdots, l_P\} \times (\Sigma \cup \{\_\})^2$. We shall use a product operator $\otimes$ which associates to two non-empty strings of Σ^+ of length smaller than or equal to l_P a word of Δ^* of length l_P as follows: $a_1 \cdots a_n \otimes a'_1 \cdots a'_m = \langle 1, a_1, a'_1 \rangle \cdots \langle l_P, a_{l_P}, a'_{l_P} \rangle$, where $a_1, \cdots, a_n, a'_1, \cdots, a'_m \in \Sigma$, $a_i = \_$ for all $i \in \{n + 1, \cdots, l_P\}$, and $a'_j = \_$ for all $j \in \{m + 1, \cdots, l_P\}$. Note that $\langle 1, \_, \_ \rangle(s)$, $\langle 1, \_, a'_1 \rangle(s)$, or $\langle 1, a_1, \_ \rangle(s)$ can not be returned by operator $\otimes$.

Example 1. Let $l_P = 4$, then $\mathsf{a} \otimes \mathsf{bab} = \langle 1, \mathsf{a}, \mathsf{b} \rangle \langle 2, \_, \mathsf{a} \rangle \langle 3, \_, \mathsf{b} \rangle \langle 4, \_, \_ \rangle$.

Let $A = (Q_A, \Delta, \delta_A, F_{Q_A}, q_A)$ and $B = (Q_B, \Sigma, \delta_B, F_{Q_B}, q_B)$ be two finite automata recognizing the respective sets $L(A) = \{u_i \otimes v_i \mid \langle u_i, v_i \rangle \in P\}^+$ and $L(B) = \Sigma^+$. We may assume that both q_A and q_B are non final. We assume that the automata A and B are clean (i.e., any state accepts some input string and is reachable from the initial state q_A(or q_B) by some input string). We associate automata A, B with the following ground TRSs T_A, T_B, respectively:

$$T_A^{(i,j)} = \{q^{(i)} \to d^{(j)}(q'^{(i)}) \mid q' \in \delta_A(q, d)\} \cup \{q^{(i)} \to e \mid q \in F_{Q_A}\}$$

$$T_B^{(i,j)} = \{q^{(i)} \to a^{(j)}(q'^{(i)}) \mid q' \in \delta_B(q, a)\} \cup \{q^{(i)} \to e \mid q \in F_{Q_B}\}$$

We assume given 13 disjoint copies of the above signatures, colored with color $i \in \{0, \cdots, 12\}$:

$$\Sigma^{(i)} = \{a^{(i)} \mid a \in \Sigma\} \qquad Q_A^{(i)} = \{q^{(i)} \mid q \in Q_A\}$$
$$\Delta^{(i)} = \{d^{(i)} \mid d \in \Delta\} \qquad Q_B^{(i)} = \{q^{(i)} \mid q \in Q_B\}$$

Let $\Theta^{012} = \Delta^{(0)} \cup \Sigma^{(1)} \cup \Sigma^{(2)}$, $\Theta^{345} = \Delta^{(3)} \cup \Delta^{(4)} \cup \Sigma^{(5)}$, and $Q = Q_A^{(6)} \cup Q_A^{(7)} \cup Q_A^{(8)} \cup Q_A^{(9)} \cup Q_B^{(10)} \cup Q_B^{(11)} \cup Q_A^{(12)}$. Let e be a constant. We assume that $\text{ar}(f) = 1$ for every $f \in \Delta \cup \Sigma$. For a ground term t built from $\Delta \cup \Sigma \cup \{e\}$, $t^{(i)}$ is defined as follows: $e^{(i)} = e$ and $(f(t_1))^{(i)} = f^{(i)}(t_1^{(i)})$ for $f \in \Delta \cup \Sigma$ and term t_1.

Now, we define the flat TRS R_1 on an extended signature $F_0 = Q \cup \{e, 0, 1\}$, $F_1 = \Theta^{012} \cup \Theta^{345}$, $F_6 = \{f\}$, and $F_7 = \{g\}$. First, we color T_A and T_B:

$$T_A^{(i,j)} = \{q^{(i)} \to d^{(j)}(q'^{(i)}) \mid q' \in \delta_A(q, d)\} \cup \{q^{(i)} \to e \mid q \in F_{Q_A}\}$$

$$T_B^{(i,j)} = \{q^{(i)} \to a^{(j)}(q'^{(i)}) \mid q' \in \delta_B(q, a)\} \cup \{q^{(i)} \to e \mid q \in F_{Q_B}\}$$

Next, we define recoloring TRSs S, P and projection TRSs Π_1, Π_2:

$$S^{(i,j)} = \{a^{(i)}(x) \to a^{(j)}(x) \mid a \in \Sigma\}$$
$$P^{(i,j)} = \{d^{(i)}(x) \to d^{(j)}(x) \mid d \in \Delta\}$$
$$\Pi_1^{(i,j)} = \{\langle n, a, a'\rangle^{(i)}(x) \to a^{(j)}(x) \mid n \in \{1, \cdots, l_P\}, a \in \Sigma, a' \in \Sigma \cup \{\_\}\}$$
$$\cup \{\langle n, \_, a'\rangle^{(i)}(x) \to x \mid n \in \{2, \cdots, l_P\}, a' \in \Sigma \cup \{\_\}\}$$
$$\Pi_2^{(i,j)} = \{\langle n, a, a'\rangle^{(i)}(x) \to a'^{(j)}(x) \mid n \in \{1, \cdots, l_P\}, a \in \Sigma \cup \{\_\}, a' \in \Sigma\}$$
$$\cup \{\langle n, a, \_\rangle^{(i)}(x) \to x \mid n \in \{2, \cdots, l_P\}, a \in \Sigma \cup \{\_\}\}$$

The flat TRS R_1 is defined as follows:

$$R_0 = T_A^{(6,3)} \cup T_A^{(7,3)} \cup T_A^{(8,4)} \cup T_A^{(9,4)} \cup T_B^{(10,5)} \cup T_B^{(11,5)}$$
$$\cup P^{(3,0)} \cup \Pi_1^{(3,1)} \cup \Pi_2^{(4,2)} \cup P^{(4,0)} \cup S^{(5,1)} \cup S^{(5,2)} \cup T_A^{(12,0)}$$

$$R_1 = R_0 \cup \left\{ \begin{array}{l} 0 \to f(q_A^{(6)}, q_A^{(7)}, q_A^{(8)}, q_A^{(9)}, q_B^{(10)}, q_B^{(11)}), \\ f(x_3, x_3, x_4, x_4, x_5, x_5) \to g(x_3, x_3, x_4, x_4, x_5, x_5, q_A^{(12)}), \\ g(x_0, x_1, x_2, x_0, x_1, x_2, x_0) \to 1 \end{array} \right\}$$

By construction of R_1, if $0 \xrightarrow{*}_{R_1} 1$ then the rules of R_1 are applied as described in the following picture.

$$0 \xrightarrow{\varepsilon} f(\ q_A^{(6)}, \quad q_A^{(7)}, \quad q_A^{(8)}, \quad q_A^{(9)}, \quad q_B^{(10)}, \quad q_B^{(11)})$$

$$\quad *\downarrow_{T_A^{(6,3)}} \quad *\downarrow_{T_A^{(7,3)}} \quad *\downarrow_{T_A^{(8,4)}} \quad *\downarrow_{T_A^{(9,4)}} \quad *\downarrow_{T_B^{(10,5)}} \quad *\downarrow_{T_B^{(11,5)}}$$

$$f(\ t_3, \quad t_3, \quad t_4, \quad t_4, \quad t_5, \quad t_5)$$
$$\varepsilon\downarrow$$
$$g(\ t_3, \quad t_3, \quad t_4, \quad t_4, \quad t_5, \quad t_5, \quad q_A^{(12)})$$

$$\quad *\downarrow_{P^{(3,0)}} \quad *\downarrow_{\Pi_1^{(3,1)}} \quad *\downarrow_{\Pi_2^{(4,2)}} \quad *\downarrow_{P^{(4,0)}} \quad *\downarrow_{S^{(5,1)}} \quad *\downarrow_{S^{(5,2)}} \quad *\downarrow_{T_A^{(12,0)}}$$

$$g(\ t_0, \quad t_1, \quad t_2, \quad t_0, \quad t_1, \quad t_2, \quad t_0\ ) \quad \xrightarrow{\varepsilon} 1$$

Indeed, each of the symbols 0 and 1 occurs in only one rewrite rule of R_1, and there is only one rule to transform the function symbol f into g.

Moreover, in the above rewrite sequence by R_1, we have a subsequence $q_A^{(12)} \xrightarrow[T_A^{(12,0)}]{*} t_0$, which means that t_0 has the form $((u_{i_1} \otimes v_{i_1}) \cdots (u_{i_m} \otimes v_{i_m})(e))^{(0)}$ for some $i_1, \cdots, i_m \in \{1, \cdots, k\}$. We will show in Lemma 1 that:

$$q_A^{(6)} \xrightarrow[T_A^{(6,3)} \cup P^{(3,0)}]{*} t_0 \xleftarrow[T_A^{(9,4)} \cup P^{(4,0)}]{*} q_A^{(9)}$$

Figure 1 shows how colors are changed by the rules of R_1.

Let G_1 be the set of ground terms built from $F_0 \cup F_1 \cup F_6 \cup F_7$.

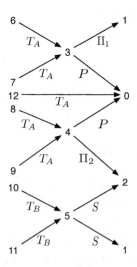

Fig. 1. Graph of the reduction of colors for R_1

Definition 3

(1) Let $\xrightarrow[R]{*}(s) = \{t \mid s \xrightarrow[R]{*} t\}$. For a subset $C \subseteq \{0, \cdots, 12\}$, let G^C be the intersection of G_1 and the set of ground terms built from e and colored function symbols in $\cup_{i \in C}(\Sigma^{(i)} \cup \Delta^{(i)} \cup Q_A^{(i)} \cup Q_B^{(i)})$.

(2) The index of an i-colored term built from $\Delta^{(i)} \cup \{e\}$ is a string of integers defined as follows: $\mathsf{index}(e) = \varepsilon$, and $\mathsf{index}(\langle n, a, a' \rangle^{(i)}(t)) = n\ \mathsf{index}(t)$.

This Lemma 1 will be useful in the proof of the following Lemma 2.

Lemma 1. Assume that $q_A^{(6)} \xrightarrow[R_0]{*} t_0$ and $q_A^{(9)} \xrightarrow[R_0]{*} t_0$ where $t_0 = ((u_{i_1} \otimes v_{i_1}) \cdots (u_{i_m} \otimes v_{i_m})(e))^{(0)}$. Then, the following properties hold:

(1) $q_A^{(6)} \xrightarrow[T_A^{(6,3)} \cup P^{(3,0)}]{*} t_0$

(2) $q_A^{(9)} \xrightarrow[T_A^{(9,4)} \cup P^{(4,0)}]{*} t_0$

Proof

(1) By definition of R_0, $q_A^{(6)} \xrightarrow[T_A^{(6,3)} \cup P^{(3,0)} \cup \Xi_1^{(3,1)}]{*} t_0$ where $\Xi_1^{(3,1)}$ is the subset of $\Pi_1^{(3,1)}$ defined by:

$$\Xi_1^{(3,1)} = \{\langle n, \_, a' \rangle^{(3)}(x) \to x \mid n \in \{2, \cdots, l_P\}, a' \in \Sigma \cup \{\_\}\}.$$

Note that $\mathsf{index}(t_0) = (1 \cdots l_P)^m$. In this rewrite sequence, if there is at least one application of some rule of $\Xi_1^{(3,1)}$, $\mathsf{index}(t_0) = (1 \cdots l_P)^m$ does not hold, since if we applied a rule $\langle n, \_, a' \rangle^{(3)}(x) \to x$, then at most $m - 1$ symbols of n would be included in $\mathsf{index}(t_0)$ whereas exactly m symbols of 1 in $\mathsf{index}(t_0)$ (since any symbol of form $\langle 1, a, a' \rangle^{(3)}$ can not be deleted). Thus, the proposition holds.

(2) Similar to (1). $\qquad\qquad\qquad\qquad\qquad\qquad\qquad\qquad\qquad\qquad\qquad\qquad\square$

Lemma 2. $0 \xrightarrow[R_1]{*} 1$ iff the PCP P has a solution.

Proof

Only if part: by definition of R_1, we have:

$$0 \xrightarrow[R_1]{} f(q_A^{(6)}, q_A^{(7)}, q_A^{(8)}, q_A^{(9)}, q_B^{(10)}, q_B^{(11)}) \xrightarrow[R_0]{\geq \varepsilon\ *} f(t_3, t_3, t_4, t_4, t_5, t_5)$$
$$\xrightarrow[R_1]{} g(t_3, t_3, t_4, t_4, t_5, t_5, q_A^{(12)}) \xrightarrow[R_0]{\geq \varepsilon\ *} g(t_0, t_1, t_2, t_0, t_1, t_2, t_0)$$
$$\xrightarrow[R_1]{} 1$$

By definition of R_0:

$$\xrightarrow[R_0]{*} (q_A^{(6)}) \subseteq G^{\{0,1,3,6\}}, \quad \xrightarrow[R_0]{*} (q_A^{(7)}) \subseteq G^{\{0,1,3,7\}},$$
$$\xrightarrow[R_0]{*} (q_A^{(8)}) \subseteq G^{\{0,2,4,8\}}, \quad \xrightarrow[R_0]{*} (q_A^{(9)}) \subseteq G^{\{0,2,4,9\}},$$
$$\xrightarrow[R_0]{*} (q_B^{(10)}) \subseteq G^{\{1,2,5,10\}}, \quad \xrightarrow[R_0]{*} (q_B^{(11)}) \subseteq G^{\{1,2,5,11\}}$$

We first show that the following condition (I) holds:

$$t_i \in G^{\{i\}} \quad \forall i \in \{0, \cdots, 5\} \tag{I}$$

Note that $t_3 \in G^{\{0,1,3\}}$ holds, since $t_3 \in \xrightarrow[R_0]{*} (q_A^{(6)}) \cap \xrightarrow[R_0]{*} (q_A^{(7)})$.
Similarly, $t_4 \in G^{\{0,2,4\}}$ and $t_5 \in G^{\{1,2,5\}}$.
Since $t_3 \xrightarrow[R_0]{*} t_0$ and $t_4 \xrightarrow[R_0]{*} t_0$, we have $t_0 \in G^{\{0,1,3\}} \cap G^{\{0,2,4\}} = G^{\{0\}}$.
Similarly, $t_1 \in G^{\{0,1,3\}} \cap G^{\{1,2,5\}} = G^{\{1\}}$ and $t_2 \in G^{\{0,2,4\}} \cap G^{\{1,2,5\}} = G^{\{2\}}$.
Hence, the condition (I) holds for $i \in \{0, 1, 2\}$.

Since $t_3 \xrightarrow[R_0]{*} t_0 \in G^{\{0\}}$ and $t_3 \xrightarrow[R_0]{*} t_1 \in G^{\{1\}}$, t_3 can not contain any symbol in $G^{\{0,1\}}$, hence $t_3 \in G^{\{3\}}$ holds.
Similarly, $t_4 \in G^{\{4\}}$ holds because $t_4 \xrightarrow[R_0]{*} t_2 \in G^{\{2\}}$ and $t_4 \xrightarrow[R_0]{*} t_0 \in G^{\{0\}}$, and $t_5 \in G^{\{5\}}$ holds because $t_5 \xrightarrow[R_0]{*} t_1 \in G^{\{1\}}$ and $t_5 \xrightarrow[R_0]{*} t_2 \in G^{\{2\}}$.
Hence, (I) holds for $i \in \{3, 4, 5\}$, as claimed.

By (I), we have: $t_3 \xrightarrow[\Pi_1^{(3,1)}]{*} t_1 \xleftarrow[S^{(5,1)}]{*} t_5$ and $t_4 \xrightarrow[\Pi_2^{(4,2)}]{*} t_2 \xleftarrow[S^{(5,2)}]{*} t_5$.

Since $q_A^{(12)} \xrightarrow[R_0]{*} t_0$, $t_0 = ((u_{i_1} \otimes v_{i_1}) \cdots (u_{i_m} \otimes v_{i_m})(e))^{(0)}$ for some $i_1, \cdots, i_m \in \{1, \cdots, k\}$. We have $m > 0$ because the initial state q_A is not final.
By Lemma 1, $t_3 \xrightarrow[P^{(3,0)}]{*} t_0 \xleftarrow[P^{(4,0)}]{*} t_4$. Thus, $t_3 = ((u_{i_1} \otimes v_{i_1}) \cdots (u_{i_m} \otimes v_{i_m})(e))^{(3)}$
and $t_4 = ((u_{i_1} \otimes v_{i_1}) \cdots (u_{i_m} \otimes v_{i_m})(e))^{(4)}$.
Since $t_3 \xrightarrow[\Pi_1^{(3,1)}]{*} t_1$, $t_1 = (u_{i_1} \cdots u_{i_m}(e))^{(1)}$,
and since $t_4 \xrightarrow[\Pi_2^{(4,2)}]{*} t_2$, $t_2 = (v_{i_1} \cdots v_{i_m}(e))^{(2)}$.
Finally, $t_1 \xleftarrow[S^{(5,1)}]{*} t_5 \xrightarrow[S^{(5,2)}]{*} t_2$, hence $t_5 = (u_{i_1} \cdots u_{i_m}(e))^{(5)} = (v_{i_1} \cdots v_{i_m}(e))^{(5)}$.
It means that the PCP P has a solution.

If part: let $i_1 \cdots i_m$ be a solution of the PCP P, and let:

$$s = (u_{i_1} \otimes v_{i_1}) \cdots (u_{i_m} \otimes v_{i_m})(e) \text{ and } t = u_{i_1} \cdots u_{i_m}(e)$$

Then, $t = v_{i_1} \cdots v_{i_m}(e)$ holds. By definition of R_1, we have:

$$0 \rightarrow f(q_A^{(6)}, q_A^{(7)}, q_A^{(8)}, q_A^{(9)}, q_B^{(10)}, q_B^{(11)}) \xrightarrow{*} f(s^{(3)}, s^{(3)}, s^{(4)}, s^{(4)}, t^{(5)}, t^{(5)})$$
$$\rightarrow g(s^{(3)}, s^{(3)}, s^{(4)}, s^{(4)}, t^{(5)}, t^{(5)}, q_A^{(12)}) \xrightarrow{*} g(s^{(0)}, t^{(1)}, t^{(2)}, s^{(0)}, t^{(1)}, t^{(2)}, s^{(0)})$$
$$\rightarrow 1.$$

Hence, $0 \xrightarrow[R_1]{*} 1$ □

As a consequence of Lemma 2, we have the following main theorem of this section.

Theorem 1. Reachability is undecidable for flat TRSs.

Since 1 is a normal form, $0 \xrightarrow[R_1]{*} 1$ iff $0 \downarrow_{R_1} 1$. Thus, the following corollary holds.

Corollary 1. Joinability is undecidable for flat TRSs.

Compared to the construction in [8] for Lemma 2, on the one hand, the above TRS R_1 is simpler and on the other hand, some rules have been added in order to permit the reduction $q_A^{(12)} \xrightarrow[T_A^{(12,0)}]{*} t_0$. This appeared to be necessary in order to fix a bug [8] where a reduction $0 \xrightarrow{*} 1$ was possible with the TRS associated to the PCP $\{\langle aa, a \rangle, \langle a, ab \rangle\}$ whereas it has no solution. The main reason for this counter-example is for lack of some consideration such as Lemma 1 above (derived from the existence of $q_A^{(12)} \xrightarrow{*} ((u_{i_1} \otimes v_{i_1}) \cdots (u_{i_m} \otimes v_{i_m})(e))^{(0)}$ and the definition of operator $\otimes$).

4   Confluence for Flat TRSs

We show that confluence for flat TRSs is undecidable by reduction of the reachability problem which has been shown undecidable in the previous section. We introduce some technical definitions in Sections 4.1.

4.1   Mapping Lemma

A mapping ϕ from T to T can be extended to TRSs as follows:

$$\phi(R) = \{\phi(\alpha) \to \phi(\beta) \mid \alpha \to \beta \in R\} \setminus \{t \to t \mid t \in T\}$$

Such a mapping ϕ can also be extended to substitutions by $\phi(\theta) = \{x \mapsto \phi(x\theta) \mid x$ in the domain of $\theta\}$. The following lemma gives a characterization of confluence for a TRS R using $\phi(R)$.

Lemma 3. A TRS R is confluent iff there exists a mapping $\phi : T \to T$ that satisfies the following conditions (1)–(4).

(1) If $s \xrightarrow{R} t$ then $\phi(s) \xrightarrow[\phi(R)]{*} \phi(t)$

(2) $\xrightarrow[\phi(R)]{} \subseteq \xrightarrow[R]{*}$

(3) $t \xrightarrow[R]{*} \phi(t)$

(4) $\phi(R)$ is confluent

Proof. Only if part: let ϕ be the identity mapping.
  If part: assume that $s \xleftarrow[R]{*} r \xrightarrow[R]{*} t$.
  By Condition (1), $\phi(s) \xleftarrow[\phi(R)]{*} \phi(r) \xrightarrow[\phi(R)]{*} \phi(t)$.
  By Condition (4), $\phi(s) \downarrow_{\phi(R)} \phi(t)$.
  By Condition (2), $\phi(s) \downarrow_R \phi(t)$.
  By Condition (3), $s \xrightarrow[R]{*} \phi(s)$ and $t \xrightarrow[R]{*} \phi(t)$. Thus, $s \downarrow_R t$.   □

This lemma is used in Section 4.2.

4.2 Proof of Undecidability

Let us introduce new function symbols $\Theta_2^{012} = \{d_2 \mid d \in \Theta^{012}\}$, where each d_2 has arity 2. We add the following rules to the TRS R_1 of Section 3:

$$R_2 = R_1 \cup \{e \to 0\} \cup \{d(x) \to d_2(0, x), d_2(1, x) \to x \mid d \in \Theta^{012}\}$$

Note that the TRS R_2 is flat. Let G_2 be the set of ground terms built from $F_0 \cup F_1 \cup F_6 \cup F_7 \cup \Theta_2^{012}$.

First, we show that $0 \xrightarrow[R_2]{*} 1$ iff $0 \xrightarrow[R_1]{*} 1$. For this purpose, we will introduce another reduction mapping ϕ and another TRS R_2' and show a technical lemma.

Let ψ be the mapping over G_2 defined as follows.

$$\begin{aligned}
\psi(h(t_1, \cdots, t_n)) &= e & \text{(if } h \in \{0, 1, f, g\}) \\
\psi(d_2(t_1, t_2)) &= d(\psi(t_2)) & \text{(if } d \in \Theta^{012}) \\
\psi(h(t_1, \cdots, t_n)) &= h(\psi(t_1), \cdots, \psi(t_n)) & \text{(otherwise)}
\end{aligned}$$

Let $R_2' = R_1 \cup \{e \to 0\} \cup \{d(x) \to d_2(0, x) \mid d \in \Theta^{012}\}$.

Lemma 4. For any $s \in G_2$, if $s \xrightarrow[R_2']{} t$ then $\psi(s) \xrightarrow[R_1]{=} \psi(t)$.

Proof. We prove this lemma by induction on the structure of s.

Base case: if $s \in Q$ then $s = \psi(s) \xrightarrow[R_1]{} \psi(t) = t$.
If $s \in \{e, 0\}$ then $\psi(s) = \psi(t) = e$.

Induction step

Case of $s \in \{f(s_1, \cdots, s_6), g(s_1, \cdots, s_7)\}$: in this case, $\psi(s) = \psi(t) = e$.
Case of $s = d(s_1)$ where $d \in \Theta^{345}$: if $t = d(t_1)$ and $s_1 \xrightarrow[R_2']{} t_1$ then $\psi(s) = d(\psi(s_1)) \xrightarrow[R_1]{} d(\psi(t_1)) = \psi(t)$ by the induction hypothesis. Otherwise either $t = d'(s_1)$ with $d' \in \Theta^{012}$ and $\psi(s) = d(\psi(s_1)) \xrightarrow[R_1]{} d'(\psi(s_1)) = \psi(t)$ or $t = s_1$ and $\psi(s) = d(\psi(s_1)) \xrightarrow[R_1]{} \psi(s_1) = \psi(t)$.
Case of $s = d(s_1)$ where $d \in \Theta^{012}$: if $t = d(t_1)$ and $s_1 \xrightarrow[R_2']{} t_1$ then $\psi(s) = d(\psi(s_1)) \xrightarrow[R_1]{=} d(\psi(t_1)) = \psi(t)$ by the induction hypothesis. Otherwise $t = d_2(0, s_1)$ and $\psi(s) = \psi(t) = d(\psi(s_1))$.
Case of $s = d_2(s_1, s_2)$ where $d \in \Theta^{012}$: in this case, $t = d_2(t_1, t_2)$ holds for some t_1, t_2 and either $s_1 \xrightarrow[R_2']{} t_1$ and $s_2 = t_2$ or $s_2 \xrightarrow[R_2']{} t_2$ and $s_1 = t_1$, hence $\psi(s) = d(\psi(s_2)) \xrightarrow[R_1]{=} d(\psi(t_2)) = \psi(t)$ by the induction hypothesis. □

Lemma 5. $0 \xrightarrow[R_2]{*} 1$ iff $0 \xrightarrow[R_1]{*} 1$.

Proof. The if part is obvious. For the only if part, by definition of R_2, if $0 \xrightarrow[R_2]{*} 1$ then there exists a shortest sequence γ that satisfies:

$$\begin{aligned}
\gamma : 0 &\xrightarrow[R_1]{} f(q_A^{(6)}, q_A^{(7)}, q_A^{(8)}, q_A^{(9)}, q_B^{(10)}, q_B^{(11)}) \xrightarrow[R_2]{> \varepsilon *} f(t_3, t_3, t_4, t_4, t_5, t_5) \\
&\xrightarrow[R_1]{} g(t_3, t_3, t_4, t_4, t_5, t_5, q_A^{(12)}) \xrightarrow[R_2]{> \varepsilon *} g(t_0, t_1, t_2, t_0, t_1, t_2, t_0) \\
&\xrightarrow[R_1]{} 1.
\end{aligned}$$

Note that $d_2(1, x) \to x$ can not be applied in γ. Indeed, if $d_2(1, x) \to x$ is applied in γ, then γ must contain a subsequence $0 \xrightarrow[R_2]{*} 1$ since 1 appears only in the right-hand side of the rule $\mathsf{g}(x_0, x_1, x_2, x_0, x_1, x_2, x_0) \to 1$, g is only generated by the rule $\mathsf{f}(x_3, x_3, x_4, x_4, x_5, x_5) \to \mathsf{g}(x_3, x_3, x_4, x_4, x_5, x_5, q_A^{(12)})$, and f is only generated by the rule $0 \to \mathsf{f}(q_A^{(6)}, q_A^{(7)}, q_A^{(8)}, q_A^{(9)}, q_B^{(10)}, q_B^{(11)})$. This contradicts the hypothesis that γ is a shortest sequence. Thus,

$$\mathsf{f}(q_A^{(6)}, q_A^{(7)}, q_A^{(8)}, q_A^{(9)}, q_B^{(10)}, q_B^{(11)}) \xrightarrow[R_2]{\geq \varepsilon *} \mathsf{f}(t_3, t_3, t_4, t_4, t_5, t_5)$$

and
$$\mathsf{g}(t_3, t_3, t_4, t_4, t_5, t_5, q_A^{(12)}) \xrightarrow[R_2]{\geq \varepsilon *} \mathsf{g}(t_0, t_1, t_2, t_0, t_1, t_2, t_0)$$

By Lemma 4 (for sake of readability, we shall write below $\underline{x}$ instead of $\psi(x)$):

$$\mathsf{f}(\underline{q_A^{(6)}}, \underline{q_A^{(7)}}, \underline{q_A^{(8)}}, \underline{q_A^{(9)}}, \underline{q_B^{(10)}}, \underline{q_B^{(11)}}) \xrightarrow[R_1]{*} \mathsf{f}(\underline{t_3}, \underline{t_3}, \underline{t_4}, \underline{t_4}, \underline{t_5}, \underline{t_5})$$

and
$$\mathsf{g}(\underline{t_3}, \underline{t_3}, \underline{t_4}, \underline{t_4}, \underline{t_5}, \underline{t_5}, \underline{q_A^{(12)}}) \xrightarrow[R_1]{*} \mathsf{g}(\underline{t_0}, \underline{t_1}, \underline{t_2}, \underline{t_0}, \underline{t_1}, \underline{t_2}, \underline{t_0})$$

Since $\psi(q) = q$ for every $q \in Q$, $0 \xrightarrow[R_1]{} \mathsf{f}(\underline{q_A^{(6)}}, \underline{q_A^{(7)}}, \underline{q_A^{(8)}}, \underline{q_A^{(9)}}, \underline{q_B^{(10)}}, \underline{q_B^{(11)}})$.

By definition of R_1, $\mathsf{f}(\underline{t_3}, \underline{t_3}, \underline{t_4}, \underline{t_4}, \underline{t_5}, \underline{t_5}) \xrightarrow[R_1]{} \mathsf{g}(\underline{t_3}, \underline{t_3}, \underline{t_4}, \underline{t_4}, \underline{t_5}, \underline{t_5}, \underline{q_A^{(12)}})$ and $\mathsf{g}(\underline{t_0}, \underline{t_1}, \underline{t_2}, \underline{t_0}, \underline{t_1}, \underline{t_2}, \underline{t_0}) \xrightarrow[R_1]{} 1$. Altogether $0 \xrightarrow[R_1]{*} 1$. $\square$

We shall show next that R_2 is confluent iff $0 \xrightarrow[R_2]{*} 1$ by using Lemma 3. We need the following lemma for that purpose.

Lemma 6. If $0 \xrightarrow[R_2]{*} 1$ then $t \xrightarrow[R_2]{*} 1$ for any $t \in G_2$.

Proof. First, we note that for any $q \in Q$, there exists $s \in G_2$ which does not contain any function symbol in Q such that $q \xrightarrow[R_2]{*} s$. Since both of the automata A and B are clean, there exists $u \in \Delta^{(3)*} \cup \Delta^{(4)*} \cup \Sigma^{(5)*} \cup \Delta^{(0)*}$ such that $q \xrightarrow[R_0]{*} u(e)$.

Thus, it suffices to show that for any $t \in G_2$ which does not contain any function symbol in Q, $t \xrightarrow[R_2]{*} 1$. We show this proposition by induction on the structure of t:

Base case: by $e \xrightarrow[R_2]{} 0 \xrightarrow[R_2]{*} 1$.

Induction step: let $t = h(t_1, \cdots, t_n)$ where $n > 0$ and $h \in \Theta^{012} \cup \Theta^{345} \cup \{\mathsf{f}, \mathsf{g}\} \cup \Theta_2^{012}$. By the induction hypothesis, $h(t_1, \cdots, t_n) \xrightarrow[R_2]{*} h(1, \cdots, 1)$.
For every $d \in \Theta^{345}$, $d(1) \xrightarrow[R_1]{} d'(1)$ for some $d' \in \Theta^{012}$ or $d(1) \xrightarrow[R_1]{} 1$.
For every $d' \in \Theta^{012}$, $d'(1) \xrightarrow[R_2]{} d'_2(0, 1) \xrightarrow[R_2]{*} d'_2(1, 1)$.
For every $d'_2 \in \Theta_2^{012}$, $d'_2(1, 1) \xrightarrow[R_2]{} 1$.
Moreover, $\mathsf{f}(1, \cdots, 1) \xrightarrow[R_1]{} \mathsf{g}(1, \cdots, 1, q_A^{(12)}) \xrightarrow[R_1]{} \mathsf{g}(1, \cdots, 1, u(e))$ $\square$
$$\xrightarrow[R_2]{*} \mathsf{g}(1, \cdots, 1, 1)$$
$$\xrightarrow[R_1]{} 1 \quad \text{where } u \in \Delta^{(0)*}$$

Let $\phi(t)$ be the term obtained from t by replacing every maximal ground subterm (w.r.t. $\geq_{\mathsf{sub}}$) by 1. Note that:

$$\phi(R_0) = P^{(3,0)} \cup \Pi_1^{(3,1)} \cup \Pi_2^{(4,2)} \cup P^{(4,0)} \cup S^{(5,1)} \cup S^{(5,2)}$$

$$\phi(R_1) = \phi(R_0) \cup \left\{ \begin{array}{l} \mathsf{f}(x_3, x_3, x_4, x_4, x_5, x_5) \to \mathsf{g}(x_3, x_3, x_4, x_4, x_5, 1), \\ \mathsf{g}(x_0, x_1, x_2, x_0, x_1, x_2, x_0) \to 1 \end{array} \right\}$$

$$\phi(R_2) = \phi(R_1) \cup \{d(x) \to d_2(1, x), d_2(1, x) \to x \mid d \in \Theta^{012}\}.$$

Note also that the rules of T_A and T_B vanish in $\phi(R_0)$. The following technical lemma is used in the proof of Lemma 8.

Lemma 7. For any non-constant function symbol $h \in \Theta^{012} \cup \Theta^{345} \cup \{\mathsf{f}, \mathsf{g}\} \cup \Theta_2^{012}$, $h(1, \cdots, 1) \xrightarrow[\phi(R_2)]{*} 1$.

Proof

For every $d \in \Theta^{345}$, $d(1) \xrightarrow[\phi(R_1)]{} d'(1)$ for some $d' \in \Theta^{012}$ or $d(1) \xrightarrow[\phi(R_1)]{} 1$.
For every $d' \in \Theta^{012}$, $d'(1) \xrightarrow[\phi(R_2)]{} d_2'(1, 1)$.
For every $d_2' \in \Theta_2^{012}$, $d_2'(1, 1) \xrightarrow[\phi(R_2)]{} 1$.
Moreover, $\mathsf{f}(1, \cdots, 1) \xrightarrow[\phi(R_1)]{} \mathsf{g}(1, \cdots, 1) \xrightarrow[\phi(R_1)]{} 1$. Thus, the lemma holds. □

We show now how the hypotheses of Lemma 3 hold for R_2 and ϕ.

Lemma 8. If $0 \xrightarrow[R_2]{*} 1$ then the following properties hold.

(1) If $s \xrightarrow[R_2]{} t$ then $\phi(s) \xrightarrow[\phi(R_2)]{*} \phi(t)$.
(2) $\xrightarrow[\phi(R_2)]{} \subseteq \xrightarrow[R_2]{*}$.
(3) $t \xrightarrow[R_2]{*} \phi(t)$.
(4) $\phi(R_2)$ is confluent.

Proof

(1) By induction on the structure of s. If s is a ground term then $\phi(s) = \phi(t) = 1$. Thus, we assume that s is not ground. Let $s \xrightarrow[R_2]{p} t$.
If $p = \varepsilon$ then $s = \alpha\theta \to \beta\theta = t$ where $\alpha \to \beta \in R_2$. Let $s = h(s_1, \cdots, s_n)$ for some $h \in \Theta^{012} \cup \Theta^{345} \cup \{\mathsf{f}, \mathsf{g}\} \cup \Theta_2^{012}$ and $s_1, \cdots, s_n$, and $\alpha = h(a_1, \cdots, a_n)$. Since R_2 is flat, $a_1 \cdots a_n \in X \cup F_0$. If a_i is a variable then $\phi(a_i\theta) = a_i\phi(\theta)$. If a_i is a constant then $\phi(s_i) = \phi(a_i) = 1$. Thus, $\phi(s) = \phi(\alpha)\phi(\theta)$. Similarly, $\phi(t) = \phi(\beta)\phi(\theta)$, so $\phi(s) \xrightarrow[\phi(R_2)]{} \phi(t)$ holds.
If $p \neq \varepsilon$ then $s = h(s_1, \cdots, s_i, \cdots, s_n)$, $t = h(s_1, \cdots, t_i, \cdots, s_n)$, and $s_i \xrightarrow[R_2]{} t_i$ where $i \in \{1, \cdots, n\}$. Since s is not ground,

$$\phi(s) = h(\phi(s_1), \cdots, \phi(s_i), \cdots, \phi(s_n)).$$

By the induction hypothesis, $\phi(s_i) \xrightarrow[\phi(R_2)]{*} \phi(t_i)$.
If t is not ground then $h(\phi(s_1), \cdots, \phi(t_i), \cdots, \phi(s_n)) = \phi(t)$.
If t is ground then $h(\phi(s_1), \cdots, \phi(t_i), \cdots, \phi(s_n)) = h(1, \cdots, 1)$.
By Lemma 7, $h(1, \cdots, 1) \xrightarrow[\phi(R_2)]{*} 1 = \phi(t)$. Thus, $\phi(s) \xrightarrow[\phi(R_2)]{*} \phi(t)$ holds.

(2) Since $\phi(R_2) \setminus \Big(\big(\{d(x) \to d_2(1,x) \mid d \in \Theta^{012}\} \cup \{f(x_3, x_3, x_4, x_4, x_5, x_5) \to$ $g(x_3, x_3, x_4, x_4, x_5, x_5, 1)\} \big) \Big) \subseteq R_2$, it suffices to show that:

$$d(x\theta) \xrightarrow[R_2]{*} d_2(1, x\theta)$$
$$\text{and } f(x_3\theta, x_3\theta, x_4\theta, x_4\theta, x_5\theta, x_5\theta) \xrightarrow[R_2]{*} g(x_3\theta, x_3\theta, x_4\theta, x_4\theta, x_5\theta, x_5\theta, 1)$$

We have that: $d(x\theta) \xrightarrow[R_2]{} d_2(0, x\theta) \xrightarrow[R_2]{*} d_2(1, x\theta)$.
By Lemma 6:

$$f(x_3\theta, x_3\theta, x_4\theta, x_4\theta, x_5\theta, x_5\theta) \xrightarrow[R_1]{} g\big(x_3\theta, x_3\theta, x_4\theta, x_4\theta, x_5\theta, x_5\theta, q_A^{(12)}\big)$$
$$\xrightarrow[R_2]{*} g(x_3\theta, x_3\theta, x_4\theta, x_4\theta, x_5\theta, x_5\theta, 1)$$

(3) By Lemma 6.

(4) We can easily show that $\phi(R_2)$ is terminating by using a lexicographic path order induced by a precedence $>$ that satisfies the following conditions: for any $d \in \Theta^{345}, d' \in \Theta^{012}, d'' \in \Theta_2^{012}, d > d' > d'' > 1$ and $f > g > 1$. Thus, it suffices to show that every critical peak of $\phi(R_2)$ is joinable.

For every $a, a' \in \Sigma$, $\langle n, a, a' \rangle^{(0)}(x) \leftarrow \langle n, a, a' \rangle^{(3)}(x) \to a^{(1)}(x)$ is joinable by: $\langle n, a, a' \rangle^{(0)}(x) \to \langle n, a, a' \rangle_2^{(0)}(1, x) \to x \leftarrow a_2^{(1)}(1, x) \leftarrow a^{(1)}(x)$.

For every $a' \in \Sigma$, $\langle n, \_, a' \rangle^{(0)}(x) \leftarrow \langle n, \_, a' \rangle^{(3)}(x) \to x$ is joinable by: $\langle n, \_, a' \rangle^{(0)}(x) \to \langle n, \_, a' \rangle_2^{(0)}(1, x) \to x$.

For every $a, a' \in \Sigma$, $\langle n, a, a' \rangle^{(0)}(x) \leftarrow \langle n, a, a' \rangle^{(4)}(x) \to a'^{(2)}(x)$ is joinable by: $\langle n, a, a' \rangle^{(0)}(x) \to \langle n, a, a' \rangle_2^{(0)}(1, x) \to x \leftarrow a_2'^{(2)}(1, x) \leftarrow a'^{(2)}(x)$.

For every $a \in \Sigma$, $\langle n, a, \_ \rangle^{(0)}(x) \leftarrow \langle n, a, \_ \rangle^{(4)}(x) \to x$ is joinable by: $\langle n, a, \_ \rangle^{(0)}(x) \to \langle n, a, \_ \rangle_2^{(0)}(1, x) \to x$.

For every $a \in \Sigma$, $a^{(1)}(x) \leftarrow a^{(5)}(x) \to a^{(2)}(x)$ is joinable by: $a^{(1)}(x) \to a_2^{(1)}(1, x) \to x \leftarrow a_2^{(2)}(1, x) \leftarrow a^{(2)}(x)$.   □

Lemma 9. R_2 is confluent iff $0 \xrightarrow[R_2]{*} 1$.

Proof. The if part follows from Lemmata 3 and 8.
For the only if part, by $\langle n, \_, a' \rangle^{(0)}(x) \xleftarrow[P^{(3,0)}]{} \langle n, \_, a' \rangle^{(3)}(x) \xrightarrow[\Pi_1^{(3,1)}]{} x$, the confluence ensures that $\langle n, \_, a' \rangle^{(0)}(x) \downarrow_{R_2} x$.
Since x is a normal form, $\langle n, \_, a' \rangle^{(0)}(x) \xrightarrow[R_2]{*} x$. Thus, there exists a sequence:

$$\langle n, \_, a' \rangle^{(0)}(x) \xrightarrow[R_2]{} \langle n, \_, a' \rangle_2^{(0)}(0, x) \xrightarrow[R_2]{*} \langle n, \_, a' \rangle_2^{(0)}(1, x) \xrightarrow[R_2]{} x$$

It follows that $0 \xrightarrow[R_2]{*} 1$ holds.   □

By Lemmata 2, 5, 9, the following theorem holds.

Theorem 2. Confluence is undecidable for flat TRSs.

The above TRS R_2 differs from the analogous one of [8]. Indeed, in some cases, with the TRS of [8] we may have $0 \xrightarrow[R_2]{*} 1$ whereas $0 \xrightarrow[R_1]{*} 1$ does not hold, which is a problem for the correctness of the reduction. This error was corrected in [9], but Lemma 9 does not hold for the TRS of this report. Therefore, the above TRS R_2 and the above proof differ from the ones of [9].

5 Concluding Remarks

We have shown that the properties of reachability, joinability and confluence are undecidable for flat TRSs. These results are negative solutions to the open problems posed in [7], and striking compared with the results that the word and unification problems for shallow TRSs are decidable [3]. The undecidability of reachability is shown by a reduction of the Post's Correspondence Problem and the case of joinability and confluence are treated both by a reduction of reachability (with a non trivial reduction in the case of confluence).

The proof techniques involved in our constructions, namely term coloring in Section 3, the criteria for confluence of Lemma 3 and the ground term mapping of Section 4 appeared to be very useful in this context and we believe that they could be of benefit to other decision problems.

Note that the only rules not linear in our TRS are $\mathsf{f}(x_3, x_3, x_4, x_4, x_5, x_5) \rightarrow \mathsf{g}(x_3, x_3, x_4, x_4, x_5, x_5, q_A^{(12)})$ and $\mathsf{g}(x_0, x_1, x_2, x_0, x_1, x_2, x_0) \rightarrow 1$ (both left and right members of the first rule are non-linear and they share variables, which is crucial in our reduction). Hence, we have narrowed dramatically the gap between known decidable and undecidable cases of confluence, reachability and joinability of TRS. All three properties are indeed decidable for TRSs whose left members of rules are flat and right members are flat and linear [17, 12, 16].

It will be a next step to find non-right-linear subclasses of flat (or shallow) TRSs with the decidable property for some of these decision problems. For example, what about the class of flat and semi-constructor TRSs? Here, a semi-constructor TRS is such a TRS that all defined symbols appearing in the right-hand side of each rewrite rule occur only in its ground subterms.

Another interesting question is: does there exist a subclass of TRSs such that exactly one of reachability and confluence is decidable? For the related question about whether there exists a subclass such that exactly one of reachability and joinability is decidable, the existence of such a confluent subclass has been shown in [10, 11].

Acknowledgements

We would like to thank Professor Masahiko Sakai of Nagoya University for his helpful comments. This work was supported in part by Grant-in-Aid for Scientific Research 15500009 from Japan Society for the Promotion of Science.

References

1. F. Baader and T. Nipkow. *Term Rewriting and All That*. Cambridge University Press, 1998.
2. H. Comon, G. Godoy, and R. Nieuwenhuis. The confluence of ground term rewrite systems is decidable in polynomial time. In *Proc. 42nd Symp. Foundations of Computer Science (FOCS'2001), Las Vegas, NV, USA, Oct. 2001*, 2001.

3. H. Comon, M. Haberstrau, and J.-P. Jouannaud. Syntacticness, cycle-syntacticness and shallow theories. *Information and Computation*, 111(1):154–191, 1994.

4. M. Dauchet, T. Heuillard, P. Lescanne, and S. Tison. Decidability of the confluence of finite ground term rewrite systems and of other related term rewrite systems. *Information and Computation*, 88:187–201, 1990.

5. N. Dershowitz and J.-P. Jouannaud. Rewrite systems. In *Handbook of Theoretical Computer Science*, volume B, pages 243–320. Elsevier Science Publishers B. V., 1990.

6. G. Godoy and A. Tiwari. Confluence of shallow right-linear rewrite systems. In *Computer Science Logic, 14th Annual Conf., CSL 2005*, pages 541–556. LNCS 3634, 2005.

7. G. Godoy, A. Tiwari, and R. Verma. On the confluence of linear shallow term rewrite systems. In *In Symp. on Theoretical Aspects of Computer Science(STACS 2003)*, pages 85–96. LNCS 2607, 2003.

8. F. Jacquemard. Reachability and confluence are undecidable for flat term rewriting systems. *Inf.Process.Lett.*, 87:265–270, 2003.

9. F. Jacquemard. Erratum to the paper: Reachability and confluence are undecidable for flat term rewriting system. Research Report LSV-05-09, 2005.

10. I. Mitsuhashi, M. Oyamaguchi, Y. Ohta, and T. Yamada. The joinability and related decision problems for semi-constructor TRSs. *Trans, IPS Japan*, 47(5), 2006. To appear.

11. I. Mitsuhashi, M. Oyamaguchi, and T. Yamada. The reachability and related decision problems for monadic and semi-constructor TRSs. To appear in Inf. Process.Lett.

12. T. Nagaya and Y. Toyama. Decidability for left-linear growing term rewriting systems. In *Proc.10th RTA*, pages 256–270. LNCS 1631, 1999.

13. M. Oyamaguchi. The Church-Rosser property for ground term-rewriting systems is decidable. *Theoretical Computer Science*, 49(1):43–79, 1987.

14. M. Oyamaguchi. The reachability and joinability problems for right-ground term-rewriting systems. *J.Inf.Process.*, 13(3):347–354, 1990.

15. K. Salomaa. Deterministic tree pushdown automata and monadic tree rewriting systems. *J.Comput.Syst.Sci.*, 37:367–394, 1988.

16. T. Takai, Y. Kaji, and H. Seki. Right-linear finite path overlapping term rewriting systems effectively preserve recognizability. In *Proc.11th RTA*, pages 246–260. LNCS 1833, 2000.

17. A. Tiwari. Deciding confluence of certain term rewriting systems in polynomial time. In *Proc. IEEE Symposium on Logic in Computer Science*, pages 447–456. IEEE Society, 2002.

Some Properties of Triangular Sets and Improvement Upon Algorithm CharSer

Yong-Bin Li

[1] Chengdu Institute of Computer Applications,
Chinese Academy of Sciences, Chengdu, Sichuan 610041, China
[2] School of Applied Mathematic,
University of Electronic Science and Technology of China,
Chengdu, Sichuan 610054, China
yongbinli@uestc.edu.cn

Abstract. We present some new properties of triangular sets, which have rather theoretical contribution to understand the structure of the affine varieties of triangular sets. Based on these results and the famous algorithm CharSet, we present two modified versions of the algorithm CharSer that can decompose any nonempty polynomial set into characteristic series. Some examples show that our improvement can efficiently avoid for redundant decompositions, and reduce the branches of the decomposition tree at times.

Keywords: Triangular sets, quasi-normal zero, characteristic set, algorithm CharSer.

1 Introduction

Let $\mathbf{K}$ be a field of characteristic 0 and $\mathbf{K}[x_1,\ldots,x_n]$ (or $\mathbf{K}[\mathbf{x}]$ for short) the ring of polynomials in the variables $(x_1,\ldots,x_n)$ with coefficients in $\mathbf{K}$. A *polynomial set* is a finite set $\mathbb{P}$ of nonzero polynomials in $\mathbf{K}[\mathbf{x}]$. For any polynomial $P \notin \mathbf{K}$, the biggest index p such that $\deg(P,x_p) > 0$ is called the *class*, x_p the *leading variable*, and $\deg(P,x_p)$ the *leading degree* of P, denoted by $\mathrm{cls}(P)$, $\mathrm{lv}(P)$ and $\mathrm{ldeg}(P)$, respectively. A finite nonempty ordered set $\mathbb{T} = [f_1,\ldots,f_s]$ of polynomials in $\mathbf{K}[\mathbf{x}]\backslash\mathbf{K}$ is called a *triangular set* if $\mathrm{cls}(f_1) < \cdots < \mathrm{cls}(f_s)$. Let a triangular set $\mathbb{T}$ be written as the following form

$$\mathbb{T} = [f_1(u_1,\ldots,u_r,y_1),\ldots,f_s(u_1,\ldots,u_r,y_1,\ldots,y_s)], \qquad (1)$$

where $(u_1,\ldots,u_r,y_1,\ldots,y_s)$ is a permutation of $(x_1,\ldots,x_n)$, and we always assume $r \geq 1$ throughout this paper. Let $F \neq 0$ be a polynomial and G any polynomial in $\mathbf{K}[\mathbf{x}]$, the *pseudo-remainder* of G with respect to F in $\mathrm{lv}(F)$ denoted by $\mathrm{prem}(G,F)$. One can see the detail definition in [18] or an alternative one in [14]. For any polynomial set $\mathbb{P} \subset \mathbf{K}[\mathbf{u},y_1,\ldots,y_s]\backslash\mathbf{K}[\mathbf{u}]$ and polynomial $P \in \mathbb{P}$, with the notation in [18], $\mathrm{prem}(P,\mathbb{T})$ stands for the *pseudo-remainder* of P with respect to $\mathbb{T}$, and $\mathrm{res}(P,\mathbb{T})$ the *resultant* of P with respect to $\mathbb{T}$, respectively. The extension field $\bar{\mathbf{K}}$ of $\mathbf{K}$ considered in this paper is the complex

J. Calmet, T. Ida, and D. Wang (Eds.): AISC 2006, LNAI 4120, pp. 82–93, 2006.

number field. We speak about the set of all zeros of $\mathbb{P}$ in $\tilde{\mathbf{K}}^n$ which is denoted by Zero($\mathbb{P}$). It is obvious that Zero($\mathbb{P}$) is a closed set in Euclidean space $\tilde{\mathbf{K}}^n$. While speaking about a *polynomial system*, we refer to a pair $[\mathbb{P}, \mathbb{Q}]$ of polynomial sets. The set all zeros of $[\mathbb{P}, \mathbb{Q}]$ is defined as

$$\text{Zero}(\mathbb{P}/\mathbb{Q}) \triangleq \{\mathbf{z} \in \tilde{\mathbf{K}}^n : P(\mathbf{z}) = 0, \ Q(\mathbf{z}) \neq 0, \ \forall P \in \mathbb{P}, \ Q \in \mathbb{Q}\}.$$

For any triangular set $\mathbb{T}$ in the form (1), we write $\text{ldeg}(\mathbb{T})$ for $\prod_{f \in \mathbb{T}} \text{ldeg}(f)$, and sat($\mathbb{T}$) the *saturation* ideal Ideal($\mathbb{T}$) : ini($\mathbb{T}$)$^\infty$, respectively. $\mathbb{C}_{f_i}$ denotes the set of all the nonzero coefficients of f_i in y_i, $I_i = \text{ini}(f_i)$ the leading coefficient of f_i in y_i for each i, and ini($\mathbb{T}$) the set of all I_i. For any $\bar{\mathbf{z}} = (\bar{\mathbf{u}}, \bar{y}_1, \ldots, \bar{y}_s) \in$ Zero($\mathbb{T}$), we write $\bar{\mathbf{z}}^{\{j\}}$ for $\bar{\mathbf{u}}, \bar{y}_1, \ldots, \bar{y}_j$ or $(\bar{\mathbf{u}}, \bar{y}_1, \ldots, \bar{y}_j)$ with $\bar{\mathbf{u}} = \bar{\mathbf{z}}^{\{0\}}$ and $\bar{\mathbf{z}} = \bar{\mathbf{z}}^{\{s\}}$.

A triangular set $\mathbb{T} = [f_1, \ldots, f_s]$ is called a *regular set* which was introduced independently by Yang and Zhang([23]) and Kalkbrener([9]), if res($I_i, \mathbb{T}$) $\neq 0$, for $i = 2, 3, \cdots, s$. In addition, the notion of *Lazard triangular sets* was introduced in [10] which are just special regular sets. Referring to [1, 2], one can find the properties of Lazard triangular sets and regular sets.

The next assertion proved by Aubry et al. in [1] (see also Theorem 6.2.4 in [18]): A triangular set $\mathbb{T}$ is a regular set if and only if sat($\mathbb{T}$) = $\{P \in \mathbf{K}[\mathbf{x}] :$ prem($P, \mathbb{T}$) = 0$\}$.

Let $\mathbb{T}$ be a regular set in $\mathbf{K}[\mathbf{x}]$. A zero $\mathbf{z}_0 \in$ Zero($\mathbb{T}$) is called a *quasi-normal zero* if $\mathbf{z}_0^{\{i-1\}} \notin$ Zero($\mathbb{C}_{f_i}$) for any $1 \leq i \leq s$, also said to be satisfying the *nondegenerate condition* (see [25] for details). $\mathbb{T}$ is called a *strong regular set* if every zero of $\mathbb{T}$ is also a quasi-normal zero. Referring to [12], we have Zero($\mathbb{T}$) = Zero(sat($\mathbb{T}$)) if $\mathbb{T}$ is a strong regular set.

Let $\mathbb{T}$ be a regular set and P a polynomial in $\mathbf{K}[\mathbf{x}]$. The following properties presented in [13] are equivalent:

- Zero($\mathbb{T}/$ini($\mathbb{T}$)) $\subseteq$ Zero($\{P\}$);
- for any quasi-normal zero z_0 of $\mathbb{T}$, $z_0 \in$ Zero($\{P\}$);
- there exists an integer $0 < d \leq \text{ldeg}(\mathbb{T})$ such that prem($P^d, \mathbb{T}$) = 0.

It is natural question to ask whether the above interesting properties can extend to triangular sets in general. This effort has rather theoretical contribution to understand the structure of the affine varieties of triangular sets.

2   Definitions and Properties

We will try to extend the theory of the weakly nondegenerate condition to triangular sets in this section.

2.1   Preliminaries

The following definition is an extension of the concept of quasi-normal zero of regular sets.

Definition 1. Let $\mathbb{T} = [f_1, \ldots, f_s]$ be a triangular set in $\mathbf{K}[\mathbf{x}]$. A zero $\mathbf{z}_0 \in$ Zero($\mathbb{T}$) is called a *quasi-normal zero* of $\mathbb{T}$ if for any $1 \leq i \leq s$, either condition holds:

  a. $I_i(\mathbf{z}_0^{\{i-1\}}) \neq 0$;
  b. $\mathrm{res}(I_i, \mathbb{T}) \neq 0$ and $\mathbf{z}_0^{\{i-1\}} \notin$ Zero($\mathbb{C}_{f_i}$).

Example 2. Let a triangular set $\mathbb{T} = [x_1^2 - u^2, x_1(x_1 + 1)x_2^2 + vx_2 + u(u - 1), (x_1 - u)x_3^2 + x_3 - x_2]$ in $\mathbf{K}[u, v, x_1, x_2, x_3]$, which is not a regular set. The zero $\mathbf{z}_1 = (u = 1, v = 1, x_1 = -1, x_2 = 0, x_3 = 0)$ is a quasi-normal zero of $\mathbb{T}$, but the zero $\mathbf{z}_2 = (u = 0, v = 1, x_1 = 0, x_2 = 0, x_3 = 0)$ is not.

Example 3. Let a triangular set $\mathbb{T} = [ux_1 - s, vx_2^2 + x_2 - u, sx_3 - u]$ in $\mathbf{K}[s, u, v, x_1, x_2, x_3]$. It is easy to see that each element of the following set

$$\mathbf{X}_0 = \{(s = 0, u = 0, v = w_1, x_1 = w_2, x_2 = 0, x_3 = w_3) : \forall w_i \in \tilde{\mathbf{K}}\}$$

is not a quasi-normal zero of $\mathbb{T}$.

For any triangular set $\mathbb{T} \subset \mathbf{K}[\mathbf{x}]$, we denote by QnZero($\mathbb{T}$) the set of all quasi-normal zeros of $\mathbb{T}$ and by $\overline{\mathrm{QnZero}(\mathbb{T})}^E$ the closure of QnZero($\mathbb{T}$) in Euclidean space $\tilde{\mathbf{K}}^n$. $\mathbf{U}(\mathbf{z}^*, \varepsilon)$ denotes the ε-neighborhood of $\mathbf{z}^* = (x_1^*, \ldots, x_n^*) \in \tilde{\mathbf{K}}^n$ defined by the set $\{\mathbf{z} = (x_1, \ldots, x_n) \in \tilde{\mathbf{K}}^n : |x_i - x_i^*| < \varepsilon, i = 1, \cdots, n\}$.

Given a positive integer number n, we define a mapping Ψ_n of $\tilde{\mathbf{K}}^{n+1}$ into $\tilde{\mathbf{K}}[y]$ as follows,

$$\Psi_n(\mathbf{a}) = a_n y^n + a_{n-1} y^{n-1} + \ldots + a_1 y + a_0 \in \tilde{\mathbf{K}}[y],$$

for any $\mathbf{a} = (a_n, a_{n-1}, \ldots, a_1, a_0) \in \tilde{\mathbf{K}}^{n+1}$.

The following result is another description of Lemma 5 in [25] or Lemma 2 in Chapter 3 in [24] with the above notation, which plays an important role in this section.

Lemma 4. Let $G = b_k y^k + \cdots + b_1 y + b_0$ be a polynomial in $\tilde{\mathbf{K}}[y]$ with $k \geq 1$ and $b_k \neq 0$. For any integer $n \geq k$, $\varepsilon > 0$ and $y^* \in$ Zero($\{G\}$), there exits a $\delta > 0$ such that

$$\mathbf{U}(y^*, \varepsilon) \cap \mathrm{Zero}(\{\Psi_n(\mathbf{b})\}) \neq \emptyset$$

for any $\mathbf{b} \in \mathbf{U}(\tilde{\mathbf{b}}, \delta)$ with $\tilde{\mathbf{b}} = (\overbrace{0, \cdots, 0}^{n-k}, b_k, b_{k-1}, \ldots, b_1, b_0) \in \tilde{\mathbf{K}}^{n+1}$.

2.2   Main Results

The following assertion is an extension of Theorem 2 in [25] or Theorem 1 in Chapter 3 in [24].

Theorem 5. For any triangular set $\mathbb{T} = [f_1(\mathbf{u}, y_1), \ldots, f_s(\mathbf{u}, y_1, \ldots, y_s)]$, we have

$$\overline{\mathrm{QnZero}(\mathbb{T})}^E \subseteq \mathrm{Zero}(\mathrm{sat}(\mathbb{T})).$$

Proof: We first claim that $\text{QnZero}(\mathbb{T}) \subseteq \text{Zero}(\text{sat}(\mathbb{T}))$. For any given $\mathbf{z}^* = (\mathbf{u}^*, y_1^*, \cdots, y_s^*) \in \text{QnZero}(\mathbb{T})$, we set $\eta = \{j : I_j(\mathbf{z}^{*\{j-1\}}) = 0 \text{ for } 1 \le j \le s\}$ with the above notation. Given any $P \in \text{sat}(\mathbb{T})$, it follows that there exists an integer $d > 0$ such that

$$(\prod_{i=1}^{s} I_i)^d P \in \text{Ideal}(\mathbb{T}).$$

If $\eta = \emptyset$, namely, $I_i(\mathbf{z}^{*\{i-1\}}) \neq 0$ for any $1 \le i \le s$, then it is obvious that $\mathbf{z}^* \in \text{Zero}(\{P\})$.

Now consider the case $\eta \neq \emptyset$, we assume without loss of generality that $\eta = \{l_1, \ldots, l_m\}$ with $1 \le l_1 \le \cdots \le l_m \le s$. It follows from Definition 1 that

$$\text{res}(I_{l_i}, \mathbb{T}) \in \mathbf{K}[\mathbf{u}] \setminus \{0\}, \text{ for } 1 \le i \le m.$$

Supposing $P(\mathbf{z}^*) \neq 0$, by the continuity of the functions determined by P and I_j for any $j \notin \eta$, there is an $\varepsilon > 0$ such that

$$P(\mathbf{z}) \neq 0, I_j(\mathbf{z}^{\{j-1\}}) \neq 0 \text{ for any } j \notin \eta \text{ and } \mathbf{z} \in \mathbf{U}(\mathbf{z}^*, \varepsilon). \tag{2}$$

According to Lemma 4, we can obtain a number $\delta_s > 0$ (with $\delta_s < \varepsilon$) such that

$$\mathbf{U}(y_s^*, \varepsilon) \cap \text{Zero}(\{f_s(\overline{\mathbf{z}}^{\{s-1\}}, y_s)\}) \neq \emptyset$$

for any $\overline{\mathbf{z}}^{\{s-1\}} \in \mathbf{U}(\mathbf{z}^{*\{s-1\}}, \delta_s)$ whether or not $s \in \eta$.

Analogously, we can find a number $\delta_{s-1} > 0$ (with $\delta_{s-1} < \delta_s$) such that

$$\mathbf{U}(y_{s-1}^*, \delta_s) \cap \text{Zero}(\{f_{s-1}(\overline{\mathbf{z}}^{\{s-2\}}, y_{s-1})\}) \neq \emptyset$$

for any $\overline{\mathbf{z}}^{\{s-2\}} \in \mathbf{U}(\mathbf{z}^{*\{s-2\}}, \delta_{s-1})$ whether or not $s - 1 \in \eta$.

By doing in this way successively, $\cdots$, the numbers $\delta_s, \delta_{s-1}, \cdots, \delta_2$ are obtained one by one.

At last, we can obtain a number $\delta_1 > 0$ (with $\delta_1 < \delta_2$) such that

$$\mathbf{U}(y_1^*, \delta_2) \cap \text{Zero}(\{f_1(\overline{\mathbf{z}}^{\{0\}}, y_1)\}) \neq \emptyset$$

for any $\overline{\mathbf{z}}^{\{0\}} \in \mathbf{U}(\mathbf{z}^{*\{0\}}, \delta_1)$.

Now set $R = \prod_{j \in \eta} \text{res}(I_j, \mathbb{T}) \in \mathbf{K}[\mathbf{u}]$, Since the following set is a dense set in $\tilde{\mathbf{K}}^r$

$$\{\mathbf{u} \in \tilde{\mathbf{K}}^r : R(\mathbf{u}) \neq 0\},$$

we can get an $\mathbf{u}_0 \in \mathbf{U}(\mathbf{z}^{*\{0\}}, \delta_1)$ with $R(\mathbf{u}_0) \neq 0$. Starting from $\mathbf{u}_0$ by above argument. It can be found successively the numbers $y_1^0, \cdots, y_s^0$ such that

$$\mathbf{z}_0 = (\mathbf{u}_0, y_1^0, \cdots, y_s^0) \in \text{Zero}(\mathbb{T}) \cap \mathbf{U}(\mathbf{z}^*, \varepsilon).$$

This implies that $I_i(\mathbf{z}_0) \neq 0$ for any $1 \le i \le s$, so $P(\mathbf{z}_0) = 0$. This reduces a contradiction to (2), hence $P(\mathbf{z}^*) = 0$. It follows that $\text{QnZero}(\mathbb{T}) \subseteq \text{Zero}(\text{sat}(\mathbb{T}))$. Since $\overline{\text{Zero}(\text{sat}(\mathbb{T}))}^E = \text{Zero}(\text{sat}(\mathbb{T}))$ we know that

$$\overline{\text{QnZero}(\mathbb{T})}^E \subseteq \text{Zero}(\text{sat}(\mathbb{T})).$$

This completes the proof of the theorem.

Starting from Theorem 5 and it's proof, the next corollaries can easily be found.

Corollary 6. Let $\mathbb{T} = [f_1(\mathbf{u}, y_1), \ldots, f_s(\mathbf{u}, y_1, \ldots, y_s)]$ be a triangular set. Given a $\check{\mathbf{z}} = (\check{\mathbf{u}}, \check{z}_1, \ldots, \check{z}_s) \in \mathrm{QnZero}(\mathbb{T})$ and $\varepsilon > 0$, there exist a $\check{\delta} > 0$ and a polynomial $R \in \mathbf{K}[\mathbf{u}] \setminus \mathbf{K}$ such that, one can get a zero

$$(\mathbf{u}^*, z_1^*, \ldots, z_s^*) \in \mathbf{U}(\check{\mathbf{z}}, \varepsilon) \bigcap \mathrm{QnZero}(\mathbb{T})$$

for any $\mathbf{u}^* \in \mathbf{U}(\check{\mathbf{u}}, \check{\delta})$ with $R(\mathbf{u}^*) \neq 0$.

Corollary 7. Let $\mathbb{T}$ be a triangular set in $\mathbf{K}[\mathbf{x}]$. If $\mathrm{prem}(P^d, \mathbb{T}) = 0$ for some integer $d > 0$, then $\overline{\mathrm{QnZero}(\mathbb{T})}^E \subseteq \mathrm{Zero}(\{P\})$.

Corollary 8. Let $\mathbb{T}$ be a triangular set in $\mathbf{K}[\mathbf{x}]$ such that $\overline{\mathrm{QnZero}(\mathbb{T})}^E = \mathrm{Zero}(\mathbb{T})$. Then, $\mathrm{Zero}(\mathbb{T}) = \mathrm{Zero}(\mathrm{sat}(\mathbb{T}))$.

Example 9. Let a triangular set $\mathbb{T} = [x_1^2 - u^2, (x_1 + u_1)x_2^2 + (x_1 + u_1)x_2 + 1, x_1 x_3^2 - x_3 + u]$ in $\mathbf{K}[u, x_1, x_2, x_3]$, it is easy to see that $\mathbb{T}$ is not a regular set and $\mathrm{QnZero}(\mathbb{T}) = \mathrm{Zero}(\mathbb{T})$. Then, we have $\mathrm{Zero}(\mathbb{T}) = \mathrm{Zero}(\mathrm{sat}(\mathbb{T}))$ by Corollary 8.

Example 10. Let a triangular set $\mathbb{T} = [x_1 - u_1, u_2 x_2^2 + u_3 x_2 + u_4]$, it is easy to see that $\mathbf{X}_0 \subseteq \mathrm{Zero}(\mathbb{T}) \setminus \mathrm{QnZero}(\mathbb{T})$ where

$$\mathbf{X}_0 = \{(u_1 = c_1, u_2 = u_3 = u_4 = 0, x_1 = c_1, x_2 = c_2) : \forall c_1, c_2 \in \tilde{\mathbf{K}}\}.$$

By the above results, it is difficult to determine the relationship of $\mathrm{Zero}(\mathbb{T})$ and $\mathrm{Zero}(\mathrm{sat}(\mathbb{T}))$. But, we will get that $\mathrm{Zero}(\mathbb{T}) = \mathrm{Zero}(\mathrm{sat}(\mathbb{T}))$ according to the following Theorem 14.

To discuss the general situation, we first introduce some notation. For each triangular set $\mathbb{T}$, we now denote $\mathbb{R}_f = \{\mathrm{res}(c, \mathbb{T}) \neq 0 : c \in \mathcal{C}_f\}$ for any $f \in \mathbb{T}$ and write $\gcd(\mathbb{R}_f)$ for the greatest common divisor of polynomials in $\mathbb{R}_f$ over $\mathbf{K}[\mathbf{u}]$ if $\mathbb{R}_f \neq \emptyset$.

Definition 11. Let $\mathbb{T}$ be a triangular set in $\mathbf{K}[\mathbf{x}]$. We establish

$$\mathbb{U}_{\mathbb{T}} \triangleq \{c : \mathrm{res}(c, \mathbb{T}) = 0, \text{ for } c \in \mathrm{ini}(\mathbb{T})\}$$
$$\cup \{\gcd(\mathbb{R}_f) : \mathrm{res}(\mathrm{ini}(f), \mathbb{T}) \neq 0, \mathbb{R}_f \cap \mathbf{K} = \emptyset, \text{ for } f \in \mathbb{T}\} \setminus \mathbf{K}.$$

One can compute $\mathbb{U}_{\mathbb{T}}$ by the following algorithm Comp for any triangular set $\mathbb{T}$.

Algorithm Comp: $\mathbb{U}_{\mathbb{T}} \leftarrow \mathrm{Comp}(\mathbb{T})$. Given a triangular set $\mathbb{T}$ in $\mathbf{K}[\mathbf{x}]$, this algorithm computes $\mathbb{U}_{\mathbb{T}}$.

C1. Set $\mathbb{U} \leftarrow \emptyset$; $\mathbb{T}^* \leftarrow \mathbb{T}$.
C2. While $\mathbb{T}^* \neq \emptyset$ do:

    C2.1. Let f be an element of $\mathbb{T}^*$ and set $\mathbb{T}^* \leftarrow \mathbb{T}^* \setminus \{f\}$.
    C2.2. Compute $\mathrm{res}(\mathrm{ini}(f), \mathbb{T})$.

C2.2.1. If $\mathrm{res}(\mathrm{ini}(f), \mathbb{T}) = 0$ then set $\mathbb{U} \leftarrow \mathbb{U} \cup \{\mathrm{ini}(f)\}$ and go to C2.
C2.2.2. If $\mathrm{res}(\mathrm{ini}(f), \mathbb{T}) \neq 0$ and $\mathbb{C}_f \cap \mathbf{K} = \emptyset$ then set

$$\mathbb{R}_f^* \leftarrow \{\mathrm{res}(c, \mathbb{T}): \text{ for } c \in \mathbb{C}_f\}.$$

C2.2.3. Set $\mathbb{R}_f \leftarrow \mathbb{R}_f^* \setminus \{0\}$ and $\mathbb{A} \leftarrow \emptyset$. If $\mathbb{R}_f \cap \mathbf{K} = \emptyset$, then compute $\gcd(\mathbb{R}_f)$.
C2.2.4. If $\gcd(\mathbb{R}_f) \in \mathbf{K}[\mathbf{u}] \setminus \mathbf{K}$, then set $\mathbb{A} \leftarrow \mathbb{A} \cup \{\gcd(\mathbb{R}_f)\}$.
C2.2.5. Set $\mathbb{U} \leftarrow \mathbb{U} \cup \mathbb{A}$.
C3. Set $\mathbb{U}_\mathbb{T} \leftarrow \mathbb{U}$.

Example 12. Let a triangular set $\mathbb{T}^* = [f_1, f_2, f_3]$ in $\mathbf{K}[x_1, x_2, x_3, x_4]$ under $x_1 \prec x_2 \prec x_3 \prec x_4$, where

$$\begin{aligned}
f_1 &= -x_2^2 + x_1, \\
f_2 &= -x_2 x_3^3 + (2x_1 - 1)x_3^2 - x_2(x_1 - 2)x_3 - x_1, \\
f_3 &= (x_2 x_3 + 1)x_4 + x_1 x_3 + x_2.
\end{aligned}$$

By the above notation, we know

$$\begin{aligned}
\mathbb{C}_{f_1} &= \{-1, x_1\}, \ \mathbb{C}_{f_2} = \{-x_2, 2x_1 - 1, -x_2(x_1 - 2), -x_1\}, \\
\mathbb{C}_{f_3} &= \{x_2 x_3 + 1, x_1 x_3 + x_2\}.
\end{aligned}$$

It is obvious that

$$\mathbb{R}_{f_1} = \mathbb{C}_{f_1} = \{-1, x_1\}, \ \mathbb{R}_{f_2} = \{-x_1, 2x_1 - 1, -x_1(x_1 - 2)^2, -x_1\},$$

and as $\mathrm{res}(\mathrm{ini}(f_3), \mathbb{T}^*) = 0$. Thus

$$\mathbb{U}_{\mathbb{T}^*} = \{\mathrm{ini}(f_3)\} = \{x_2 x_3 + 1\}.$$

Similarly, one can compute that

$$\mathbb{U}_{\mathbb{T}_1^*} = \{u\}; \ \mathbb{U}_{\mathbb{T}_2^*} = \emptyset;$$

where

$$\begin{aligned}
\mathbb{T}_1^*[1] &= x(27x^4 + 27x^3 + 9x^2 + 36ux^2 + 18x^2 u^2 \\
&\quad + 12u^2 x + x + 24ux + 4u + 2u^2), \\
\mathbb{T}_1^*[2] &= uy - 3x - 1, \\
\mathbb{T}_1^*[3] &= (2u^2 + 3x^2 + x)z - 6x - 2
\end{aligned}$$

under $u \prec x \prec y \prec z$ and $\mathbb{T}_2^* = [ux_1 - s, vx_2^2 + x_2 - u, sx_3 - u]$ under $s \prec u \prec v \prec x_1 \prec x_2 \prec x_3$.

Furthermore, we will get an interesting result that

$$\mathrm{Zero}(\mathbb{T}/\mathbb{U}_\mathbb{T}) \subseteq \mathrm{Zero}(\mathrm{sat}(\mathbb{T}))$$

for any triangular set $\mathbb{T}$. In order to prove it, the next proposition plays a crucial role.

Proposition 13. Let $\mathbb{T} = [f_1(\mathbf{u}, y_1), \ldots, f_s(\mathbf{u}, y_1, \ldots, y_s)]$ be a triangular set and a zero $\mathbf{z}_0 = (\dot{\mathbf{u}}, \dot{z}_1, \ldots, \dot{z}_s) \in \text{Zero}(\mathbb{T}/U_{\mathbb{T}})$ as the above notation. If $\mathbf{z}_0^{\{k-1\}} \in \text{QnZero}(\mathbb{T}^{\{k-1\}})$, but $\mathbf{z}_0^{\{k\}} \notin \text{QnZero}(\mathbb{T}^{\{k\}})$ for some $1 \leq k \leq s$, then, given $\varepsilon > 0$, there exist a $\delta > 0$ and a polynomial $R \in \mathbf{K}[\mathbf{u}] \setminus \mathbf{K}$ such that, one can obtain a zero

$$(\mathbf{u}^*, z_1^*, \ldots, z_k^*) \in \mathbf{U}(\mathbf{z}_0^{\{k\}}, \varepsilon) \bigcap \text{QnZero}(\mathbb{T}^{\{k\}})$$

for any $\mathbf{u}^* \in \mathbf{U}(\dot{\mathbf{u}}, \delta)$ with $R(\mathbf{u}^*) \neq 0$.

Proof. Suppose that

$$f_k = \sum_{c \in \mathbb{C}_{f_k}} c y_k^{n_c}.$$

We first claim $\text{res}(c_0, \mathbb{T}^{\{k-1\}}) \neq 0$ with $c_0 = \text{ini}(f_k)$. In fact, if $\text{res}(c_0, \mathbb{T}^{\{k-1\}}) = 0$, this means $c_0 \in U_{\mathbb{T}}$, then $\mathbf{z}_0^{\{k\}} \in \text{QnZero}(\mathbb{T}^{\{k\}})$, it reduces a contradiction. Furthermore, it follows from $\mathbf{z}_0^{\{k\}} \notin \text{QnZero}(\mathbb{T}^{\{k\}})$ that $c(\mathbf{z}_0^{\{k-1\}}) = 0$ for any $c \in \mathbb{C}_{f_k}$. In addition, we have that $\mathbb{R}_{f_k} \cap \mathbf{K} = \emptyset$ and $r_c(\dot{\mathbf{u}}) = 0$ for any $r_c \in \mathbb{R}_{f_k}$.

Given $\varepsilon > 0$, as $\mathbf{z}_0^{\{k-1\}} \in \text{QnZero}(\mathbb{T}^{\{k-1\}})$, one can obtain a $\delta > 0$ and a polynomial $R_0 \in \mathbf{K}[\mathbf{u}] \setminus \mathbf{K}$ by Corollary 6 such that, there is a zero

$$(\mathbf{u}^*, z_1^*, \ldots, z_{k-1}^*) \in \mathbf{U}(\mathbf{z}_0^{\{k-1\}}, \varepsilon) \bigcap \text{QnZero}(\mathbb{T}^{\{k-1\}}) \tag{3}$$

for any $\mathbf{u}^* \in \mathbf{U}(\dot{\mathbf{u}}, \delta)$ with $R_0(\mathbf{u}^*) \neq 0$.
Now set $R = R_0 \text{res}(c_0, \mathbb{T}^{\{k-1\}})$ and

$$\mathbf{S}^* = \mathbf{U}(\dot{\mathbf{u}}, \delta) \bigcap \{\mathbf{u} \in \mathbf{K}^r : R(\mathbf{u}) \neq 0\}.$$

For any $\bar{\mathbf{u}} \in \mathbf{S}^*$, there is a zero

$$(\bar{\mathbf{u}}, \bar{z}_1, \ldots, \bar{z}_{k-1}) \in \mathbf{U}(\mathbf{z}_0^{\{k-1\}}, \varepsilon) \bigcap \text{QnZero}(\mathbb{T}^{\{k-1\}})$$

by the form (3), it implies that $c_0(\bar{\mathbf{u}}, \bar{z}_1, \ldots, \bar{z}_{k-1}) \neq 0$. Set

$$\mathbf{Z} = \{(\bar{\mathbf{u}}, \bar{z}_1, \ldots, \bar{z}_{k-1}) \in \mathbf{U}(\mathbf{z}_0^{\{k-1\}}, \varepsilon) \bigcap \text{QnZero}(\mathbb{T}^{\{k-1\}}) : \bar{\mathbf{u}} \in \mathbf{S}^*\}.$$

We proceed to prove that

$$\mathbf{U}(\mathbf{z}_0^{\{k\}}, \varepsilon) \bigcap \text{QnZero}(\mathbb{T}^{\{k\}}) \neq \emptyset.$$

It follows from the fact $c(\mathbf{z}_0^{\{k-1\}}) = 0$ and $r_c(\dot{\mathbf{u}}) \neq 0$ for any $c \in \mathbb{C}_{f_k}$ that

$$\inf(\{f_k(\bar{\mathbf{u}}, \bar{z}_1, \ldots, \bar{z}_{k-1}, z_k) : (\bar{\mathbf{u}}, \bar{z}_1, \ldots, \bar{z}_{k-1}) \in \mathbf{Z}\}) = 0$$

for any $z_k \in (\dot{z}_k - \varepsilon, \dot{z}_k + \varepsilon)$.
Given a $\bar{z}_k \in (\dot{z}_k - \varepsilon/2, \dot{z}_k + \varepsilon/2)$, the above fact implies that there is a $(\bar{\mathbf{u}}, \bar{z}_1, \ldots, \bar{z}_{k-1}) \in \mathbf{Z}$ and δ_0 such that $f_k(\bar{\mathbf{u}}, \bar{z}_1, \ldots, \bar{z}_{k-1}, z_k) = \delta_0$ with $|\delta_0| < \bar{\delta}$ for any $\bar{\delta} > 0$. Let $\bar{f}_k = f_k(\bar{\mathbf{u}}, \bar{z}_1, \ldots, \bar{z}_{k-1}, y_k) \in \mathbf{K}[y_k]$ and $g_k = \bar{f}_k -$

$\bar{\delta}_0$ where $g_k(\bar{z}_k) = 0$. According to Lemma 4, one can prove that there is a $(\mathbf{u}^\star, z_1^\star, \ldots, z_{k-1}^\star) \in \mathbf{Z}$ and $z_k^\star \in (\dot{z}_k - \varepsilon, \dot{z}_k + \varepsilon)$ such that f_k vanishes at $(\mathbf{u}^\star, z_1^\star, \ldots, z_{k-1}^\star, z_k^\star)$. Hence

$$(\mathbf{u}^\star, z_1^\star, \ldots, z_{k-1}^\star, z_k^\star) \in \mathbf{U}(\mathbf{z}_0^{\{k\}}, \varepsilon) \bigcap \mathrm{QnZero}(\mathbb{T}^{\{k\}}).$$

Since ε is given arbitrarily, this completes the proof the proposition.

The next result follows by an argument analogous to the proof of Theorem 5, we omit the details.

Theorem 14. For any triangular set $\mathbb{T}$, we have

$$\mathrm{Zero}(\mathbb{T}/\mathbb{U}_\mathbb{T}) \subseteq \overline{\mathrm{QnZero}(\mathbb{T})}^E \subseteq \mathrm{Zero}(\mathrm{sat}(\mathbb{T})).$$

Corollary 15. Let $\mathbb{T}$ be a triangular set in $\mathbf{K}[\mathbf{x}]$ with $\mathbb{U}_\mathbb{T} = \emptyset$. Then,

$$\mathrm{Zero}(\mathbb{T}) = \mathrm{Zero}(\mathrm{sat}(\mathbb{T})).$$

Example 16. (Continued from Example 10). Since $\mathbb{U}_\mathbb{T} = \emptyset$, we have that $\mathrm{Zero}(\mathbb{T}) = \mathrm{Zero}(\mathrm{sat}(\mathbb{T}))$.

3 Improvement Upon Algorithm CharSer

Let $\mathbb{T}$ be a triangular set as (1) and P any polynomial. P is said to be *reduced* with respect to $\mathbb{T}$ if $\deg(P, y_i) < \deg(f_i, y_i)$ for all i. $\mathbb{T}$ is said to be *noncontradictory ascending set* if every $f \in \mathbb{T} \cup \mathrm{ini}(\mathbb{T})$ is reduced with respect to $\mathbb{T} \setminus \{f\}$. The following concept of *characteristic sets* was introduced by Wu in [20].

Definition 17. An ascending set $\mathbb{T}$ is called a *characteristic set* of nonempty polynomial set $\mathbb{P} \subset \mathbf{K}[\mathbf{x}]$ if

$$\mathbb{T} \subset \mathrm{Ideal}(\mathbb{P}), \quad \mathrm{prem}(\mathbb{P}, \mathbb{T}) = \{0\}.$$

Definition 18. A finite set or sequence Ψ of ascending sets $\mathbb{T}_1, \ldots, \mathbb{T}_e$ is called a *characteristic series* of polynomial set $\mathbb{P}$ in $\mathbf{K}[\mathbf{x}]$ if the following zero decomposition holds

$$\mathrm{Zero}(\mathbb{P}) = \bigcup_{i=1}^e \mathrm{Zero}(\mathbb{T}_i/\mathrm{ini}(\mathbb{T}_i))$$

and $\mathrm{prem}(\mathbb{P}, \mathbb{T}_i) = \{0\}$ for every i.

Given a nonempty polynomial set $\mathbb{P} \subset \mathbf{K}[\mathbf{x}]$, two algorithms CharSet and CharSer[1] which were developed by Wu in [20, 21, 22] and described by Wang in [18] compute a characteristic set $\mathbb{T}$ of $\mathbb{P}$ (denoted by CharSet($\mathbb{P}$)) and characteristic series $\{\mathbb{T}_1, \ldots, \mathbb{T}_e\}$ of $\mathbb{P}$. Several authors continued and improved Wu's

---

[1] One can obtain their respective implementations in the Epsilon library at: http://www-calfor.lip6.fr/~wang/epsilon.

approach, mainly Chou and Gao([3, 4, 5]), Dahan et al.([6]), Gallo and Mishra ([7, 8]), Li([11]) and Wang([15, 16, 17]). As an application of these new properties of triangular sets, we proceed to improve algorithm CharSer. The next result which follows from Theorem 14 plays an important in this section.

Proposition 19. Let $\mathbb{P}$ a nonempty polynomial set in $\mathbf{K}[\mathbf{x}]$. If $\mathbb{U}_{\mathbb{T}} = \emptyset$ with $\mathbb{T} = \text{CharSet}(\mathbb{P})$, then $\text{Zero}(\mathbb{T}) = \text{Zero}(\mathbb{P})$.

Based algorithm CharSet, we next to present two modified versions of algorithm CharSer according to the above results, and omit the details of the proofs. The following algorithm CharSerA is only different to in step qC2.5 of CharSer in [18] where ini($\mathbb{T}$) is replaced by $\mathbb{U}_{\mathbb{T}}$.

Algorithm qCharSerA: $\Psi \leftarrow \text{CharSerA}(\mathbb{P})$. Given a nonempty polynomial set $\mathbb{P}$ in $\mathbf{K}[\mathbf{x}]$, this algorithm computes a finite set triangular sets Ψ such that

$$\text{Zero}(\mathbb{P}) = \bigcup_{\mathbb{T} \in \Psi} \text{Zero}(\mathbb{T}/\mathbb{U}_{\mathbb{T}}).$$

qC1. Set $\Phi \leftarrow \{\mathbb{P}\}$, $\Psi \leftarrow \emptyset$.
qC2. While $\Phi \neq \emptyset$ do:

    qC2.1. Let $\mathbb{F}$ be an element of Φ and set $\Phi \leftarrow \Phi \backslash \{\mathbb{F}\}$.
    qC2.2. Compute $\mathbb{T} \leftarrow \text{CharSet}(\mathbb{F})$.
    qC2.3. If $\mathbb{T}$ is noncontradictory, then compute $\mathbb{U}_{\mathbb{T}}$ by algorithm Comp.
    qC2.4. If $\mathbb{U}_{\mathbb{T}} = \emptyset$, then set $\Psi \leftarrow \Psi \cup \{\mathbb{T}\}$.
    qC2.5. If $\mathbb{U}_{\mathbb{T}} \neq \emptyset$, then set

$$\Psi \leftarrow \Psi \cup \{\mathbb{T}\}, \quad \Phi \leftarrow \Phi \cup \{\mathbb{F} \cup \mathbb{T} \cup \{I\} : I \in \mathbb{U}_{\mathbb{T}}\}.$$

In order to avoid producing superfluous triangular sets at the utmost, we proceed to improve further algorithm QuasiCharSerA. Note that we always treat elements of $\mathbb{U}_{\mathbb{T}}$ in the above algorithm equally without discrimination. More delicate considering elements of $\mathbb{U}_{\mathbb{T}}$ and the above results lead to the following algorithm.

Algorithm QuasiCharSerB: $\Xi \leftarrow \text{QuasiCharSerB}(\mathbb{P})$. Given a nonempty polynomial set $\mathbb{P}$ in $\mathbf{K}[\mathbf{x}]$, this algorithm computes a finite set of triangular systems Ξ such that

$$\text{Zero}(\mathbb{P}) = \bigcup_{[\mathbb{T}, \mathbb{U}] \in \Xi} \text{Zero}(\mathbb{T}/\mathbb{U}).$$

qC1. Set $\Phi \leftarrow \emptyset$, $\Xi \leftarrow \emptyset$.
qC2. Compute $\mathbb{T} \leftarrow \text{CharSet}(\mathbb{P})$.
qC3. If $\mathbb{T}$ is non-contradictory, then set $\Phi \leftarrow \{< \mathbb{P}, \mathbb{T} >\}$.
qC4. While $\Phi \neq \emptyset$ do:

    qC4.1. Let $< \mathbb{F}_1, \mathbb{F}_2 >$ be an element of Φ and set $\Phi \leftarrow \Phi \backslash \{< \mathbb{F}_1, \mathbb{F}_2 >\}$.
    qC4.2. Compute $\mathbb{U}_{\mathbb{T}} \leftarrow \text{Comp}(\mathbb{F}_2)$ and set $\mathbb{U} \leftarrow \mathbb{U}_{\mathbb{T}}$.
    qC4.3. If $\mathbb{U} = \emptyset$, then set $\Xi \leftarrow \Xi \cup \{[\mathbb{F}_2, \emptyset]\}$ and go to qC4.

qC4.4. Set $\mathbb{U}_0 \leftarrow \mathbb{U}$, $\mathbb{U}^* \leftarrow \emptyset$.

qC4.5. While $\mathbb{U}_0 \neq \emptyset$ do:

    qC4.5.1. Let I be an element of $\mathbb{U}_0$ and set $\mathbb{U}_0 \leftarrow \mathbb{U}_0 \backslash \{I\}$.

    qC4.5.2. Compute $\mathbb{T}_0 \leftarrow \text{CharSet}(\mathbb{F}_2 \cup \{I\})$ and $\mathbb{U}_{\mathbb{T}_0} \leftarrow \text{Comp}(\mathbb{T}_0)$. If $\mathbb{U}_{\mathbb{T}_0} = \emptyset$ and $\text{prem}(\mathbb{P}, \mathbb{T}_0) = \{0\}$, then go to qC4.5.

    qC4.5.3. Compute $\mathbb{T} \leftarrow \text{CharSet}(\mathbb{F}_1 \cup \mathbb{F}_2 \cup \{I\})$ and set

$$\Phi \leftarrow \Phi \cup \{< \mathbb{F}_1 \cup \mathbb{F}_2 \cup \{I\}, \mathbb{T} >\}.$$

    qC4.5.4. Set $\mathbb{U}^* \leftarrow \mathbb{U}^* \cup \{I\}$.

qC4.6. Set $\Xi \leftarrow \Xi \cup \{[\mathbb{F}_2, \mathbb{U}^*]\}$.

Example 20. Let $\mathbb{P} = \{P_1, P_2, P_3\}$ with

$$P_1 = -x_2 x_3 x_4 - x_4 - x_1 x_3 + x_2^2 - x_2 - x_1,$$
$$P_2 = x_2 x_3^3 - 2x_1 x_3^2 + x_3^2 + x_1 x_2 x_3 - 2x_2 x_3 + x_1,$$
$$P_3 = 2x_2 x_3 x_4 + 2x_4 + x_2 x_3^3 - 2x_1 x_3^2 + x_3^2 + x_1 x_2 x_3$$
$$\quad - 2x_2 x_3 + 2x_1 x_3 + 2x_2 + x_1.$$

Under the variable ordering $x_1 \prec x_2 \prec x_3 \prec x_4$, one can compute $\text{Charset}(\mathbb{P}) = \mathbb{T}^*$ which has already been given in Example 12. By the above description, one can easily get $\text{QuasiCharSerA}(\mathbb{P}) = \{\mathbb{T}^*, \mathbb{T}_2\}$, where $\mathbb{T}_2 = [-x_1 + x_2^2, x_2 x_3 + 1]$. It is easy to see that $\mathbb{U}_{\mathbb{T}_2} = \emptyset$, this implies that

$$\text{Zero}(\mathbb{P}) = \text{Zero}(\mathbb{T}^*/\mathbb{U}_{\mathbb{T}^*}) \cup \text{Zero}(\mathbb{T}_2/\mathbb{U}_{\mathbb{T}_2}) = \text{Zero}(\mathbb{T}^*/\{x_2 x_3 + 1\}) \cup \text{Zero}(\mathbb{T}_2).$$

One can check that $\text{prem}(\mathbb{P}, \mathbb{T}_2) = \{0\}$ and $\mathbb{T}_2 = \text{CharSet}(\mathbb{T}^* \cup \{x_2 x_3 + 1\})$, this means $\mathbb{T}_2$ is a redundant one. We can compute $\text{QuasiCharSerB}(\mathbb{P}) = \{[\mathbb{T}^*, \emptyset]\}$. Namely,

$$\text{Zero}(\mathbb{P}) = \text{Zero}(\mathbb{T}^*).$$

Compared with algorithm CharSer, we get $\text{CharSer}(\mathbb{P}) = \{\mathbb{T}^*, \mathbb{T}_2, \mathbb{T}_3\}$, where $\mathbb{T}_3 = [x_1, x_2, x_3, x_4]$, and

$$\text{Zero}(\mathbb{P}) = \text{Zero}(\mathbb{T}^*/\{x_2 x_3 + 1\}) \cup \text{Zero}(\mathbb{T}_2/\{x_2\}) \cup \text{Zero}(\mathbb{T}_3).$$

Example 21. Let $\mathbb{P} = \{uy - 3x - 1, -2zu - yxz + 2y, xz^2 - zu + y\}$. This set of polynomials has been considered by Wang in [19]. Under the variable ordering $u \prec x \prec y \prec z$, one can compute $\text{Charset}(\mathbb{P}) = \mathbb{T}_1^*$ which is given in Example 12. By the above description, one can easily get $\text{QuasiCharSerA}(\mathbb{P}) = \{\mathbb{T}_1^*, \mathbb{T}_2, \mathbb{T}_3\}$ and $\text{QuasiCharSerB}(\mathbb{P}) = \{[\mathbb{T}_1^*, \{u\}], [\mathbb{T}_2, \{y\}], [\mathbb{T}_3, \emptyset]\}$, where $\mathbb{T}_2 = [u, 3x + 1, y(y - 12), y(6 + z)]$, $\mathbb{T}_3 = [u, 3x + 1, y, z^2]$. It follows that

$$\text{Zero}(\mathbb{P}) = \text{Zero}(\mathbb{T}_1^*/\{u\}) \cup \text{Zero}(\mathbb{T}_2/\{y\}) \cup \text{Zero}(\mathbb{T}_3).$$

Compared with algorithm CharSer, we have that $\text{CharSer}(\mathbb{P}) = \{\mathbb{T}_1^*, \mathbb{T}_2, \mathbb{T}_3\}$, and

$$\text{Zero}(\mathbb{P}) = \text{Zero}(\mathbb{T}_1^*/\{2u^2 + 3x^2 + x, u\}) \cup \text{Zero}(\mathbb{T}_2/\{y\}) \cup \text{Zero}(\mathbb{T}_3).$$

Example 22. Let $\mathbb{P} = \{ux_1 - s, vx_2^2 + x_2 - u, sx_3 - u\}$ under the variable ordering $s \prec u \prec v \prec x_1 \prec x_2 \prec x_4$. It is obvious that $\mathrm{CharSet}(\mathbb{P}) = \mathbb{T}_2^*$ which is given in Example 12 or the same as $\mathbb{T}$ in Example 3. We know that $\mathbb{U}_{\mathbb{T}_2^*} = \emptyset$. This implies that

$$\mathrm{QuasiCharSerA}(\mathbb{P}) = \{\mathbb{T}_2^*\}; \ \mathrm{QuasiCharSerB}(\mathbb{P}) = \{[\mathbb{T}_2^*, \emptyset]\}.$$

It follows that $\mathrm{Zero}(\mathbb{P}) = \mathrm{Zero}(\mathbb{T}_2^*)$.

Compared with algorithm CharSer, we have that

$$\mathrm{CharSer}(\mathbb{P}) = \{\mathbb{T}_2^*, \mathbb{T}_2, \mathbb{T}_3, \mathbb{T}_4\}$$

where

$$\mathbb{T}_2 = [v, ux_1 - s, -x_2 + u, -sx_3 + u], \ \mathbb{T}_3 = [s, u, x_2(vx_2 + 1)], \mathbb{T}_4 = [s, u, v, x_2]$$

and

$$\mathrm{Zero}(\mathbb{P}) = \mathrm{Zero}(\mathbb{T}_2^*/\{s, u, v\}) \cup \mathrm{Zero}(\mathbb{T}_2/\{u, s\}) \cup \mathrm{Zero}(\mathbb{T}_3/\{v\}) \cup \mathrm{Zero}(\mathbb{T}_4).$$

Acknowledgments. The author is very grateful to Prof. Lu Yang and Prof. Jing-Zhong Zhang who taught him their work and thoughts, and have kept advising and helping him, and to the referees for their suggestions and corrections in this paper.

References

1. Aubry, P., Lazard, D., Moreno Maza, M. On the theories of triangular sets. *J. Symb. Comput.*, 1999, **28**: 105-124.
2. Aubry, P., Moreno Maza, M. Triangular sets for solving polynomial systems: a comparative implementation of four methods. *J. Symb. Comput.*, 1999, **28**: 125-154.
3. Chou, S.-C., Gao, X.-S. Ritt-Wu's decomposition algorithm and geometry theorem proving. In *Springer's LNCS 449*, 1990, 207-220.
4. Chou, S.-C., Gao, X.-S. Computations with parametric equations. In *Proceedings ISAAC'91*, 1991, 122-127.
5. Chou, S.-C., Gao, X.-S. Solving parametric algebraic systems. In *Proceedings ISAAC'92*, 1992, 335-341.
6. Dahan, X., Moreno Maza, M., Schost, É., Wu, W., Xie, Y. Lifting techniques for triangular decomposition. In *ISSAC'05*, 2005, 108-115.
7. Gallo, G., Mishra, B. Efficient algorithms and bounds for Wu-Ritt characteristic sets. In *Proceedings MEGA'90*, 1990, 119-142.
8. Gallo, G., Mishra, B. Wu-Ritt characteristic sets and their complexity. In Goodman, J.E., Pollack, R. and Steiger, W., eds, Discrete and Computational Geometry: Papers from the DIMACS Special Year, volume 6, *Dimacs Series in Discrete Mathematics and Theoretical Computer Science*, 1991, 111-136.
9. Kalkbrener, M. A generalized Euclidean algorithm for computing triangular representations of algebraic varieties. *J. Symb. Comput.*, 1993, **15**: 143-167.

10. Lazard, D. A new method for solving algebraic systems of positive dimension. *Discrete Appl. Math.*, 1991, **33**: 147-160.

11. Li, Z.-M. Determinant polynomial sequences. *Chinese Sci. Bull.*, 1989, **34**: 1595-1599.

12. Li, Y.-B. Zhang, J.-Z., Yang, L. Decomposing polynomial systems into strong regular sets. In *Proceedings ICMS 2002*, 2002, 361-371.

13. Li, Y.-B. Applications of the theory of weakly nondegenerate conditions to zero decomposition for polynomial systems. *J. Symb. Comput.*, 2004, **38**: 815-832.

14. Li, Y.-B. An alternative algorithm for computing the pseudo-remainder of multivariate polynomials. *Applied Mathematics and Computation*, 2006, **173**: 484-492.

15. Wang, D. Some improvements on Wu's method for solving systems of algebraic equations. In Wen-Tsün, W. and Min-De, C., eds, *Proc. of the Int. Workshop on Math. Mechanisation, Beijing, China.* 1992.

16. Wang, D. An elimination method for polynomial systems. *J. Symb. Comput.*, 1993, **16**, 83-114.

17. Wang, D. An implementation of the characteristic set method in Maple. In Pfalzgraf, J. and Wang, D., eds, *Automated Practical Reasoning: Algebraic Approaches.* Springer, Wien, 1995, 187-201.

18. Wang, D. Elimination methods. Springer, Wien/New York, 2001.

19. Wang, D. Elimination Practice. Imperial College Press, London, 2003.

20. Wu, W.-T. On the decision problem and the mechanization of theorem-proving in elementary geometry. *Scientia Sinica*, 1978, **21**: 159-172.

21. Wu, W.-T. On zeros of algebraic equations–an application of Ritt principle. *Kexue Tongbao*, 1986, **31**, 1-5.

22. Wu, W.-T. A zero structure theorem for polynomial equations solving. *MM Research Preprints*, 1987, **1**, 2-12.

23. Yang, L., Zhang, J.-Z. Search dependency between algebraic equations: An algorithm applied to automated reasoning. Technical Report ICTP/91/6, International Center For Theoretical Physics, International Atomic Energy Agency, Miramare, Trieste, 1991.

24. Yang L., Zhang J.-Z., Hou, X.-R. Non-Linear equation systems and automated theorem proving, Shanghai: Shanghai Sci & Tech Education Publ. House, 1996 (in Chinese).

25. Zhang, J.-Z., Yang, L., Hou, X.-R. A note on Wu Wen-Tsün's nondegenerate condition. Technical Report ICTP/91/160, International Center For Theoretical Physics, International Atomic Energy Agency, Miramare, Trieste, 1991; Also in *Chinese Science Bulletin*, 1993, **38**:1, 86-87.

A New Definition for Passivity and Its Relation to Coherence*

Moritz Minzlaff[1] and Jacques Calmet[2]

[1] Universität Karlsruhe
moritz.minzlaff@stud.uni-karlsruhe.de
[2] Fakultät für Informatik, Universität Karlsruhe
calmet@ira.uka.de

Abstract. It is an essential step in decomposition algorithms for radical differential ideals to satisfy the so-called Rosenfeld property. Basically all approaches to achieve this step are based on one of two concepts: *Coherence* or *passivity*.

In this paper we will give a modern treatment of passivity. Our focus is on questions regarding the different definitions of passivity and their relation to coherence. The theorem by Li and Wang stating that passivity in Wu's sense implies coherence is extended to a broader setting. A new definition for passivity is suggested and it is shown to allow for a converse statement so that coherence and passivity are seen to be equivalent.

1 Introduction

The Rosenfeld property provides a link between differential algebra and algebra. It allows to solve questions about a system of differential polynomials in a purely algebraic fashion. For example, if the given system of differential polynomials is known to have the Rosenfeld property, its radical differential ideal may be decomposed into prime differential ideals using tools in a finite polynomial algebra [1].

To satisfy the Rosenfeld property, decomposition algorithms follow one of two methods: Either the concept of *coherence*, which appeared in Rosenfeld's original work [12] and which bears resemblance to using S-polynomials in Gröbner basis theory, is used, or involutive ideas are applied to guarantee *passivity* of the given differential system.

While the definition of coherence is standardized, the same cannot be said about passivity. Not only have different involutive divisions been used, often passivity is introduced in correspondence with a previously fixed method for involutive completion of the given system which again is based on a certain involutive reduction. While these definitions and coherence are believed to be linked concepts, little has been proven about their relation. This naturally leads to the following three questions which are at the centre of this work:

---

* Work partially supported by the EU GIFT project (NEST- Adventure Project no. 5006).

J. Calmet, T. Ida, and D. Wang (Eds.): AISC 2006, LNAI 4120, pp. 94–105, 2006.

1. Is passivity independent of the chosen involutive completion?
2. Every passive system (with respect to Wu's definition) is coherent. [9] Is the converse also true?
3. Is passivity independent of the chosen involutive division?

In the following, we will show that the answer to all three questions is "Yes". We first introduce the notions that will be used in this paper in Sect. 2. In Sect. 3, a new definition for passivity is suggested. We show this new definition to be independent of the way that completions are computed and reduction is performed. Section 4 contains a comparison of our definition with passivity in the sense of other authors. We show that the new definition is equivalent to the one provided by Wu and that it is closely related to passivity in the sense of Ritt and Chen & Gao. Section 5 is devoted to an equivalence statement between coherence and passivity. In the final Sect. 6 we summarize our main results.

2   Basic Notions

We use Kolchin's "Differential Algebra and Algebraic Groups" [8] and a paper by Seiler [13] as the main references for differential algebra and involutive divisions, respectively. If not mentioned otherwise, their notation will be used.

Differential Algebra. Let $\mathbb{K}$ be a differential field of characteristic zero with set of m pairwise commuting derivations $\Delta = \{\delta_1, \ldots, \delta_m\}$. We adopt the convention to write "Δ-" instead of "differential". The commutative monoid generated by Δ is denoted by Θ; Θ^+ be the set Θ without its identity. The elements of Θ are called *derivative operators*. As the monoids Θ and $\mathbb{N}^m$ are isomorphic, we identify derivative operators with m-tuples of non-negative integers; this also implies that we write the addition of two elements of $\mathbb{N}^m$ multiplicatively. We consider the Δ-polynomial ring $\mathbb{K}[\![\mathcal{Y}]\!]$ in n Δ-indeterminates $\mathcal{Y} = \{Y_1, \ldots, Y_n\}$.[1] The elements of $\Theta\mathcal{Y}$ are called *derivatives*. Given a subset $A \subseteq \mathbb{K}[\![\mathcal{Y}]\!]$, denote by $\langle A \rangle$ and $[A]$ the *ideal* and the Δ-*ideal* generated by A, respectively. Furthermore, let ΘA and $\Theta^+ A$ be the union of all θa, where $a \in A, \theta \in \Theta$ and $\theta \in \Theta^+$, respectively. The *saturation* of an ideal I by a finite set $H \subseteq \mathbb{K}[\![\mathcal{Y}]\!]$ is defined as

$$I : H^\infty := \{p \in \mathbb{K}[\![\mathcal{Y}]\!] \mid \exists h \in H^\infty : hp \in I\},$$

where H^∞ is the smallest multiplicatively closed subset of $\mathbb{K}[\![\mathcal{Y}]\!]$ containing H. If I is a Δ-ideal, then so is $I : H^\infty$.

We choose a fixed ranking on $\Theta\mathcal{Y}$. Hence, we may speak of the leader, initial, separant, and degree of a given $p \in \mathbb{K}[\![\mathcal{Y}]\!] \setminus \mathbb{K}$. These shall be denoted by $\mathrm{LEAD}(p), \mathrm{INIT}(p), \mathrm{SEP}(p)$, and $\mathrm{DEG}(p)$, respectively. For a subset $A \subseteq \mathbb{K}[\![\mathcal{Y}]\!] \setminus \mathbb{K}$, let $\mathrm{LEAD}(A), I_A$, and S_A be the set of all leaders, initials, and separants of elements of A, respectively. We put $H_A := I_A \cup S_A$. Given a subset $A \subseteq \mathbb{K}[\![\mathcal{Y}]\!] \setminus \mathbb{K}$

---

[1] Note that, in contrast to Kolchin, we use the notation of [6] to refer to rings of Δ-polynomial.

and $u \in \Theta \mathcal{Y}$, $A_{<u}$ and $A_{\leq u}$ denote the union of all elements in A whose leaders rank lower than u and lower than or equal to u, respectively; likewise for $A_{\geq u}$ and $A_{>u}$. As we will often work with Noetherian induction, we recall that rankings are well-orderings.

Definition 1. *A Δ-system is an ordered pair (A, H) of finite, non-empty subsets $A, H \subseteq \mathbb{K}[\![\mathcal{Y}]\!]$ such that $A \cap \mathbb{K} = \emptyset$ and $H_A \subseteq H^\infty$.*

Whenever (A, H) is a Δ-system we assume that A is triangular and partially autoreduced and that H is partially reduced w.r.t. A.

Definition 2. *Let $p, q \in R$ with $\mathrm{LEAD}(p) = v = \mathrm{LEAD}(q)$, $d = \mathrm{DEG}(p)$, and $e = \mathrm{DEG}(q)$. The pseudo-S-polynomial $S(p, q)$ of p and q is defined as*

$$S(p, q) := \frac{\mathrm{INIT}(q)v^e p - \mathrm{INIT}(p)v^d q}{\mathrm{GCD}(v^d, v^e)} .$$

Definition 3. *A Δ-system (A, H) is coherent if for every pair $(p, q) \in \Theta^+ A \times \Theta^+ A$ such that $\mathrm{LEAD}(p) = v = \mathrm{LEAD}(q)$ the following holds:*

$$S(p, q) \in \langle \Theta A_{<v} \rangle : H^\infty .$$

Coherence can be decided algorithmically: For all $Y \in \mathcal{Y}$, consider all pairs $(p, q) \in A \times A$ with $\mathrm{LEAD}(p) = \alpha Y$ and $\mathrm{LEAD}(q) = \beta Y$, some $\alpha, \beta \in \mathbb{N}^m$. Given such a pair, define $\gamma := \mathrm{LCM}(\alpha, \beta)$. Then (A, H) is coherent if the above condition is satisfied by the finitely many pairs $(\frac{\gamma}{\alpha}p, \frac{\gamma}{\beta}q)$.

Theorem 1 (Rosenfeld Lemma). *Let (A, H) be a coherent Δ-system. In that case, every $p \in [A] : H^\infty$ that is partially reduced w.r.t. A belongs to $\langle A \rangle : H^\infty$.*

A proof has been given by Boulier et al. [1] The latter property in the above theorem is called the *Rosenfeld property*; examples of its applications can be found in many parts of the literature. [1,5,7] Recall that coherence is only a sufficient condition for the Rosenfeld property. [14]

Involutive Divisions and Δ-Polynomials. Let $\mathcal{N} \subseteq \mathbb{N}^m$ be finite and $\alpha \in \mathcal{N}$. The *cone* of $\alpha \in \mathcal{N}$ is given by $\mathcal{C}(\alpha) := \{\beta\alpha \mid \beta \in \mathbb{N}^m\}$. The *span* $\langle \mathcal{N} \rangle$ of $\mathcal{N}$ is the union of all cones of its members, i.e. $\langle \mathcal{N} \rangle = \cup_{\alpha \in \mathcal{N}} \mathcal{C}(\alpha)$. Given an involutive division $\mathcal{L}$, we denote by $N_{\mathcal{N},\alpha} \subseteq \{1, \ldots, m\}$ the set of *multiplicative indices* of α relative to $\mathcal{N}$ (and $\mathcal{L}$). The *involutive cone* $\mathcal{C}_{\mathcal{L},\mathcal{N}}(\alpha)$ of α relative to $\mathcal{N}$ is given by the set

$$\{\beta\alpha \mid (\beta_1, \ldots, \beta_m) \in \mathbb{N}^m, \beta_i \neq 0 \text{ only if } i \in N_{\mathcal{N},\alpha}\} .$$

The *involutive span* $\langle \mathcal{N} \rangle_{\mathcal{L}}$ of $\mathcal{N}$ is the union of all involutive cones of its members. A finite set $\mathcal{N} \subseteq \mathcal{N}' \subseteq \mathbb{N}^m$ is an *involutive completion* of $\mathcal{N}$, if $\langle \mathcal{N}' \rangle_{\mathcal{L}} = \langle \mathcal{N} \rangle$. It is *minimal* if $\mathcal{N}'$ is contained in every involutive completion of $\mathcal{N}$. In general, (minimal) involutive completions need not exist. However, if there are no

$\alpha, \beta \in \mathcal{N}$ such that α is a divisor of β, then there is always a minimal involutive completion relative to Thomas and Janet division; see [4].

For a subset $A \subseteq \mathbb{K}[\![\mathcal{Y}]\!] \setminus \mathbb{K}$, define for each $Y \in \mathcal{Y}$ a set $\mathcal{N}_{A,Y} := \{\alpha \in \mathbb{N}^m \mid \alpha Y \in \text{LEAD}(A)\}$. Given a $p \in A$ with $\text{LEAD}(p) = \alpha Y$, the *involutive cone of p* is defined as

$$\mathcal{C}_{\mathcal{L},A}(p) := \{\beta p \mid \beta \alpha \in \mathcal{C}_{\mathcal{L},\mathcal{N}_{A,Y}}(\alpha)\} \ .$$

An element of $\Theta\{p\}$ is called a *multiplicative prolongation (of $p \in A$)* if it belongs to $\mathcal{C}_{\mathcal{L},A}(p)$. It is a non-multiplicative prolongation otherwise. A derivative $u \in \Theta\mathcal{Y}$ is said to be a *proper multiplicative derivative of A (or more precisely of* $\text{LEAD}(p)$, $p \in A$) if $u \in \text{LEAD}(\mathcal{C}_{\mathcal{L},A}(p))\setminus\{\text{LEAD}(p)\}$. Finally, a Δ-polynomial that does not contain any proper multiplicative derivative of A is said to be *partially involutively reduced* w.r.t. A.

3   Passive Differential Systems

Throughout this section, let $\mathcal{L}$ be a fixed but arbitrary involutive division.

Definition 4. *Let $A \subseteq \mathbb{K}[\![\mathcal{Y}]\!] \setminus \mathbb{K}$ be finite. The* involutive span $[A]_{\mathcal{L}}$ *of A is given by*

$$[A]_{\mathcal{L}} := \left\langle \bigcup_{p \in A} \mathcal{C}_{\mathcal{L},A}(p) \right\rangle \ .$$

For every $u \in \Theta\mathcal{Y}$, define a subideal $[A]_{\mathcal{L}}^u$ of $[A]_{\mathcal{L}}$ by

$$[A]_{\mathcal{L}}^u := \left\langle \bigcup_{p \in A} \mathcal{C}_{\mathcal{L},A}(p) \cap \mathbb{K}[\Theta\mathcal{Y}_{\leq u}] \right\rangle \ .$$

Definition 5. *Let $A \subseteq \mathbb{K}[\![\mathcal{Y}]\!] \setminus \mathbb{K}$ be triangular. A set $A^{\mathcal{L}}$ containing A is an* involutive completion *of A, if it is minimal with the following properties.*

1. *For all $Y \in \mathcal{Y}$, $\mathcal{N}_{A^{\mathcal{L}},Y}$ is a minimal involutive completion of $\mathcal{N}_{A,Y}$.*
2. *$H_A^\infty = H_{A^{\mathcal{L}}}^\infty$.*
3. *$p \in A^{\mathcal{L}} \setminus A$ implies $\text{DEG}(p) = 1$.*
4. *$p \in A^{\mathcal{L}} \setminus A$ with $\text{LEAD}(p) = v$ implies $p \in \langle \Theta A_{\leq v} \rangle$.*

If a minimal involutive completion $\mathcal{N}_{A,Y}^{\mathcal{L}}$ of $\mathcal{N}_{A,Y}$ is given for every $Y \in \mathcal{Y}$, an involutive completion of A can be obtained in the following way: For every $Y \in \mathcal{Y}$ and $\alpha \in \mathcal{N}_{A,Y}^{\mathcal{L}} \setminus \mathcal{N}_{A,Y}$, choose a $p \in A$ and $\beta \in \mathbb{N}^m$ such that $\text{LEAD}(\beta p) = \alpha Y$. As $\alpha \in \langle \mathcal{N}_{A,Y} \rangle$, such a pair (β, p) exists. The union of all these βp and all $p \in A$ is an involutive completion of A. We denote this set by $A_*^{\mathcal{L}}$. Further examples of involutive completions can be found in [11, 15, 3].

Definition 6. *Let (A, H) be a Δ-system and $A^{\mathcal{L}}$ be an involutive completion of A. Then (A, H) is* passive *w.r.t. $A^{\mathcal{L}}$ if $\langle \Theta A_{\leq u} \rangle : H^\infty = [A^{\mathcal{L}}]_{\mathcal{L}}^u : H^\infty$ for all $u \in \Theta\mathcal{Y}$.*

We are now in the position to provide an answer to the first question posed in the introduction: The following three statements show that passivity does not depend on the chosen involutive completion.

Proposition 1. *Let (A, H) be a Δ-system such that an involutive completion $A^{\mathcal{L}}$ of A exists. Every $p \in [A^{\mathcal{L}}]_{\mathcal{L}}^{u} : H^{\infty}$, when multiplied by a suitable $h \in H^{\infty}$, may be written as linear combination*

$$hp = \sum_{i=1}^{k} \lambda_i m_i,$$

where each m_i is a finite product of multiplicative prolongations of $A^{\mathcal{L}}$, all of which rank lower than or equal to u, and where all λ_i are reduced w.r.t. A at all derivatives $v \in \Theta \mathcal{Y}_{\leq u}$. Moreover, if $\mathrm{LEAD}(p) \leq u$, then each λ_i is partially reduced w.r.t. A.

Proof. By definition, there is for every $p \in [A]_{\mathcal{L}}^{u} : H^{\infty}$ an $h \in H^{\infty}$ such that hp is a linear combination of multiplicative prolongations lower than or equal to u, say

$$hp = \sum_{i=1}^{k} \lambda_{\beta,q}(\beta q) \ . \tag{1}$$

Assume some λ_i is not reduced w.r.t. A at some derivative ranking lower than or equal to u. We call any such derivative a *nuisance* and proceed to remove them by Noetherian induction. If a given representation of the form (1) contains no nuisances, then there is nothing to show. Otherwise, let v be the highest ranking nuisance. Without loss of generality, v shall appear with degree e in λ_1 and we may assume there is no occurrence of v as a nuisance of higher degree. We find a multiplicative prolongation γr of $r \in A^{\mathcal{L}}$ such that $\mathrm{LEAD}(\gamma r) = v$ and $\mathrm{DEG}(r)$ is less than or equal to the degree of λ_1 in v: If $v \in \Theta^{+}\mathrm{LEAD}(A)$, then $\mathrm{DEG}(\gamma r) = 1$ since A is partially autoreduced. Otherwise, the claim is clear since λ_1 is not reduced at v w.r.t. A. In any case, we may pseudo-divide λ_1 by γr at v to obtain

$$h'\lambda_1 = \lambda(\gamma r) + \lambda_1',$$

where $h' \in \mathrm{INIT}(\gamma r)^{\infty}$. Note that λ_1' neither contains v nor any nuisances ranking higher than v. Moreover, λ contains v in a strictly lower degree than e. We multiply (1) by h' and plug the result into the above equation. In this new equation, there is one less occurrence of v as a nuisance of degree e. Since $\mathrm{INIT}(\gamma r) < v$, multiplying by h' did not introduce any new nuisances ranking as high as v or higher. We may repeat the argument to eventually completely remove v as a nuisance and are done by induction hypothesis.

If $\mathrm{LEAD}(p) \leq u$, then only the λ_i may contain derivatives $v \in \mathrm{LEAD}(\Theta^{+}A)_{>u}$. Hence, it may be assumed that all λ_i are free of these derivatives. This proves the claim. $\qquad\square$

Corollary 1. *Let (A, H) and $A^{\mathcal{L}}$ be as above. Any $p \in [A^{\mathcal{L}}]^u_{\mathcal{L}} : H^\infty$, when multiplied by a suitable $h \in H^\infty$, may be written as*

$$hp = \lambda(\gamma r) + s,$$

where $\lambda \in \mathbb{K}[\![\mathcal{Y}]\!]$, $s \in [A^{\mathcal{L}}]^v_{\mathcal{L}}$ for some $v < u$ and γr is a multiplicative prolongation of $r \in A^{\mathcal{L}}$, uniquely determined by $\mathrm{LEAD}(\gamma r) = u$. If $\mathrm{LEAD}(p) < u$ and $u \in \mathrm{LEAD}(\Theta^+ A)$, then it may be assumed that $\lambda = 0$.

Proof. The first claim follows directly from Proposition 1. If $\mathrm{LEAD}(p) < u \in \mathrm{LEAD}(\Theta^+ A)$, then no $u' \in \mathrm{LEAD}(\Theta^+ A)_{\geq u}$ appears on the left hand side, so the same must be true for the right hand side. But by Proposition 1, there is an equation such that the only possibility for an element of $\mathrm{LEAD}(\Theta^+ A)_{\geq u}$ to appear is as leader of γr. The claim follows.   □

Proposition 2. *Let (A, H) be a Δ-system and $A^{\mathcal{L}}_1$, $A^{\mathcal{L}}_2$ be involutive completions of A. Then (A, H) is passive w.r.t. $A^{\mathcal{L}}_1$ if and only if (A, H) is passive w.r.t. $A^{\mathcal{L}}_2$*

Proof. Due to symmetry reasons, it is enough to show that whenever (A, H) is passive w.r.t. $A^{\mathcal{L}}_1$, then for all $u \in \Theta\mathcal{Y}$ we have $\alpha p \in [A^{\mathcal{L}}_2]^u_{\mathcal{L}} : H^\infty$ for all $p \in A$, $\alpha \in \mathbb{N}^m$ with $\mathrm{LEAD}(\alpha p) = u$.

Therefore, assume that (A, H) is passive w.r.t. $A^{\mathcal{L}}_1$ and consider $p \in A$, $\alpha \in \mathbb{N}^m$ with $u = \mathrm{LEAD}(\alpha p)$. We proceed by Noetherian induction on the rank of u. If there is no $\beta q \in \Theta A$ such that $\mathrm{LEAD}(\beta q) < u$, then $\alpha = 0$ and $p \in [A^{\mathcal{L}}_2]^u_{\mathcal{L}} : H^\infty$. Otherwise, $\alpha p \in [A^{\mathcal{L}}_1]^u_{\mathcal{L}} : H^\infty$ by assumption. So, Corollary 1 implies

$$h(\alpha p) = \lambda_1(\gamma_1 r_1) + s, \qquad (2)$$

where $\gamma_1 r_1$ is a multiplicative prolongation of $r_1 \in A^{\mathcal{L}}_1$, uniquely determined by $\mathrm{LEAD}(\gamma_1 r_1) = v$, and $s \in [A^{\mathcal{L}}_1]^v_{\mathcal{L}}$ for some $v < u$. If $\lambda_1 = 0$, then note that every $\beta q \in \Theta A^{\mathcal{L}}$ with $\mathrm{LEAD}(\beta q) \leq v$ belongs to $\Theta A_{\leq v}$ by property (iv) of involutive completions. Thus, we are done by the induction hypothesis.

Otherwise, since $A^{\mathcal{L}}_2$ is an involutive completion of A and A is partially autoreduced, there is a unique pair (γ_2, r_2) so that $\gamma_2 r_2$ is a multiplicative prolongation of $r_2 \in A^{\mathcal{L}}_2$ with $\mathrm{LEAD}(\gamma_2 r_2) = u$ and $\mathrm{DEG}(\gamma_2 r_2) = 1 = \mathrm{DEG}(\gamma_1 r_1)$. By definition of involutive completions $\gamma_2 r_2 \in \langle \Theta A_{\leq u} \rangle : H^\infty$ and by assumption $\langle \Theta A_{\leq u} \rangle : H^\infty = [A^{\mathcal{L}}_1]^u_{\mathcal{L}} : H^\infty$. Pseudo-dividing $\gamma_1 r_1$ by $\gamma_2 r_2$ at u, we obtain

$$\mathrm{INIT}(\gamma_2 r_2)(\gamma_1 r_1) = \mathrm{INIT}(\gamma_1 r_1)(\gamma_2 r_2) + s',$$

where $s' \in [A^{\mathcal{L}}_1]^u_{\mathcal{L}} : H^\infty$. We find some $h' \in H^\infty$ such that $h's' = \sum_{i=1}^k \lambda_i m_i$ as in Proposition 1. Since $u = \mathrm{LEAD}(\alpha p) \in \mathrm{LEAD}(\Theta^+ A)$, u does not appear in any λ_i and can only appear in m_i if $\gamma_1 r_1$ is part of the product. Without loss of generality, this is the case for $i = k'+1, \ldots, k$ and we shift all those $\lambda_i m_i$ to the left hand side of the above equation. Hence, we obtain an equation of the form

$$(h'\mathrm{INIT}(\gamma_2 r_2) + \lambda)(\gamma_1 r_1) = h'\mathrm{INIT}(\gamma_1 r_1)(\gamma_2 r_2) + \sum_{i=1}^{k'} \lambda_i m_i,$$

where $m_i < u$ $(i = 1, \ldots, k')$. Comparing the coefficient of u on both sides, it follows that $h'' := h' \text{INIT}(\gamma_2 r_2) + \lambda$ belongs to H^∞. Therefore, if (2) is multiplied by h'' and if we plug the above equation into the result, then we obtain

$$h'' h(\alpha p) = \lambda_1 \left(h' \text{INIT}(\gamma_1 r_1)(\gamma_2 r_2) + \sum_{i=1}^{k'} \lambda_i m_i \right) + h'' s$$

$$= \lambda_1 h' \text{INIT}(\gamma_1 r_1)(\gamma_2 r_2) + s'',$$

where $s'' \in [A_1^{\mathcal{L}}]_{\mathcal{L}}^{u'} \subseteq \langle \Theta A_{\leq u'} \rangle$ for some $u' < u$. By induction hypothesis, the proof is done. $\square$

In light of this result, we simply say that a Δ-system is *passive* without reference to an involutive completion. We will end this section with a finite criterion for passivity. It is an analogue of local involutivity; hence, the involutive division has to be continuous. See, e.g., [13], Def. 5.2 and Prop. 5.3.

Proposition 3 (A Finite Criterion). *Let $\mathcal{L}$ be continuous, (A, H) be a Δ-system, and $A^{\mathcal{L}}$ be an involutive completion of A. If $\delta_i p \in [A^{\mathcal{L}}]_{\mathcal{L}}^u : H^\infty$ whenever $p \in A^{\mathcal{L}}$, $\delta_i \in \Delta$, and $\text{LEAD}(\delta_i p) = u$, then (A, H) is passive.*

Note that the above is actually only a condition on those $\delta_i p$ that are non-multiplicative prolongations of $p \in A^{\mathcal{L}}$.

Proof. Let $p \in A$ and $\alpha \in \mathbb{N}^m$ and $u = \text{LEAD}(\alpha p)$. To show that $\alpha p \in [A^{\mathcal{L}}]_{\mathcal{L}}^u : H^\infty$ we proceed by Noetherian induction on the rank of u. If there is no $\beta q \in \Theta A$ with $\text{LEAD}(\beta q) < \text{LEAD}(\alpha p)$, then $\alpha = 0$ and $p \in A \subseteq [A^{\mathcal{L}}]_{\mathcal{L}}^u : H^\infty$.

If $\alpha p \in \mathcal{C}_{\mathcal{L},A^{\mathcal{L}}}(p)$, there is nothing to show. Otherwise, we may assume that $\alpha \neq 0$ and $\alpha = \alpha' \delta_i$ where $\delta_i p \notin \mathcal{C}_{\mathcal{L},A^{\mathcal{L}}}(p)$. By assumption, Corollary 1 may be applied to yield

$$h_1(\delta_i p) = \lambda_1(\gamma_1 r_1) + s_1,$$

where $s_1 \in [A^{\mathcal{L}}]_{\mathcal{L}}^v$ for some $v < \text{LEAD}(\delta_i p)$ and $\text{LEAD}(\gamma_1 r_1) = \text{LEAD}(\delta_i p)$. We apply α' to both sides and shift all terms except $h_1(\alpha' \delta_i p)$ from the left to the right, i.e.

$$h_1(\alpha p) = \lambda_1(\alpha' \gamma_1 r_1) + \sum_{\substack{\beta_1, \beta_2 \in \mathbb{N}^m, \\ \beta_1 \neq 0, \beta_1 \beta_2 = \alpha}} (\beta_1 \lambda_1)(\beta_2 \gamma_1 r_1)$$

$$+ \alpha' s_1 - \sum_{\substack{\beta_1, \beta_2 \in \mathbb{N}^m, \\ \beta_1 \neq 0, \beta_1 \beta_2 = \alpha}} (\beta_1 h_1)(\beta_2 \delta_i p) \ .$$

Since $s_1 \in [A^{\mathcal{L}}]_{\mathcal{L}}^v \subseteq \langle \Theta A_{\leq v} \rangle$ and $\alpha' v < u$, we have $\alpha' s_1 \in \langle \Theta A_{<u} \rangle$. Also, $(\beta_2 \gamma_1 r_1)$, $(\beta_2 \delta_i p) \in \langle \Theta A_{<u} \rangle$, the former by definition of involutive completions. By induction hypothesis, we obtain

$$h_1(\alpha p) = \lambda_1(\alpha' \gamma_1 r_1) + s_1', \tag{3}$$

for a suitable $s_1 \in [A^{\mathcal{L}}]^{v_1}$, $v_1 < u$ and $\text{LEAD}(\alpha' \gamma_1 r_1) = u$. Hence, if $\alpha' \gamma_1 r_1$ is a multiplicative prolongation of r_1, then the claim is proven. Otherwise, we may write $\alpha'\gamma_1 = \gamma'_1\delta_{(i_1)}$, where $\delta_{(i_1)}r_1 \notin \mathcal{C}_{\mathcal{L},A^\mathcal{L}}(r_1)$. We may use Corollary 1 again and repeat the above described argument to find

$$h_2(\alpha'\gamma_1 r_1) = \lambda_2(\gamma'_1\gamma_2 r_2) + s_2, \tag{4}$$

where $s_2 \in [A^\mathcal{L}]^{v_2}$, $v_2 < u$ and $\text{LEAD}(\gamma'_1\gamma_2 r_2) = u$. If we multiply (3) by h_2 and substitute (4) into the resulting equation, then we arrive at

$$\begin{aligned} h_2 h_1(\alpha p) &= \lambda_1\left(\lambda_2(\gamma'_1\gamma_2 r_2) + s_2\right) + h_2 s'_1 \\ &= \lambda_1\lambda_2(\gamma'_1\gamma_2 r_2) + \lambda_1 s_2 + h_2 s'_1 \\ &= \lambda'_2(\gamma'_1\gamma_2 r_2) + s'_2, \end{aligned}$$

for some $s'_2 \in [A^\mathcal{L}]^{v_2}_\mathcal{L}$. If again $\gamma'_1\gamma_2 r_2$ is not a multiplicative prolongation of r_2, we may continue the process and construct a sequence $r_1, r_2, r_3, \ldots$ such that the derivative operator appearing in $\text{LEAD}(\delta_{(i_j)}r_j)$ is in the involutive cone of the derivative operator in $\text{LEAD}(r_{j+1})$ for $j = 1, 2, 3 \ldots$ and so on. In other words, the derivative operators $\beta_1, \beta_2, \beta_3, \ldots$ appearing in the leaders of $r_1, r_2, r_3, \ldots$ form a sequence satisfying [13], Def. 5.2. By continuity of $\mathcal{L}$, the elements in the sequence $\beta_1, \beta_2, \beta_3, \ldots$ are pairwise distinct. Hence, the elements in the sequence $r_1, r_2, r_3, \ldots$ are pairwise distinct and therefore, this sequence is finite. After a finite number of steps, we eventually find an $h \in H^\infty$ such that $h(\alpha p) = \lambda(\gamma r) + s$, where γr is a multiplicative prolongation of $r \in A^\mathcal{L}$ and $s \in [A^\mathcal{L}]^v_\mathcal{L}$ for some $v < u$. By induction hypothesis, the proof is done. □

4   Comparison of Definitions

Let us compare Definition 6 to Ritt's definition of passivity, which is based upon Riquier's work, and two more definitions proposed by Wu and Chen & Gao.

Definition 7 ([11], p. 161). *Let $\mathcal{L}$ be the Janet division and $(A, \{1\})$ be a Δ-system such that A is autoreduced.[2] Let $A^\mathcal{L}$ be an involutive completion of A. Then $(A, \{1\})$ is passive if every non-multiplicative prolongation $\delta_i p$ of a $p \in A^\mathcal{L}$ may be represented as*

$$\delta_i p = \gamma r + \sum_{i=1}^{k} m_i,$$

where γr is a multiplicative prolongation of $r \in A^\mathcal{L}$, uniquely determined by $\text{LEAD}(\gamma r) = \text{LEAD}(\delta_i p)$ and each m_i is a finite product of multiplicative prolongations all of which rank lower than $\text{LEAD}(\delta_i p)$.

Definition 8 ([15], p. 300). *Let $\mathcal{L}$ be the Thomas division and (A, H_A) be a Δ-system such that A is autoreduced. Let $A^\mathcal{L}$ be an involutive completion of A.*

---

[2] This type of Δ-system is also called *orthonomic*.

Then (A, H_A) is passive *if every non-multiplicative prolongation $\delta_i p$ of a $p \in A^{\mathcal{L}}$ may be represented as*

$$h\text{INIT}(\gamma r)(\delta_i p) = h\text{INIT}(\delta_i p)(\gamma r) + \sum_{i=1}^{k} \lambda_i m_i,$$

where $h \in H^{\infty}$, γr is a multiplicative prolongation of $r \in A^{\mathcal{L}}$, uniquely determined by $\text{LEAD}(\gamma r) = \text{LEAD}(\delta_i p)$, each m_i is a finite product of multiplicative prolongations all of which rank lower than $\text{LEAD}(\delta_i p)$, and all λ_i are partially reduced w.r.t. A.

Definition 9 ([3], p. 481). *Let $\mathcal{L}$ be an involutive division and (A, H_A) be a Δ-system such that A is autoreduced. Let $A^{\mathcal{L}}$ be an involutive completion of A.[3] Then (A, H_A) is* passive *if every αp ($\alpha \in \mathbb{N}^m$, $p \in A^{\mathcal{L}}$) can be represented as*

$$h(\alpha p) = \sum_{i=1}^{k} \lambda_i (\beta_i q_i),$$

where $h \in H_A^{\infty}$, each $\beta_i q_i$ is a multiplicative prolongation of $q_i \in A$ ranking lower than or equal to $\text{LEAD}(\alpha p)$,[4] and the following holds.

- *If $\beta_i \neq 0$, then λ_i is partially involutively reduced at all derivatives $u > \text{LEAD}(\beta_i q_i)$.*
- *If $\beta_i = 0$, then λ_i is partially involutively reduced w.r.t. $A^{\mathcal{L}}$ and reduced w.r.t. all $q \in A^{\mathcal{L}}$ such that $q_i < q$.*

Due to Proposition 3, it is not difficult to see that each of these three definitions implies Definition 6. Conversely, for a given αp with $\alpha \neq 0$ and $p \in A^{\mathcal{L}}$ there is a multiplicative prolongation γr of $r \in A^{\mathcal{L}}$, uniquely determined by $\text{LEAD}(\gamma r) = \text{LEAD}(\alpha p)$, with $S(\alpha p, \gamma r) = \text{INIT}(\gamma r)(\alpha p) - \text{INIT}(\alpha p)(\gamma r)$. Applying Corollary 1 and Proposition 1 to this pseudo-S-polynomial, this leads to an equation of the form

$$h\text{INIT}(\gamma r)(\alpha p) = h\text{INIT}(\alpha p)(\gamma r) + \sum_{i=1}^{k} \lambda_i m_i, \tag{5}$$

where $h \in H^{\infty}$, all λ_i are partially reduced w.r.t. A, and all m_i are products of multiplicative prolongations of $A^{\mathcal{L}}$ all of which rank lower than $\text{LEAD}(\alpha p)$. This can be rewritten so that it satisfies Chen & Gao's definition except for the very last item. We think this is an artefact of the reduction algorithm Chen & Gao are using. Taking $\alpha = \delta_i$, we see that (5) implies passivity in Wu's sense. If in addition $H = \{1\}$, then $h = \text{INIT}(\gamma r) = \text{INIT}(\alpha p) = 1$ and we arrive at a representation very similar to the one in Ritt's definition. Unlike Ritt, who arrives at the above representation by a series of reductions which must return a zero remainder, we cannot make the stronger assumption that the λ_i equal 1.

---

[3] In Chen & Gao's definition, a weaker notion than involutive completion is used. However, in the case which is of algorithmic interest, i.e. where a statement similar to Proposition 3 holds, they have to assume that $A^{\mathcal{L}}$ is an involutive completion.

[4] The latter requirement on the rank of the multiplicative prolongations is not found in Chen & Gao's original definition. However, it is meant to be included [2].

5 An Equivalence Statement

The following is a generalization of Theorem 3 in [9] to arbitrary involutive divisions and Δ-systems (A, H) where A is triangular, partially autoreduced and where H is partially reduced w.r.t. A. It can be seen as a differential algebra, pseudo-division analogue of the well known fact that any involutive basis is a Gröbner basis, see, e.g., [4].

Theorem 2. *Let $\mathcal{L}$ be an involutive division and (A, H) be a Δ-system. If (A, H) is passive, then (A, H) is coherent.*

Proof. Let $p'q' \in \Theta^+ A$ such that $\text{LEAD}(p') = u = \text{LEAD}(q')$. We claim that $S(p', q') \in \langle \Theta A_{<u} \rangle : H^\infty$. Assume $p' = \alpha p$ and $q' = \beta q$ for some $p, q \in A$. By passivity, we may apply Corollary 1 to αp and βq and obtain

$$h_p \alpha p = \lambda_p(\gamma r) + s_p, \; h_q \beta q = \lambda_q(\gamma r) + s_q,$$

where $s_p, s_q \in [A^\mathcal{L}]_\mathcal{L}^v$, $v < u$. Since $u \in \text{LEAD}(\Theta^+ A)$ and A is partially autoreduced, γr is linear. Furthermore,

$$\begin{aligned} h_p h_q S(p', q') &= h_q \text{INIT}(q')(\lambda_p(\gamma r) + s_p) - h_p \text{INIT}(p')(\lambda_q(\gamma r) + s_q) \\ &= (h_q \text{INIT}(q')\lambda_p - h_p \text{INIT}(p')\lambda_q)(\gamma r) \qquad (6) \\ &\quad + (h_q \text{INIT}(q')s_p - h_p \text{INIT}(p')s_q) \;. \end{aligned}$$

The derivative u cannot appear in h_p or h_q, since H is partially reduced w.r.t. A. Since $p', q' \in \Theta A$, we have $\text{INIT}(p'), \text{INIT}(q') \in H$, so u cannot be contained in either one of them. Furthermore, $\text{LEAD}(S(p', q')) < u$ by definition and likewise, u can only appear in the coefficients of $s_p, s_q \in [A^\mathcal{L}]_\mathcal{L}^v$ when both Δ-polynomials are represented as linear combinations of multiplicative prolongations ranking lower than or equal to v.

Let $\text{TAIL}(\gamma r) := \gamma r - \text{INIT}(\gamma r)\text{LEAD}(\gamma r)$. If $-\text{TAIL}(\gamma r)/\text{INIT}(\gamma r)$ is substituted for u in (6) and if the resulting equation is multiplied by a suitable power of $\text{INIT}(\gamma r)$ to get rid of denominators, then we obtain an equation of the form

$$\text{INIT}(\gamma r)^k h_p h_q S(p, q) = \text{INIT}(\gamma r)^{k_1} h_q \text{INIT}(q')\tilde{s}_p - \text{INIT}(\gamma r)^{k_2} h_p \text{INIT}(p')\tilde{s}_q \;.$$

where $\tilde{s}'_p, \tilde{s}'_q$ are elements of $[A^\mathcal{L}]_\mathcal{L}^v$. Thus, all of the right hand side belongs to $[A^\mathcal{L}]_\mathcal{L}^v \subseteq \langle \Theta A_{<u} \rangle : H^\infty$. Moreover, $\text{INIT}(\gamma r) \in H^\infty$ since $\text{INIT}(\gamma r) = \text{SEP}(r)$ or $\text{INIT}(\gamma r) = \text{INIT}(r)$. In both cases, $\text{INIT}(\gamma r) \in H_A^\infty$ by definition of involutive completions and $H_A^\infty \subseteq H^\infty$ by assumption. The claim follows. $\qquad \square$

We now turn to the second question stated in the introduction, namely whether the converse to the above statement also holds. The next theorem shows that this is true. The answer to the third question is an immediate corollary: Indeed, if passivity is equivalent to coherence then it has to be independent of the chosen involutive division as coherence obviously has this property. The key ingredient to the following proof is that passivity is independent of the chosen involutive completion.

Theorem 3. *Let $\mathcal{L}$ be an involutive division and (A, H) be a Δ-system. Assume that an involutive completion of A w.r.t. $\mathcal{L}$ exists. If (A, H) is coherent, then (A, H) is passive.*

Proof. Since passivity is independent of the chosen involutive completion, we may choose $A_*^{\mathcal{L}}$. By definition of $A_*^{\mathcal{L}}$, $\Theta A_{* <u}^{\mathcal{L}} = \Theta A_{<u}$. In conjunction with property (ii) of involutive completions, it follows that $(A_*^{\mathcal{L}}, H)$ is coherent.

Let $p \in A$ and $\alpha \in \mathbb{N}^m$ be arbitrary and put $u = \text{LEAD}(\alpha p)$. We proceed by Noetherian induction on the rank of $\text{LEAD}(\alpha p)$ to show that $\alpha p \in [A_*^{\mathcal{L}}]_{\mathcal{L}}^u : H^\infty$. If there is no $\beta q \in \Theta A$ such that $\text{LEAD}(\beta q) < \text{LEAD}(\alpha p)$, then $\alpha = 0$ and p belongs to $[A_*^{\mathcal{L}}]_{\mathcal{L}}^u : H^\infty$.

There is also nothing to show if $\alpha p \in A_*^{\mathcal{L}}$. So we may assume that $\alpha p \in (\Theta A \setminus A) \cap (\Theta A_*^{\mathcal{L}} \setminus A_*^{\mathcal{L}})$. Since $A_*^{\mathcal{L}}$ is an involutive completion, there is a multiplicative prolongation γr of $r \in A_*^{\mathcal{L}}$ with $\text{LEAD}(\gamma r) = u$. Since A is partially autoreduced, γr linear. Hence,

$$S(\alpha p, \gamma r) = \text{INIT}(\gamma r)\alpha p - \text{INIT}(\alpha p)\gamma r \ .$$

Coherence of $(A_*^{\mathcal{L}}, H)$ implies that $S(\alpha p, \gamma r) \in \langle \Theta A_{* <u}^{\mathcal{L}} \rangle : H^\infty$. Thus by the above equation and the fact that $\text{INIT}(\gamma r) \in H_{A_*^{\mathcal{L}}}^\infty = H_A^\infty \subseteq H^\infty$, the claim is proven by induction hypothesis. $\square$

Corollary 2. *Let (A, H) be a Δ-system. Consider only those involutive divisions for which A has an involutive completion. The following are equivalent:*

1. (A, H) *is passive w.r.t. to all involutive divisions.*
2. (A, H) *is passive w.r.t. to one involutive division.*
3. (A, H) *is coherent.*

6   Conclusions

We have suggested a new approach to passivity in differential algebra. Our definition is a generalization of the one used by Wu. Moreover, it is closely related to passivity in the sense of Ritt and Chen & Gao. Using the new definition, we have provided a positive answer to all three questions posed in the introduction: The corresponding statements are Proposition 2, Theorem 3 and Corollary 2. We remark that the proofs can be used together with assumptions that are less restrictive than the ones made in this work, see [10].

Hopefully, the insights on passivity presented here lead to new ideas combining the properties of coherence and passivity to be used in decomposition algorithms in differential algebra. It would also be of interest to explore the connection between passivity as described in this work and passivity as it is treated in the formal theory of partial differential equations.

Acknowledgements

The first author would like to thank Dongming Wang for introducing him to the theory of coherence and passivity. We also thank Vladimir Gerdt and Werner Seiler for general comments regarding a previous version of this paper.

References

1. Boulier, F., Lazard, D., Ollivier, F., Petitot, M.: Computing representations for radicals of finitely generated differential ideals. Technical report, no. IT306, LIFL (1997)
2. Chen, Y.: Private communication (2005)
3. Chen, Y., Gao, X.: Involutive Characteristic Sets of Algebraic Partial Differential Equation Systems. Science in China (Series A) **46:4** (2003) 469–487
4. Gerdt, V.P., and Blinkov, Y.A.: Involutive Bases of Polynomial Ideals. Mathematics and Computers in Simulation **45** (1998) 519ff
5. Hubert, E.: Factorization Free Decomposition Algorithms in Differential Algebra. Journal of Symbolic Computation **29:4–5** (2000) 641–622
6. Hubert, E.: Notes on Triangular Sets and Triangulation-Decomposition Algorithms II: Differential Systems. Lecture Notes In Computer Science **2630** (2003) 40-87
7. Hubert, E., Le Roux, N.: Computing power series solutions of a nonlinear PDE system. ISSAC '03: Proceedings of the 2003 International Symposium on Symbolic and Algebraic Computation (2003) 148-155
8. Kolchin, E.R.: Differential Algebra and Algebraic Groups. Pure and Applied Mathematics **54** (1973)
9. Li, Z., Wang, D.: Coherent, Regular and Simple Systems in Zero Decompositions of Partial Differential Systems. Systems Science and Mathematical Sciences **12:5** (1999) 43–60
10. Minzlaff, M.: On The Decomposition of Radical Differential Ideals. Diplomarbeit, Universität Karlsruhe (TH) (2006)
11. Ritt, J.F.: Differential Algebra. Amer. Math. Soc., Colloquium Publications **33** (1950)
12. Rosenfeld, A.: Specializations in Differential Algebra. Trans. Amer. Math. Soc. **90** (1959) 394–407
13. Seiler, W.M.: A Combinatorial Approach to Involution and Delta-Regularity I: Involutive Bases in Polynomial Algebras of Solvable Type. Preprint Universität Mannheim (2002)
14. Sit, W.: The Ritt-Kolchin theory for differential polynomials. Differential Algebra and Related Topics (2002) 1–70
15. Wu, W.T.: On The Foundation of Algebraic Differential Geometry. Systems Science and Mathematical Sciences **2:4** (1989) 289–312

A Full System of Invariants
for Third-Order Linear Partial Differential
Operators

Ekaterina Shemyakova

Research Institute for Symbolic Computation (RISC),
J.Kepler University,
Altenbergerstr. 69, A-4040 Linz, Austria
kath@risc.uni-linz.ac.at
http://www.risc.uni-linz.ac.at

Abstract. A full system of invariants for a third-order bivariate hyperbolic linear partial differential operator L is found under the gauge transformation $g(x_1, x_2)^{-1} L g(x_1, x_2)$. That is, all other invariants can be obtained from this full system, and two operators are equivalent with respect to the gauge transformations if and only if their full systems of invariants are equal. To obtain the invariants, we generalize the notion of Laplace invariants from the case of order two to that of arbitrary order. This is done through the notion of common obstacles to factorizations into first-order factors. Explicit formulae for the invariants of a general operator are given in terms of the coefficients of the operator.
The majority of the results were obtained using Maple 9.5.

Keywords: Laplace invariants, partial differential operators, Maple.

1 Introduction

The idea of looking for the invariants of a differential operator is very popular. As early as 1769/1770, Laplace and Euler found two invariants for a second-order linear bivariate hyperbolic operator L under the gauge transformations $L \mapsto g(x_1, x_2)^{-1} L g(x_1, x_2)$. Thus, the operator

$$L = D_1 D_2 + a D_1 + b D_2 + c$$

has the invariants

$$h = \partial_1(a) + ab - c, \quad k = \partial_2(b) + ab - c.$$

Using the invariants, the operator may be rewritten in the form

$$L = (D_1 + b)(D_2 + a) + h = (D_2 + a)(D_1 + b) + k.$$

If $h = 0$ or $k = 0$, the operator L is factorable and one can find the general solution of the corresponding equation. Darboux proved [3] that h and k together

J. Calmet, T. Ida, and D. Wang (Eds.): AISC 2006, LNAI 4120, pp. 106–115, 2006.

form a full system of invariants, in other words a basis, which provides an easy way to judge whether one operator can be obtained as a gauge transformation of another.

Possible ways of generalizing these invariants include considering a broader group of transformations, considering nonlinear operators, or considering non-commuting derivations [1, 2, 4, 7]. For instance, in [1, 2], the non-commutative case for a hyperbolic second-order operator was done.

Another possibility is to extend the ideas to the case of higher-order operators. Thus, some invariants were found in [10] as a by-product of his generalized Laplace transformations theory. Also some invariants were found in [5] using the idea of incomplete factorization. However, a full system of invariants was not found.

In the present paper we introduce a complete set of invariants for a third-order hyperbolic linear partial differential operator L in two independent variables under gauge transformations. We obtain explicit formulae in special variables for which the symbol of the operator has the simplest form $D_1 D_2(D_1 + D_2)$.

Algorithmically it is not an easy task to find such a change of variables, since one has to solve a general first-order linear partial differential equation. So one prefers to have explicit formulae in the general case, that is for the symbol $\mathrm{Sym}_L = s_{30}D_1^3 + s_{21}D_1^2 D_2 + s_{12}D_1 D_2^2 + s_{03}D_2^3$. In fact, one can find the formulae even in this case. For example, the simplest invariant

$$I_1 = -2a_{20} + a_{11} - 2a_{02}$$

from Theorem 4 (given below) has the form

$$-6s_{21}s_{03}a_{20} + 2s_{12}^2 a_{20} + 9s_{30}s_{03}a_{11} - s_{21}s_{12}a_{11} - 6s_{12}s_{30}a_{02} + 2s_{21}^2 a_{02}$$

where a_{ij} denotes the coefficient of $D_i D_j$ in L. However, the general form of the next simplest invariant

$$I_2 = \partial_1(a_{20}) - \partial_2(a_{02})$$

is huge: its length in MAPLE [6] (as measured by the function `length`) is 27083. However, one can obtain the formulae for the general case using our method, large though it be.

One may note that in Theorem 4, we give the invariants in short and compact forms, meaning that one can prove that they form a complete system even working by hand. However, the discovery of these forms of the invariants, and the proof of their correctness, relies on extensive computations, which were performed using Maple.

2   Notation

Let K be a commutative ring with 1. Let $\Delta = \{\partial_1, \ldots, \partial_n\}$ be commuting derivations acting on K. That is, for every $\partial \in \Delta$ and for every $a, b \in K$, we have $\partial(a + b) = \partial(a) + \partial(b)$ and $\partial(ab) = a\partial(b) + \partial(a)b$. Consider the ring of linear differential operators

$$K[D] = K[D_1, \ldots, D_n] \, ,$$

where $D_1, \ldots, D_n$ correspond to the derivations $\partial_1, \ldots, \partial_n$ respectively. Thus $K[D]$ is the ring of non-commutative polynomials in $D_1, \ldots, D_n$ with the following multiplication rules:

$$D_i D_j = D_j D_i , \quad D_i(a) = a D_i + D_i(a)$$

for all $a \in K$ and any $i, j \in \{1, \ldots, n\}$. We use the notation

$$D^{(i_1, \ldots, i_n)} := D_1^{i_1} \ldots D_n^{i_n} ,$$

and define the order as follows:

$$|D^{(i_1, \ldots, i_n)}| = \operatorname{ord}(D^{(i_1, \ldots, i_n)}) := i_1 + \ldots + i_n ,$$

and in addition the order of the zero operator is $-\infty$.

Now every operator $L \in K[D]$ has the form

$$L = \sum_{|J| \le d} a_J D^J ,$$

where $d = \operatorname{ord}(L)$ and $a_J \in K$, $J \in \mathbb{N}^n$. We define $L_i = \sum_{|J|=i} a_J D^J$. Then one may rewrite L in the form

$$L = \sum_{i=0}^{d} L_i .$$

The operator L_d is called the symbol of L and is denoted by Sym_L. Finally we introduce the notation $K_i[D] = \{L \in K[D] | L \equiv L_{\operatorname{ord}(L)}\}$ for all i.

3   Common Obstacles to Factorizations

In this section we briefly recapitulate some results from [8,9], because they are essential to the next sections.

Definition 1. *Let $L \in K[D]$ and suppose that its symbol has a decomposition $\operatorname{Sym}_L = S_1 \ldots S_k$. Then we say that the factorization*

$$L = F_1 \circ \ldots \circ F_k, \quad \text{where} \quad \operatorname{Sym}_{F_i} = S_i , \ \forall i \in \{1, \ldots, k\},$$

is of the factorization type $(S_1)(S_2) \ldots (S_k)$.

Definition 2. *Let $L \in K[D]$, $\operatorname{Sym}_L = S_1 \ldots S_k$. An operator $R \in K[D]$ is called a common obstacle to factorization of the type $(S_1)(S_2) \ldots (S_k)$ if there exists a factorization of this type for the operator $L - R$ and R has minimal possible order.*

Let R be such a common obstacle of order t, and $d_i = \operatorname{ord}(S_i)$, $i = 1, \ldots, k$. Then

$$R = L - F_1 \circ \ldots \circ F_k ,$$

where

$$F_j = S_j + \sum_{i=0}^{d_j-1} G_i^j ,$$

and $G_i^j \in K_i[D]$ for all i and j.

One can change any term whose order is less than or equal to $t - (d - d_j)$ in any factor F_j, $j \in \{1, \ldots, k\}$ and get new factors F_j' as a result. Then one may see that

$$R' = L - F_1' \circ \ldots \circ F_k'$$

is again a common obstacle of order t.

Definition 3. *The main (common) obstacle is defined to be $R = L - F_1' \circ \ldots \circ F_k'$, where each factor F_j' is obtained from the factor F_j by replacing all terms of order less than or equal to $t - (d - d_j)$ in F_j with zero.*

Example 1. Consider the hyperbolic operator

$$L = D_1 D_2 - a D_1 - b D_2 - c,$$

where $a, b, c \in K$. Then P_1 is a common obstacle to factorizations of the type $(D_1)(D_2)$ if there exist $g_0, h_0 \in K$ such that

$$L - P_1 = (D_1 - g_0) \circ (D_2 - h_0).$$

Comparing the terms on the two sides of the equation, one gets

$$g_0 = b, \ h_0 = a , \quad P_1 = \partial_x(a) - ab - c.$$

Analogously, we get a common obstacle to factorization of the type $(D_2)(D_1)$:

$$P_2 = \partial_y(b) - ab - c,$$

and the corresponding factorization for $L - P_2$:

$$L - P_2 = (D_1 - a) \circ (D_2 - b).$$

Thus, the obtained common obstacles P_1 and P_2 are the famous Laplace invariants [10].

Theorem 1. *Let $n = 2$, $L \in K[D]$, $\mathrm{ord}(L) = d$, and $\mathrm{Sym}_L = S_1 \cdot S_2 \ldots S_k$, where S_i, $i \in \{1, \ldots, k\}$ are pairwise coprime. Then the order of common obstacles is less then $d - 1$.*

Let us recall that a second-order partial differential operator is called (strictly) hyperbolic if its symbol has exactly two different factors. By analogy, a partial differential operator of order d is (strictly) hyperbolic if its symbol has exactly d different factors.

Theorem 2. *Let $n = 2$, and let $L \in K[D]$ be a strictly hyperbolic operator of some order d. Let the factorization type of the factorization of L into exactly d factors be fixed. Then a common obstacle is unique.*

Theorem 3. *Let P be a common obstacle of $L \in K[D]$, then $g^{-1}Pg$ is a common obstacle of $g^{-1}Lg$, where $g \in K$.*

4   Generalization of the Laplace Invariants

First of all, we recall the definition of an invariant of an operator $L \in K[D]$.

Definition 4. *Let $L \in K[D]$. An algebraic differential expression I in coefficients of L is an invariant under gauge transformations $L_1 = g(x, y)^{-1}Lg(x, y)$ if it does not change under these transformations.*

For some fixed $L \in K[D]$, the set of all invariants forms a ring, meaning that the sum, the difference and the product of two invariants are also invariants. Trivial examples of an invariant are coefficients of the symbol.

In the following proposition, we collect known information about the common obstacles of a bivariate, strictly hyperbolic operator and show that the symbol of a common obstacle can be considered as a generalization of the Laplace invariants.

Proposition 1. *Let $n = 2$. Fix a strictly hyperbolic operator $L \in K[D]$, and denote the order of L by d. Consider factorizations of L into first-order factors. Then*

1. *the order of common obstacles less than or equal to $d - 2$;*
2. *a common obstacle is unique for each factorization type;*
3. *there are $d!$ common obstacles;*
4. *if $d = 2$, then the common obstacles of order 0 are the Laplace invariants;*
5. *the symbol of a common obstacle is an invariant.*

Proof. The first statement follows from Theorem 1, while the second one is the result of Theorem 2.

There are $d!$ different types of factorizations of L into d factors, and so $d!$ common obstacles (not necessarily different). This proves the third statement.

The fourth statement follows from Example 1.

To prove the last statement we use Theorem 3 and the fact that the symbol of an operator in $K[D]$ is invariant under the gauge transformations. So, the symbol of a common obstacle of L equals the symbol of a common obstacle of $g^{-1}Lg$, $g \in K^*$. On the other hand, the second statement implies that common obstacles are unique. So, we are done.

Note that for a second order operator, by Theorem 1, all common obstacles have order 0. Thus, the symbol of a common obstacle is just the common obstacle itself. That is why a common obstacle is invariant for a second-order hyperbolic operator. In general, that is not the case. Moreover, for a common obstacle P, none of the operators $P_i, i \in \{0, \dots, \mathrm{ord}(L) - 1\}$ is an invariant in general.

5   A Full System of Invariants for Third Order LPDOs

Let $n = 2$. We consider a third-order strictly hyperbolic operator $L \in K[D]$. Using an appropriate change of variables, one can find a form for L in which the symbol is

$$\mathrm{Sym}_L = D_1 D_2 (D_1 + D_2).$$

So, without loss of generality it is enough to consider only this case.

For convenience we denote the derivations ∂_1 and ∂_2 by ∂_x and ∂_y respectively.

Theorem 4. *Let* $n = 2$. *Consider an operator* $L \in K[D]$:

$$L = D_1 D_2 (D_1 + D_2) + a_{20} D_1^2 + a_{11} D_1 D_2 + a_{02} D_2^2 + a_{10} D_1 + a_{01} D_2 + a_{00}.$$

Then the following is a full system of invariants of L.

$$
\begin{aligned}
I_1 &= -2a_{20} + a_{11} - 2a_{02}, \\
I_2 &= \partial_x(a_{20}) - \partial_y(a_{02}), \\
I_3 &= a_{10} + a_{20}(a_{20} - a_{11}) + \partial_y(a_{20} - a_{11}), \\
I_4 &= a_{01} + a_{02}(-a_{11} + a_{02}) + \partial_x(-a_{11} + a_{02}), \\
I_5 &= a_{00} - a_{01}a_{20} - a_{10}a_{02} + a_{02}a_{20}a_{11} + \\
&\quad + (2a_{02} - a_{11} + 2a_{20})\partial_x(a_{20}) + \partial_{xy}(a_{20} - a_{11} + a_{02}).
\end{aligned}
\tag{1}
$$

Thus, an operator $L' \in K[D]$

$$L' = D_1 D_2 (D_1 + D_2) + b_{20} D_1^2 + b_{11} D_1 D_2 + b_{02} D_2^2 + b_{10} D_1 + b_{01} D_2 + b_{00}$$

is equivalent to L *(w.r.t. the gauge transformations) if and only if their corresponding invariants* I_1, I_2, I_3, I_4, I_5 *are equal.*

Proof. Suppose the corresponding systems of invariants $\{I_1, I_2, I_3, I_4, I_5\}$ of L and $\{I_1', I_2', I_3', I_4', I_5'\}$ of L' are equal. We look for a function $f \in K$, such that

$$f^{-1}Lf = L'. \tag{2}$$

We equate the coefficients of D_1^2, D_2^2 on both sides of (2), and get

$$b_{20} = a_{20} + \partial_y(g), \quad b_{02} = a_{02} + \partial_x(g). \tag{3}$$

Consider $f \in K$, such that

$$g = \ln f.$$

Now, $I_2 = I_2'$ implies

$$\partial_x(b_{20} - a_{20}) = \partial_y(b_{02} - a_{02}). \tag{4}$$

Therefore, there is only one (up to a multiplicative constant) function f, which satisfies the conditions (3). We fix this function f, and check whether the others coefficients on the both sides of (2) are equal.

We introduce the following notation:

$$P = \partial_x(g), \quad Q = \partial_y(g).$$

We will use the following two lemmas.

Lemma 1. *The following equalities hold:*

$$\frac{\partial_{xy}(f)}{f} = \partial_x(g)\partial_y(g) + \partial_{xy}(g) = PQ + \partial_x(Q) \; ,$$

$$\frac{\partial_{yy}(f)}{f} = \partial_y^2(g) + \partial_{yy}(g) = Q^2 + \partial_y(Q) \; ,$$

$$\frac{\partial_{xxy}(f)}{f} = Q\partial_x(P) + 2P\partial_x(Q) + P^2Q + \partial_{xx}(Q) \; ,$$

$$\frac{\partial_{xyy}(f)}{f} = P\partial_y(Q) + 2Q\partial_y(P) + PQ^2 + \partial_{xy}(Q) \; .$$

Proof. Use definitions of g, P, Q, equalities (3), and $\partial_y(P) = \partial_x(Q)$.
For instance, the proof of the first statement is as follows. Since

$$\partial_x(g) = \frac{\partial_x(f)}{f} \; ,$$

we have

$$\frac{\partial_{xy}(f)}{f} = \frac{\partial_y(f\partial_x(g))}{f} = \frac{\partial_y(f)\partial_x(g) + f\partial_{xy}(g)}{f} = \partial_x(g)\partial_y(g) + \partial_{xy}(g) \; .$$

To get a form of the expression in P and Q one may just substitute

$$\partial_x(g) = P \; , \quad \partial_y(g) = Q \; .$$

Lemma 2. *The following equalities hold:*

1. $b_{20} = Q + a_{20}$,
2. $b_{02} = P + a_{02}$,
3. $b_{11} = 2P + 2Q + a_{11}$,
4. $b_{10} = a_{10} + 2PQ + a_{11}Q + Q^2 + 2a_{20}P + \partial_y(2P + Q)$,
5. $b_{01} = a_{01} + 2PQ + a_{11}P + P^2 + 2a_{02}Q + \partial_x(P + 2Q)$.

Proof. The equalities $1, 2$ follow from (3). Equality 3 is implied by $I_1 = I_1'$, while
equalities $4, 5$ are consequences of $I_3 = I_3'$ and $I_4 = I_4'$ respectively.

To finish the proof it is enough to prove the following lemma.

Lemma 3. *Let the coefficients of $D_1 D_2$, D_1 and D_2 be c_{11}, c_{10} and c_{01} and let
the free coefficient in $(f^{-1}L * f - L')$ be c_{00}. Then $c_{11} = c_{10} = c_{01} = c_{00} = 0$.*

Proof. From the equality $I_1 = I_1'$, we have $c_{11} = 0$.
Now, by the lemma 1, we compute

$$c_{10} = a_{10} - b_{10} + a_{11}\partial_y(g) + \frac{\partial_{yy}(f) + 2\partial_{xy}(f)}{f} + 2a_{20}\partial_x(g)$$

$$= a_{10} - b_{10} + a_{11}Q + Q^2 + 2PQ + \partial_y(Q) + 2\partial_x(Q) \; , \tag{5}$$

which is 0, by equality 4 from lemma 2.

Analogously, the coefficient

$$c_{01} = a_{01} - b_{01} + a_{11}\partial_x(g) + \frac{\partial_{xx}(f) + 2\partial_{xy}(f)}{f} + 2a_{02}\partial_y(g)$$

is 0, by equality 5 from lemma 2.

Now, let us consider the free coefficient

$$c_{00} = a_{00} - b_{00} + a_{10}\partial_x(g) + a_{01}\partial_y(g)$$
$$+\frac{a_{20}\partial_{xx}(f) + a_{11}\partial_{xy}(f) + a_{02}\partial_{yy}(f)}{f} + \frac{\partial_{xxy}(f) + \partial_{xyy}(f)}{f} . \qquad (6)$$

We may use lemma 1 together with the equalities $1, 2$ from lemma 2 to obtain

$$c_{00} = a_{00} - b_{00} + a_{10}P + a_{01}Q + a_{20}P^2 + a_{11}PQ + a_{02}Q^2$$
$$+P^2Q + PQ^2 + (a_{20} + Q)\partial_x(P) + (a_{11} + 2(P + Q))\partial_y(P)$$
$$+(a_{02} + P)\partial_y(Q) + \partial_{xy}(P + Q). \qquad (7)$$

From $I_5 = I_5'$, and using $I_1 = I_1'$, we get

$$0 = a_{00} - b_{00} - a_{01}a_{20} + b_{01}b_{20} - a_{10}a_{02} + a_{02}a_{20}a_{11} + b_{10}b_{02}$$
$$-b_{02}b_{20}b_{11} + (-2a_{02} + a_{11} - 2a_{20})\partial_y(P) + \partial_{xy}(P + Q). \qquad (8)$$

After subtracting this equation from equation (7), and using all the substitutions from lemma 2, one gets

$$c_{00} = 0 .$$

So, we have finished the proof in one direction, more precisely, we have just proved that if the five invariants are equal, then there is $f \in K$ such that $f^{-1}Lf = L'$.

Now, suppose that L and L' are equivalent w.r.t. the gauge transformations. Then one can check that the respective invariants are equal. This requires some extended but straightforward computations of the invariants, which were done by means of Maple 9.5.

Let $n = 2$. Consider a strictly hyperbolic third order operator $L \in K[D]$. By Proposition 1, there are 6 common obstacles to factorizations of L into first order factors. By the same proposition, the symbols of these common obstacles are invariants of L.

Now, we have the full system of invariants $\{I_1, I_2, I_3, I_4, I_5\}$, which is a basis for all invariants. That is, the coefficients of the symbols of these 6 common obstacles can be expressed in terms of these five invariants. The free coefficients of common obstacles are not invariants in general, but nevertheless they can be "almost" expressed in terms of these basic invariants. More precisely, we prove the following theorem.

Theorem 5. *Let $n = 2$, and $L \in K[D]$ be a third order operator with the symbol $\mathrm{Sym}_L = S_1 \cdot S_2 \cdot S_3$, where $S_1 = D_1$, $S_2 = D_2$, $S_3 = D_1 + D_2$.*

Then for every permutation (i, j, k) of $(1, 2, 3)$ denote the main common obstacle to factorization of the type $(S_i)(S_j)(S_k)$ by Obst_{ijk}, and the symbol of Obst_{ijk} by Sym_{ijk}.

Then all Sym_{ijk} and Obst_{ijk} can be expressed using the five invariants of the full system (1) as follows:

$$
\begin{aligned}
\mathrm{Sym}_{123} &= (I_3 + I_2)D_1 &&+ I_4 D_2, \\
\mathrm{Sym}_{132} &= (I_3 + \partial_y(I_1))D_1 &&+ I_4 D_2, \\
\mathrm{Sym}_{213} &= I_3 D_1 &&+ (I_4 + I_2)D_2, \\
\mathrm{Sym}_{231} &= I_3 D_1 &&+ (I_4 + 2I_2 + \partial_x(I_1))D_2, \\
\mathrm{Sym}_{312} &= (I_3 + \partial_y(I_1))D_1 &&+ (I_4 + I_2 + \partial_x(I_1))D_2, \\
\mathrm{Sym}_{321} &= (I_3 + I_2 + \partial_y(I_1))D_1 &&+ (I_4 + 2I_2 + \partial_x(I_1))D_2.
\end{aligned}
$$

While (the main) common obstacles have the forms

$$
\begin{aligned}
\mathrm{Obst}_{123} &= T_{123}, \\
\mathrm{Obst}_{132} &= T_{132} - \partial_{xy}(I_1) + \partial_x(I_2), \\
\mathrm{Obst}_{213} &= T_{213} - I_1 I_2, \\
\mathrm{Obst}_{231} &= T_{231} - I_1 I_2 - \partial_{xy}(I_1) - \partial_y(I_2), \\
\mathrm{Obst}_{312} &= T_{321} - \partial_{xy}(I_1) + \partial_x(I_2), \\
\mathrm{Obst}_{321} &= T_{321} - I_1 I_2 - \partial_{xy}(I_1) - \partial_y(I_2),
\end{aligned}
$$

where a_{20} and a_{02} are the coefficients at ∂_{xx} and ∂_{yy} in L, and

$$
T_{ijk} = \mathrm{Sym}_{ijk} + I_5 - a_{02}[D_1](\mathrm{Sym}_{ijk}) - a_{20}[D_2](\mathrm{Sym}_{ijk}),
$$

where $[D_i](S)$ denotes the coefficient of D_i in S.

Proof. Let

$$
L = \mathrm{Sym}_L + a_{20}D_1^2 + a_{11}D_1 D_2 + a_{02}D_2^2 + a_{10}D_1 + a_{01}D_2 + a_{00}.
$$

Then

$$
\begin{aligned}
\mathrm{Sym}_{123} &= (u - \partial_x(a_{20}) + \partial_y(a_{02} - a_{11} + a_{20}))D_1 &&+ (v + \partial_x(a_{02} - a_{11}))D_2, \\
\mathrm{Sym}_{132} &= (u - 2\partial_x(a_{20}) - \partial_y(a_{20}))D_1 &&+ (v + \partial_x(a_{02} - a_{11}))D_2, \\
\mathrm{Sym}_{213} &= (u + \partial_y(a_{20} - a_{11}))D_1 &&+ (v + \partial_x(a_{02} - a_{11} + a_{20}) \\
& && \quad - \partial_y(a_{02}))D_2, \\
\mathrm{Sym}_{231} &= (u + \partial_y(a_{20} - a_{11}))D_1 &&+ (v - \partial_x(a_{02}) - 2\partial_y(a_{02}))D_2, \\
\mathrm{Sym}_{312} &= (u - 2\partial_x(a_{20}) - \partial_y(a_{20}))D_1 &&+ (v - \partial_x(a_{20} + a_{02}) \\
& && \quad - \partial_y(a_{02}))D_2, \\
\mathrm{Sym}_{321} &= (u - \partial_x(a_{20}) - \partial_y(a_{20} + a_{02}))D_1 &&+ (v - \partial_x(a_{02}) - 2\partial_y(a_{02}))D_2,
\end{aligned}
$$

where

$$
u = a_{10} - a_{20}a_{11} + a_{20}^2 \quad \text{and} \quad v = a_{01} - a_{02}a_{11} + a_{02}^2.
$$

Then direct computations prove the first statement of the theorem.

We prove the formulae for the Obst_{ijk} in a similar way, that is, we compute the formulae for Obst_{ijk} as some expressions (they are large, in fact) in the coefficients of L, and then finish the proof of the theorem by means of direct computations.

Remark 1. The invariants I_1, I_2, I_3, I_4 were obtained as a consequence of Proposition 1, that is as generalizations of Laplace invariants. More precisely, we have considered all six types of factorizations into first order factors and the corresponding common obstacles, which are uniquely determined for these types. Thus, we got twelve invariants as the coefficients of these common obstacles. However, these invariants can be expressed in terms of four invariants. Then one has to determine the last invariant I_5, which should be dependent on the coefficient a_{00} of the operator L.

Acknowledgments. This work was supported by Austrian Science Foundation (FWF) under the project SFB F013/F1304.

References

1. Anderson, I., Juras, M.: Generalized Laplace Invariants and the Method of Darboux. In: Duke J. Math., 89(1997),351–375.
2. Anderson, I., Kamran, N.: The Variational Bicomplex for Hyperbolic Second-Order Scalar Partial Differential Equations in the Plane. In: Duke J. Math., 87(1997), 265–319.
3. Darboux, G.: Leçons sur la théorie générale des surfaces et les applications géométriques du calcul infinitésimal, Vol.2, (1889), Gauthier-Villars.
4. Ibragimov, N.:Invariants of hyperbolic equations: Solution of the Laplace problem. In: Prikladnaya Mekhanika i Tekhnicheskaya Fizika, Vol. 45, No. 2, (2004), 11-21. English Translation in Journal of Applied Mechanics and Technical Physics, Vol. 45, No. 2, (2004), 158-166.
5. Kartashova, E.:Hierarchy of general invariants for bivariate LPDOs. In: J. Theoretical and Mathematical Physics, (2006), 1–8.
6. www.maple.com.
7. Morozov, O.: Contact Equivalence Problem for Linear Hyperbolic Equations. In: arxiv.org preprint math-ph/0406004.
8. Shemyakova, E.,Winkler, F.:Obstacle to Factorization of LPDOs. In: Proc. Transgressive Computing, Granada, Spain, (2006).
9. Shemyakova, E.,Winkler, F.: Obstacles to factorizations of partial differential operators into several factors. In: to be published in J. Programming and Computing Software.
10. Tsarev, S.P.: Generalized Laplace Transformations and Integration of Hyperbolic Systems of Linear Partial Differential Equations. In: Proc. ISSAC'05,(2005).

An Algorithm for Computing the Complete Root Classification of a Parametric Polynomial

Songxin Liang and David J. Jeffrey

Department of Applied Mathematics
The University of Western Ontario
London, Ontario, Canada
{sliang22, djeffrey}@uwo.ca

Abstract. The Complete Root Classification for a univariate polynomial with symbolic coefficients is the collection of all the possible cases of its root classification, together with the conditions its coefficients should satisfy for each case. Here an algorithm is given for the automatic computation of the complete root classification of a polynomial with complex symbolic coefficients. The application of complete root classifications to some real quantifier elimination problems is also described.

Keywords: Complete discrimination system, complete root classification, root classification, parametric polynomial.

1 Introduction

Complete discrimination systems and complete root classifications have been applied to questions concerning the positive definiteness of polynomials [7], to ordinary differential equations [12], to integral equations [13] and to mechanics [6]. In section 3 of this paper, we show applications to some real quantifier elimination problems.

Consider for a start a polynomial $p \in K[x]$, where K is typically $\mathbb{R}$ or $\mathbb{C}$, whose coefficients are known (not symbolic). By a Root Classification (RC), we mean a list giving the number of roots in each of several categories. We list the number of roots, including multiplicities, that are real, the number that form complex conjugate pairs, and the number that are complex but which are not part of a conjugate pair. The number of roots listed must sum to $\deg(p)$. If a polynomial has known coefficients, then its RC is unique, but if the polynomial has symbolic coefficients, then it could have several RCs, each one being valid for a range of coefficient values. The collection of all such RCs, together with the ranges of the coefficients over which they are valid, is then the Complete Root Classification (CRC) of the polynomial. We denote the complex unit by I and the complex conjugate of $r \in \mathbb{C}$ by $\bar{r}$.

Definition 1. *A Root Classification (RC) of $p \in \mathbb{C}[x]$ consists of a list of lists: $[L_1, L_2, L_3]$, where L_1, L_2, L_3 are defined below.*

J. Calmet, T. Ida, and D. Wang (Eds.): AISC 2006, LNAI 4120, pp. 116–130, 2006.

The list L_1 describes all roots $r_k^{(1)}$ of p such that $r_k^{(1)} \in \mathbb{R}$. It is the list $L_1 := [n_1^{(1)}, n_2^{(1)}, \ldots]$ where each $n_k^{(1)}$ is the multiplicity of root $r_k^{(1)}$. The list L_2 describes all roots $r_k^{(2)}$ of p such that $r_k^{(2)} \in \mathbb{C} \backslash \mathbb{R}$ and $\overline{r_k^{(2)}}$ is also a root of p. It is the list $L_2 := [n_1^{(2)}, -n_1^{(2)}, n_2^{(2)}, -n_2^{(2)}, \ldots]$ where each $n_k^{(2)}$ is the multiplicity of root $r_k^{(2)}$. The list L_3 describes all roots $r_k^{(3)}$ of p such that $r_k^{(3)}$ has not been counted in L_1 and L_2. It is the list $L_3 := [n_1^{(3)}, n_2^{(3)}, \ldots]$, where each $n_k^{(3)}$ is the multiplicity of root $r_k^{(3)}$.

Remark. Notice that in the list L_2, by listing each $n_k^{(2)}$ twice, the sum of all the absolute values of all entries equals $\deg(p)$. The negative sign in front of the second $n_k^{(2)}$ in the list L_2 is merely a mnemonic of the fact that the pair of integers describes a complex-conjugate pair.

Definition 2. *A Complete Discrimination System (CDS) of a parametric polynomial $p(x)$ consists of a set of expressions in the symbolic coefficients of $p(x)$, such that only these expressions need to be tested in order to decide a root classification.*

Definition 3. *A Complete Root Classification (CRC) of a parametric polynomial $p(x)$ is a collection of statements, each of which consists of an RC coupled with a sequence of conditions for which it is valid. The sequence of conditions uses the elements of the CDS.*

As an introductory example, the root classification of $p(x) = x^4 + ax^2 + bIx$ (where a, b are real parameters) is completely determined by a, b and D where $D = -4a^3 + 27b^2$. Therefore, the CDS is $\{a, b, D\}$. The CRC consists of the statements

- $[[4], [], []]$ if and only if $a = 0 \wedge b = 0$,
- $[[1, 1, 2], [], []]$ if and only if $a < 0 \wedge b = 0$,
- $[[2], [1, -1], []]$ if and only if $a > 0 \wedge b = 0$,
- $[[1], [], [1, 1, 1]]$ if and only if $b \neq 0 \wedge D \neq 0$,
- $[[1], [], [1, 2]]$ if and only if $b \neq 0 \wedge a \neq 0 \wedge D = 0$.

Progress on finding CDS and CRC for polynomials has been slow. The CRC for a quartic polynomial with real symbolic coefficients was found by Arnon [1] only in 1985. Arnon [2] used that result to derive the positive semidefinite conditions on polynomial $x^4 + px^2 + qx + r$. Significant progress on CDS and CRC for polynomials with degrees greater than 4 was not made until 1996 (see [2], [9]). In 1996, Yang et al. ([9], [10]) proposed a generic method for establishing the CDS for a real polynomial of any degree. Furthermore, they gave an algorithm for generating the root classification of a constant coefficient polynomial. However, they did not give an algorithm for computing a CRC. This drawback prevented the CDS and CRC from extensive applications.

As the degree of polynomial increases, human analysis will founder on the greater complication and the possibility of error. In fact, it is not easy to discover accurately how many possible cases of root classification there are for a polynomial of degree 9 with real symbolic coefficients. Thus, it is a necessary and non-trivial task to establish a generic algorithm to determine the CRC for any polynomial by computer. Moreover, compared with the case of real parametric polynomials, the analysis of complex parametric polynomials is much more complicated. If R_n and C_n denote the numbers of possible cases of RC for a polynomial of degree n with real symbolic coefficients and complex symbolic coefficients respectively, then we can see from Table 1 (an explanation will be given in section 2.2) that the ratio of C_n to R_n increases as the degree n increases. In 1999, Liang and Zhang [7] showed how to solve the problems of CDS and

Table 1. The Ratio of C_n to R_n

n	2	3	4	5	6	10	15	20	25	30
R_n	3	4	9	12	23	118	651	3177	12584	46092
C_n	6	12	27	50	98	888	9072	69545	433054	2324844
C_n/R_n	2	3	3	4	4	8	14	22	34	50

CRC for polynomials with complex coefficients, but they did not give explicit and systematic algorithms.

This paper gives a complete and explicit algorithmic method for generating the CDS and CRC of a polynomial with complex symbolic coefficients. Furthermore, this paper discusses how to remove some of those RCs which are not realizable for a sparse parametric polynomial (see Definition 10). It is interesting and also important since most polynomials arising in AI applications are sparse. In section 2, the algorithms for generating CDS and CRC for polynomials with complex symbolic coefficients are presented. In section 3, some applications of CRC to problems in real quantifier elimination and the advantages of the new method are discussed.

2   CDS and CRC

In this section, $f(x)$ is a complex (parametric) polynomial of degree n. $f(x) = f_1(x) + I * f_2(x)$, where $f_1(x)$ and $f_2(x)$ are the real part and the imaginary part of $f(x)$ respectively. By $p(x)$ and $g(x)$ we always refer to the greatest common divisor of $f_1(x)$ and $f_2(x)$ (determined by proposition 3 below) and the pseudo quotient of $f(x)$ divided by $p(x)$ respectively. A polynomial is called *primitive*, if its real part and its imaginary part are co-prime. It is easy to see that $g(x)$ is primitive. We first review some basic concepts and results needed for the algorithms in subsection 2.1. Then in subsections 2.2 to 2.4 we will present the subalgorithms which are the components of the main algorithm. In subsection 2.5, we will present the algorithm for CRC.

2.1   Review of Basic Concepts and Results

In order to make this paper self-contained, we review some definitions and propositions which can be found in [7], [9] or [10]. Notations are as above.

Proposition 1. *The set of all roots of $p(x)$ exactly consists of all real roots and all conjugate imaginary roots of $f(x)$.*

A proof can be found in [7]. Then from proposition 1, we can easily get

Proposition 2. *$g(x)$ has at most non-conjugate imaginary roots.*

Let $f(x) = a_n x^n + a_{n-1} x^{n-1} + \cdots + a_0$ and $h(x) = b_m x^m + b_{m-1} x^{m-1} + \cdots + b_0$ be two polynomials with complex coefficients, and let $s_i(f, h, x)$ and $p_i(f, h, x)$ be the principal subresultant sequence and subresultant polynomial sequence of $f(x)$ and $h(x)$ respectively. We have [11]

Proposition 3. *Suppose $a_n \neq 0$ or $b_m \neq 0$. If $s_0(f, h, x) = \cdots = s_{k-1}(f, h, x) = 0$ and $s_k(f, h, x) \neq 0$, then $\gcd(f, h) = p_k(f, h, x)$.*

Definition 4. *Let $f(x) = a_n x^n + a_{n-1} x^{n-1} + \cdots + a_0$. The following $2n \times 2n$ matrix in terms of the coefficients,*

$$
\begin{pmatrix}
a_n & a_{n-1} & a_{n-2} & \cdots & a_0 & & & & \\
0 & n a_n & (n-1)a_{n-1} & \cdots & a_1 & & & & \\
 & a_n & a_{n-1} & \cdots & a_1 & a_0 & & & \\
 & 0 & n a_n & \cdots & 2a_2 & a_1 & & & \\
 & & & \cdots & \cdots & & & & \\
 & & & \cdots & \cdots & & & & \\
 & & & & a_n & a_{n-1} & a_{n-2} & \cdots & a_0 \\
 & & & & 0 & n a_n & (n-1)a_{n-1} & \cdots & a_1
\end{pmatrix}
$$

is called the discrimination matrix of $f(x)$, and is denoted by $\mathrm{Discr}(f)$. D_k denotes the determinant of the submatrix of $\mathrm{Discr}(f)$, formed by the first $2k$ rows and the first $2k$ columns, for $k = 1, 2, \ldots, n$.

Definition 5. *The n-tuple $[D_1, D_2, \ldots, D_n]$ is called the discriminant sequence of $f(x)$.*

Definition 6. *Let $D = [D_1, D_2, \ldots, D_n]$ be the discriminant sequence of a real polynomial $p(x)$, we call the list $[\mathrm{sign}(D_1), \mathrm{sign}(D_2), \ldots, \mathrm{sign}(D_n)]$ the sign list of D.*

Definition 7. *Let D be the discriminant sequence of $p(x)$. The revised sign list $[e_1, e_2, \ldots, e_n]$ of $p(x)$ is constructed from the sign list $[s_1, s_2, \ldots, s_n]$ of D as follows, and is denoted by $rsl(p)$. If $[s_i, s_{i+1}, \ldots, s_{i+j}]$ is a section of the given list, where $s_i \neq 0$, $s_{i+1} = s_{i+2} = \ldots = s_{i+j-1} = 0$ and $s_{i+j} \neq 0$, then we replace the subsection $[s_{i+1}, \ldots, s_{i+j-1}]$ by*

$$[-s_i, -s_i, s_i, s_i, -s_i, -s_i, s_i, s_i, \ldots],$$

i.e., let $e_{i+r} = (-1)^{\lfloor (r+1)/2 \rfloor} s_i$, for $r = 1, 2, \ldots, j-1$, where $\lfloor (r+1)/2 \rfloor$ is the floor function of $(r+1)/2$. Otherwise, let $e_k = s_k$, i.e., there are no changes for other terms.

Definition 8. *Let $[d_1, d_2, \ldots, d_t]$ be the discriminant sequence of a complex polynomial $f(x)$, and k is the maximal subscript such that $d_k \neq 0$. We define the revised sign list of $f(x)$ to be the list $[1, 1, \ldots, 1, 0, 0, \ldots, 0]$ (there are k continuous 1's followed by $t - k$ continuous 0's) and denote it by $rsl(f)$.*

Letting $h(x) = f'(x)$ in proposition 3, we have

Proposition 4. *If the number of the 0's in the $rsl(f)$ is k, then $\gcd(f(x), f'(x)) = p_k(f, f', x)$.*

If we only want to know the number of the distinct real roots or imaginary roots, then the following propositions are sufficient ([9], [10]).

Proposition 5. *Given a polynomial $p(x)$ over reals, if the number of sign changes and the number of non-vanishing members of the $rsl(p)$ are v and s respectively, then the number of pairs of the distinct conjugate imaginary roots of $p(x)$ is v, and the number of the distinct real roots is $s - 2v$.*

By proposition 2 and proposition 4, we have [7]

Proposition 6. *Let $g(x)$ be a primitive polynomial of degree m. If the number of non-vanishing members of the $rsl(g)$ is r, then the number of the distinct non-conjugate imaginary roots of $g(x)$ is r.*

If we want to know not only the number of the distinct roots of a polynomial $f(x)$, but also the multiplicity of every root, then we need to consider the root classification of the "repeated part" of $f(x)$, i.e., $\gcd(f(x), f'(x))$.

Definition 9. *For convenience, let $\Delta(f)$ denote $\gcd(f(x), f'(x))$. If the number of the 0's in the $rsl(f)$ is k, then by proposition 4, $\Delta(f) = p_k(f, f', x)$. Let $\Delta^0(f) = f(x)$, $\Delta^j(f) = \Delta(\Delta^{j-1}(f))$, for $j = 1, 2, \ldots$. Then we call $\Delta^0(f), \Delta^1(f), \Delta^2(f), \ldots$ the Δ-sequence of $f(x)$.*

Remark. Let notations be as before. Let $s_i(f_1, f_2, x)$ be the principal subresultant sequence of f_1 and f_2, U and V be the Δ-sequences of $p(x)$ and $g(x)$ respectively. Let Ω be the set of all expressions in the discriminant sequences of all polynomials in W. Then the CDS of $f(x)$ is a subset of $\Omega \cup \{s_i(f_1, f_2, x)\}$.

Finally, we give two important propositions for the algorithms. The proofs are easy.

Proposition 7. *Let $f(x)$ be a polynomial over complexes. If $\Delta^j(f)$ has k distinct roots with respective multiplicities $n_1, n_2, \ldots, n_k$, then the "repeated part" $\Delta^{j+1}(f)$ of $\Delta^j(f)$ has at most k distinct roots with respective multiplicities $n_1 - 1, n_2 - 1 \ldots, n_k - 1$ (if the multiplicity of a root is 0, it means the root does not exist).*

Proposition 8. *Let $f(x)$ be a polynomial over complexes. If $\Delta^j(f)$ has k distinct roots with respective multiplicities $n_1, n_2, \ldots, n_k$, and $\Delta^{j-1}(f)$ has m distinct roots, then $m \geq k$, and the multiplicities of these m distinct roots are $n_1 + 1, n_2 + 1 \ldots, n_k + 1, 1, \ldots, 1$ (there are $m - k$ continuous 1's) respectively.*

2.2 All Possible Cases of Root Classification for a Parametric Polynomial

Given a polynomial of degree n with symbolic coefficients, from the theoretical point of view, how many possible cases of RC may it have?

The set of all possible RCs for a polynomial of degree n over reals can be expressed as: {all possible RCs with 0 pair of conjugate imaginary roots} $\cup$ {all possible RCs with 1 pair of conjugate imaginary roots} $\cup \cdots \cup$ {all possible RCs with $\lfloor n/2 \rfloor$ pairs of conjugate imaginary roots}. Therefore, the algorithm for generating the set of all possible RCs for a polynomial of degree n over reals is

AllListsReal
Input: $n \in \mathbb{N}$.
Output: the set of all possible RCs for a real polynomial of degree n.

$L := \{\}$
for k from 0 to $\lfloor n/2 \rfloor$ do

- compute the set A of all partitions of $n - 2k$
- compute the set B of all partitions of k
- change B into C as follows. $\forall b \in B$, apply mapping: $\forall i \in b$, $i \mapsto i, -i$. For example, if $B = \{[3], [1, 2]\}$, then $C = \{[3, -3], [1, -1, 2, -2]\}$
- combine the elements of A with the elements of C by distribution law to form a set U of new lists
- L:=append(L, U)

return L

For complex polynomial $f(x)$ of degree n, let notations be as before and let $\deg(g(x), x) = k$. Then the set of all possible RCs for $f(x)$ can be expressed as: {all possible RCs when $k = 0$} $\cup$ {all possible RCs when $k = 1$} $\cup \cdots \cup$ {all possible RCs when $k = n$}.

For $0 \leq k \leq n$, the algorithm to compute the set of all possible RCs with k non-conjugate imaginary roots is

ListsRealImag
Input: $n \in \mathbb{N}$, $k \in \mathbb{N}$.
Output: the set of all possible RCs for a complex polynomial of degree n with k non-conjugate imaginary roots.

Since the algorithm is similar to *AllListsReal*, we omit the details. The following example is generated by computer.

```
> ListsRealImag(5, 2)
{[[3],[],[1,1]], [[1,2],[],[1,1]], [[1,1,1],[],[1,1]],
[[1],[1,-1],[1,1]], [[3],[],[2]], [[1,2],[],[2]],[[1,1,1],[],[2]],
[[1],[1,-1],[2]]}
```

At this point, we would like to give an explanation for Table 1. The number of all partitions of n is $p(n)$, the well-known partition function (see [19], for example). Some values of $p(n)$ are as follows: $p(1) = 1, p(2) = 2, p(3) = 3, p(4) = 5, p(5) = 7, p(10) = 42, p(100) = 190569292$. Using the argument above, we can easily get the Table 1.

2.3   Conditions for Each Case of Root Classification

Let notations be as above. Suppose $b = [b_1, b_2, b_3]$ is an RC for $f(x)$. What conditions should $p(x)$ and $g(x)$ satisfy? We can express this problem exactly. In order to make the RC of $p(x)$ to be exactly $[b_1, b_2, []]$, what conditions should they satisfy for the polynomials in the Δ-sequence of $p(x)$? In order to make the RC of $g(x)$ to be exactly $[[], [], b_3]$, what conditions should they satisfy for the polynomials in the Δ-sequence of $g(x)$?

First of all, we would like to introduce some mini subalgorithms. Their implementations are easy.

1. **ImagLtInfo.** Input: an RC for a primitive polynomial. Output: a list consisting of the degree and the number of distinct non-conjugate imaginary roots of the polynomial.
2. **RealLtInfo.** Input: an RC for a polynomial over reals. Output: a list consisting of the degree, the number of distinct roots and the number of pairs of distinct conjugate imaginary roots of the given polynomial.
3. **GenRealRSL.** Input: $m, n, k \in \mathbb{N}$. Output: a sequence of all possible (revised sign) lists of which the length, the number of non-vanishing members and the number of sign changes are m, n and k respectively.
4. **GenImagRSL.** Input: $m, n \in \mathbb{N}$. Output: a list of length $m + n$ which contains m continuous 1's followed by n continuous 0's.
5. **MinusOne.** Input: an RC b. Output: an RC generated from b by decreasing the absolute values of all numbers in b by 1, and then erasing all numbers of value 0, and keeping empty lists unchanged.

Suppose $b = [b_1, b_2, []]$ is an RC of $p(x)$, and let $t = RealLtInfo(b)$. Then $p(x)$ should be a polynomial of degree $t[1]$, with $t[2]$ distinct roots and $t[3]$ pairs of distinct conjugate imaginary roots. Thus, by proposition 5, all possible $rsl(p)$ should be $rsl_0 = GenRealRSL(t[1], t[2], t[3])$. By definition 9, $\Delta^1(p)$ should be the subresultant polynomial of p and p' of index $t[1] - t[2]$, that is, $p_{t[1]-t[2]}(p, p', x)$. For the following five cases, we can determine just by rsl_0 that the RC of $p(x)$ is $b = [b_1, b_2, []]$ without further computation for the revised sign lists of other polynomials in the Δ-sequence of $p(x)$.

1. $t[1] = t[2]$. Then the $rsl(p)$ contains no 0, then there is only one polynomial in the Δ-sequence of $p(x)$, $p(x)$ itself.
2. $t[2] = 1$. Then $p(x)$ has only one distinct real root, so do the other polynomials in the Δ-sequence of $p(x)$ by proposition 7. Thus any revised sign list of the polynomials in the Δ-sequence of $p(x)$ also contains only one 1.
3. $t[2] = 2$ and $t[3] = 1$. Then $p(x)$ has only one pair of distinct conjugate imaginary root, so do the other polynomials in the Δ-sequence of $p(x)$ by proposition 7. Thus any revised sign list of the polynomials in the Δ-sequence of $p(x)$ also contains only two non-vanishing members: 1 and -1.

4. $t[1] - t[2] = 1$. Then the "repeated part" $\Delta^1(p)$ is a polynomial of degree 1, and the revised sign list of it is $[1]$.

5. $t[2] = 2\,t[3]$ and $t[1] - t[2] = 2$. Then all roots of $p(x)$ should be conjugate imaginary roots and the "repeated part" $\Delta^1(p)$ of $p(x)$ should be a polynomial of degree 2. Then by proposition 7, $\Delta^1(p)$ exactly has a pair of conjugate imaginary roots, and the revised sign list of it should be $[1, -1]$.

Noticing the definition of *MinusOne* and proposition 4, the discussion above has actually proved the correctness of the following algorithm. The algorithm for generating the conditions for $p(x)$ having $b = [b_1, b_2, []]$ as its RC is

RealCond0

Input: a real polynomial $p(x)$; an RC $b = [b_1, b_2, []]$; the variable x.
Output: a sequence of mix lists consisting of polynomials in the Δ-sequence of $p(x)$ and their all possible revised sign lists.

$t := RealLtInfo(b)$
$rsl := GenRealRSL(t[1], t[2], t[3])$
if b meets one of the five cases
      return $[p, rsl]$
else
      return cons($[p, rsl], RealCond0(\Delta^1(p), MinusOne(b), x)$)

Similarly, we can discuss the conditions for $g(x)$ having $b = [[], [], b_3]$ as its RC. Let $t := ImagLtInfo(b)$. Then the three cases $(1)\, t[1] = t[2], (2)\, t[1] = 1$ and $(3)\, t[1] - t[2] = 1$ correspond to the five cases in *RealCond0*. By propositions 4, 6, 7 and a similar discussion as in *RealCond0*, we can easily prove the correctness of the following algorithm. The algorithm for generating the conditions for $g(x)$ having $[[], [], b_3]$ as its RC is

ImagCond0

Input: a primitive polynomial $g(x)$; an RC $b = [[], [], b_3]$; the variable x.
Output: a sequence of mix lists consisting of polynomials in the Δ-sequence of $g(x)$ and their possible revised sign lists.

$t := ImagLtInfo(b)$
$rsl := GenImagRSL(t[2], t[1] - t[2])$
if b meets one of the three cases
      return $[g, rsl]$
else
      return cons($[g, rsl], ImagCond0(\Delta^1(g), MinusOne(b), x)$)

Finally, we come to the point to give the algorithm for generating the conditions for $f(x)$ having $b = [b_1, b_2, b_3]$ as its RC. Let notations be as above.

AllCond0

Input: a real polynomial $p(x)$; a primitive polynomial $g(x)$; an RC $b = [b_1, b_2, b_3]$; the variable x.
Output: b and the conditions that the Δ-sequences of $p(x)$ and $g(x)$ should satisfy.

if $\deg(g, x) = 0$

    return $[b_1, b_2, []]$ and $RealCond0(p, [b_1, b_2, []], x)$

else

    return b, $RealCond0(p, [b_1, b_2, []], x)$ and $ImagCond0(g, [[], [], b_3], x)$.

2.4 Realization of Root Classification

For a given polynomial $f(x)$ with complex symbolic coefficients, we have given algorithms for generating the set of its all possible RCs, together with the conditions it should satisfy for each case. A question arises natually: is each case of RC realizable?

For a real polynomial $p(x)$, a case of RC is realizable if and only if the conditions associated with the case are realizable. The latter holds if and only if the semi-algebraic set defined by the conditions is non-empty. Similarly, for a primitive polynomial $g(x)$, a case of RC is realizable if and only if the algebraic set defined by the associated conditions is non-empty. Deciding the emptiness of a semi-algebraic set is a difficult subject and some references are available, for example, [17] and [18]. Lemma 1.14 of [20] gives a method for deciding the emptiness of an algebraic set over $\mathbb{C}$.

Here, we choose a more direct approach to the question. We remove those RCs that can be proved to be not realizable. If a polynomial $p(x)$ of degree n has a general form, that is, $p(x) = x^n + a_{n-1}x^{n-1} + \cdots + a_1 x + a_0$, where $a_i(i = 0, 1, \ldots, n-1)$ are distinct real parameters, then it is easy to see that all the RCs outputted by *AllListsReal* are realizable. On the other hand, for sparse parametric polynomials (see the definition below), some of the RCs outputted by *AllListsReal* and *ListsRealImag* are not realizable.

Definition 10. *A parametric polynomial $f(x)$ of degree n is sparse, if at least one $D_k(2 \leq k \leq n)$ in its discriminant sequence D contains no symbols.*

Definition 11. *Let notations be as before. The RSL realization of the discriminant sequence D of $p(x)$ ($g(x)$ respectively) is a collection of all the (revised sign) lists obtained by the following process.*

If $p(x)$ has not more than 3 parameters, first we find out a sample of D-invariant decomposition for the parametric space by CAD ([3], [8]), then substitute the sample points into D one by one. If $p(x)$ has more than 3 parameters, then 3 values -1, 0 and 1 are assigned to each D_k in D which contains symbols. For $g(x)$, 2 values 0 and 1 are assigned to each D_k in D which contains symbols. Finally, we compute the revised sign lists using definition 7 for all the resulting lists.

Remark. For a given real parametric polynomial $p(x)$, the set of all possible $rsl(p)$ should be a subset of the set U_1: the RSL realization of the discriminant sequence of $p(x)$ (the two sets are the same if U_1 is obtained by CAD). Therefore, for an RC $b = [b_1, b_2, []]$, let $t = RealLtInfo(b)$, and $U_2 = GenRealRSL(t[1], t[2], t[3])$. If $U_1 \cap U_2 = \phi$ then b is not realizable for $p(x)$. On the other hand, $U_1 \cap U_2$ is the conditions that $p(x)$ should satisfy if b is realizable. Similar conclusions can be drawn for a primitive parametric polynomial $g(x)$.

For example, for $p(x) = x^5 + x^4 + a$, where a is a real parameter, its discriminant sequence is $D = [1, 1, 0, -a^2, 256a^3 + 3125a^4]$. Therefore, $p(x)$ is a sparse parametric polynomial. We claim that $b = [[1, 2, 2], [], []]$ is not realizable for $p(x)$. By CAD, we can easily obtain the RSL realization of D. It is $U_1 = \{[1, 1, -1, -1, 0], [1, 1, -1, -1, 1], [1, 1, 0, 0, 0], [1, 1, -1, -1, -1]\}$. On the other hand, $RealLtInfo(b) = [5, 3, 0]$, and $U_2 := GenRealRSL(5, 3, 0) = \{[1, 1, 1, 0, 0]\}$. Because $U_1 \cap U_2 = \phi$, b is not realizable for $p(x)$.

Based on the discussion above, we can update the algorithms in subsection 2.3 as follows.

RealCond
Input: a real polynomial $p(x)$; an RC $b = [b_1, b_2, []]$; the variable x.
Output: the conditions that the Δ-sequence of $p(x)$ should satisfy (if it is NULL, then b is not realizable).

$t := RealLtInfo(b)$
$rsl := GenRealRSL(t[1], t[2], t[3])$
if p is sparse or p has not more than 3 parameters
    $rsl := rsl \cap \{\text{the RSL realization of the discriminant sequence of } p\}$
if $rsl = \phi$
    return NULL
else if b meets one of the five cases
    return $[p, rsl]$
else
    $L := RealCond(\Delta^1(p), MinusOne(b), x)$
    if $L = $ NULL
        return NULL
    else
        return $cons([p, rsl], L)$

ImagCond
Input: a primitive polynomial $g(x)$; an RC $b = [[], [], b_3]$; the variable x.
Output: the conditions that the Δ-sequence of $g(x)$ should satisfy (if it is NULL, then b is not realizable).

$t := ImagLtInfo(b)$
$rsl := GenImagRSL(t[2], t[1] - t[2])$
if g is sparse
    $rsl := rsl \cap \{\text{the RSL realization of the discriminant sequence of } g\}$
if $rsl = \phi$
    return NULL
else if b meets one of the three cases
    return $[g, rsl]$
else
    $L := ImagCond(\Delta^1(g), MinusOne(b), x)$

if L = NULL
    return NULL
else
    return cons($[g, rsl], L$)

AllCond

Input: a real polynomial $p(x)$; a primitive polynomial $g(x)$; an RC $b = [b_1, b_2, b_3]$; the variable x.
Output: b and the conditions that the Δ-sequences of $p(x)$ and $g(x)$ should satisfy (if it is NULL, then b is not realizable).

if $\deg(g, x) = 0$
    $L := RealCond(p, b, x)$
    if L = NULL
        return NULL
    else
        return b and L
else
    $L_1 := RealCond(p, [b_1, b_2, []], x)$
    $L_2 := ImagCond(g, [[], [], b_3], x)$
    if L_1 = NULL or L_2 = NULL
        return NULL
    else
        return b, L_1 and L_2

2.5   The Algorithm for CRC

Based on the subalgorithms in the preceding subsections, now it is time to give the algorithm for generating the CRC of a complex parametric polynomial.

CRC

Input: A complex parametric polynomial $f(x)$; the variable x.
Output: the CRC of $f(x)$.

$n := \deg(f, x)$
if f is real
    $L := ListsRealImag(n, 0)$
    for b in L do
        $AllCond(f, 1 + I, b, x)$
else
    f_1=realPart(f), f_2=imagPart(f)
    compute the principal subresultant sequence $\{s_k\}$ and
    subresultant polynomial sequence $\{p_k\}, 0 \le k \le m \le n$
    for k from 0 to m do
        if $s_0 = \cdots = s_{k-1} = 0$ and $s_k \ne 0$
            $p := p_k$
            $g :=$pseudoQuotient(f, p)

$L := ListsRealImag(n, n - k)$
for b in L do
    $AllCond(p, g, b, x)$

The algorithm above has been made into a generic program in Maple, which enables the CRC to be automatically generated by computer. In [7], the CRC of a sextic polynomial over reals and the CRC of a quartic polynomial over complexes are generated by computer automatically. Notice that there is also a CRC for a sextic polynomial over reals in [9] and [10]. However, the one in [9] and [10] is accomplished by artificial analysis which needs higher skills. Furthermore, the information about the imaginary roots is not complete. In the following section, more examples will be given.

Remark. By the discussion in subsection 2.4, we can see that for real parametric polynomials which have general forms or which have not more than 3 parameters, especially for those in the examples in section 3, all the RCs in the CRC are realizable. If we use the improved CAD (see [4] and [16] for example) at the *RSL realization* within *RealCond*, we can expect that all the RCs in the CRC are realizable for real polynomials with more parameters instead of just 1, 2, or 3 parameters.

3   Applications of CRC to Real Quantifier Elimination

This section presents some applications of CRC method to real quantifier elimination. It concerns specific input formulas of the form $\forall x[\psi]$, where ψ is a single polynomial inequality. All the computations have been performed on a Pentium IV PC with 3.2 GHz CPU and 1 GB RAM. For CRC method, we use Maple 10. For CAD method, we use QEPCAD B [5].

For a real parametric polynomial $p(x)$ with positive leading coefficient, the routine of the CRC method for the specific real quantifier elimination is:

1. Compute the CRC of $p(x)$ by computer automatically.
2. Pick out the cases of RC in the CRC which don't have real roots (for positive definite) and the cases of which every real root has an even multiplicity (for positive semidefinite).
3. Interpret the conditions for each case in step 2. Denote the mapping from the set of sign lists to the set of revised sign lists in definition 7 by Φ. Let L be the set of revised sign lists in the conditions for some case in step 2. Then the conditions for that case is $\Phi^{-1}(L)$.

Example 1. Find the conditions on a, b, c such that $(\forall x)[x^4 + ax^2 + bx + c \geq 0]$.
    First, we compute the CRC of $p_4 = x^4 + ax^2 + bx + c$ by computer.

```
(*) p4:=x^4+ax^2+bx+c
(1) [[4],[],[]], iff [p4,[1,0,0,0]]
(2) [[2,2],[],[]], iff [p4,[1,1,0,0]],[p42,[1,1]]
(3) [[1,3],[],[]], iff [p4,[1,1,0,0]],[p42,[1,0]]
```

(4) `[[1,1,2],[],[]]`, iff `[p4,[1,1,1,0]]`
(5) `[[1,1,1,1],[],[]]`, iff `[p4,[1,1,1,1]]`
(6) `[[1,1],[1,-1],[]]`, iff
    `[p4,[1,1,1,-1],[1,1,-1,-1],[1,-1,-1,-1]]`
(7) `[[2],[1,-1],[]]`,iff `[p4,[1,1,-1,0],[1,-1,-1,0]]`
(8) `[[],[2,-2],[]]`, iff `[p4,[1,-1,0,0]]`
(9) `[[],[1,-1,1,-1],[]]`, iff
    `[p4,[1,-1,1,1],[1,1,-1,1],[1,-1,-1,1]]`
where,
(#1) `p4:=x^4+a*x^2+b*x+c`, and its discriminant sequence is:
 `D:=[1, -a, -2*a^3+8*c*a-9*b^2, -4*b^2*a^3+16*a^4*c+144*a*c*b^2`
 `-128*c^2*a^2-27*b^4+256*c^3]`
(#2) `p42:=-2*a*x^2-3*b*x-4*c`, and its discriminant sequence is:
`E:=[1, 9*b^2-32*c*a]`

We denote the discriminant sequences of p_4 and p_{42} by $[1, -a, D_3, D_4]$ and $[1, E_2]$ respectively. It is easy to see that $(\forall x)[x^4 + ax^2 + bx + c \geq 0] \iff$ one of the following cases holds: cases (1), (2), (7), (8), (9).

But case (1) holds $\Leftrightarrow$ the $rsl(p_4)$ be $[1,0,0,0] \Leftrightarrow a = 0 \wedge D_3 = 0 \wedge D_4 = 0$. Case (2) holds $\Leftrightarrow$ the $rsl(p_4)$ be $[1,1,0,0]$ and the $rsl(p_{42})$ be $[1,1] \Leftrightarrow a < 0 \wedge D_3 = 0 \wedge D_4 = 0 \wedge E_2 > 0$. Case (7) holds $\Leftrightarrow$ the $rsl(p_4)$ be one of $[1,1,-1,0]$ and $[1,-1,-1,0] \Leftrightarrow D_3 < 0 \wedge D_4 = 0$. Cases (8) holds $\Leftrightarrow$ the $rsl(p_4)$ be $[1,-1,0,0] \Leftrightarrow a > 0 \wedge D_3 = 0 \wedge D_4 = 0$. Case (9) holds $\Leftrightarrow$ the $rsl(p_4)$ be one of $[1,-1,1,1], [1,1,-1,1]$ and $[1,-1,-1,1] \Leftrightarrow [a \geq 0 \wedge D_4 > 0] \vee [D_3 \leq 0 \wedge D_4 > 0]$

Therefore, the solution by CRC method is
$(\forall x)[x^4 + ax^2 + bx + c \geq 0] \iff [D_4 > 0 \wedge [a \geq 0 \vee D_3 \leq 0]] \vee [D_4 = 0 \wedge D_3 \leq 0 \wedge [D_3 < 0 \vee a \geq 0 \vee E_2 > 0]]$

On the other hand, using QEPCAD, we get the following result:
$(\forall x)[x^4 + ax^2 + bx + c \geq 0] \iff F_1 \geq 0 \wedge [F_2 > 0 \vee [F_3 \geq 0 \wedge F_4 \geq 0]]$
where
$$F_1 := 256c^3 - 128a^2c^2 + 144ab^2c + 16a^4c - 27b^4 - 4a^3b^2;$$
$$F_2 := 27b^2 + 8a^3;$$
$$F_3 := 48c^2 - 16a^2c + 9ab^2 + a^4;$$
$$F_4 := 6c - a^2.$$

Example 1 is a benchmark QE problem. It takes 0.16 second to generate the CRC while it takes 0.058 second by QEPCAD to get the solution. The solution by CRC method is a little bit longer than the one by QEPCAD and the one by Lazard [15] which has 3 atomic formulas and the one by Arnon [2] which has 4 distinct atomic formulas, but is much shorter than the one by Weispfenning [14] which has 29 distinct atomic formulas. Furthermore, as the degree of the polynomial increases, the advantage of CRC method is obvious when compared with that of CAD method.

Example 2. Find the conditions on a, b, c such that $(\forall x)[x^6 + ax^2 + bx + c \geq 0]$.
The result has been given in [9]. It takes 0.47 second to generate the CRC of the given polynomial while it takes 12.9 seconds by QEPCAD to get the solution.

This is a sparse parametric polynomial. It has only 10 cases of RC in the CRC while there are 23 possible cases of RC for a general real polynomial of degree 6 (see Table 1). The solution by CRC method has 4 distinct atomic formulas while the one by QEPCAD has 5 distinct atomic formulas. The sizes of the two solutions are about the same.

Example 3. Find the conditions on a, b, c such that $(\forall x)[x^{10} + ax^2 + bx + c > 0]$.

It takes 4.41 seconds to generate the CRC. This is also a sparse parametric polynomial. It has only 10 cases of RC in the CRC while there are 118 possible cases of RC for a general real polynomial of degree 10 (see Table 1). On the other hand, CAD method failed when we tried to solve the problem.

The solution by CRC method is given by

$$(\forall x)[x^{10} + ax^2 + bx + c > 0] \iff [a > 0 \wedge D_9 = 0 \wedge D_{10} = 0]$$
$$\vee [D_9 \geq 0 \wedge D_{10} < 0] \vee [a \geq 0 \wedge D_{10} < 0], \text{ where}$$

$$D_9 = 16777216a^9 + 204800000a^4c^4 - 2073600000a^3b^2c^3$$
$$+3149280000a^2b^4c^2 - 1488034800ab^6c + 215233605b^8;$$
$$D_{10} = -67108864a^{10}c + 16777216a^9b^2 - 1638400000a^5c^5$$
$$+16588800000a^4b^2c^4 - 27993600000a^3b^4c^3 + 17006112000a^2b^6c^2$$
$$-4304672100ab^8c - 10000000000c^9 + 387420489b^{10}.$$

References

1. Arnon, D.S.: On Mechanical Quantifier Elimination for Elementary Algebra and Geometry: Automatic Solution of a Nontrivial Problem. Lecture Notes in Computer Science **204** (1985) 270–271
2. Arnon, D.S.: Geometric Reasoning with Logic and Algebra. Artificial Intelligence **37** (1988) 37–60
3. Collins, G.E.: Quantifier Elimination for the Elementary Theory of Real Closed Fields by Cylindrical Algebraic Decomposition. Lecture Notes in Computer Science **33** (1975) 134–183
4. Collins, G.E., Hong, H.: Partial Cylindrical Algebraic Decomposition for Quantifier Elimination. Journal of Symbolic Computation **12** (1991) 299–328
5. Brown, C.W.: QEPCAD B: a Program for Computing with Semi-algebraic Sets Using CADs. ACM SIGSAM Bulletin **37** (2003) 97–108
6. Lan, Y., Wang, L., Zhang, L.: The Finite Checking Method of the Largest Stability Bound for a Class of Uncertain Interval Systems. Journal of Peking University **40** (2004) 29–36
7. Liang, S., Zhang, J.: A Complete Discrimination System for Polynomials with Complex Coefficients and Its Automatic Generation. Science in China (Series E) **42** (1999) 113–128
8. Winkler, F.: Polynomial Algorithms in Computer Algebra. Springer Wien, New York (1996)
9. Yang, L.: Recent Advances on Determining the Number of Real Roots of Parametric Polynomials. Journal of Symbolic Computation **28** (1999) 225–242
10. Yang, L., Hou, X., Zeng, Z.: Complete Discrimination System for Polynomials. Science in China (Series E) **39** (1996) 628–646

11. Yang, L., Zhang, J., Hou, X.: Non-linear Equation Systems and Automated Theorem Proving. Shanghai Press of Science, Technology and Education, Shanghai (1996)
12. Yang, C., Zhu, S., Liang, Z.: Complete Discrimination of the Roots of Polynomials and Its Applications. Acta Scientiarum Naturalium Universitatis Sunyatseni **42** (2003) 5–8
13. Wang, L., Yu, W., Zhang, L.: On the Number of Positive Solutions to a Class of Integral Equations. Preprint in IMA at the University of Minnesota (2004)
14. Weispfenning, V.: Quantifier Elimination for Real Algebra - the Cubic Case. ISSAC (1994) 258–263
15. Lazard, D.: Quantifier Elimination: Optimal Solution for Two Classical Examples. Journal of Symbolic Computation **5** (1988) 261–266
16. Brown, C.W.: Simple CAD Construction and Its Applications. Journal of Symbolic Computation **31** (2001) 521–547
17. Grigor'ev, D., Vorobjov, N.: Solving Systems of Polynomial Inequalities in Subexponential Time. Journal of Symbolic Computation **5** (1988) 37–64
18. Aubry, P., Rouillier, F., Safey El Din, M.: Real Solving for Positive Dimensional Systems. Journal of Symbolic Computation **34** (2002) 543–560
19. Rademacher, H.: On the Partition Function p(n). Proc. London Math. Soc. **43** (1937) 241–254
20. Basu, S., Pollack, R., Roy, M.-F.: Algorithms in Real Algebraic Geometry. Springer-Verlag (2003)

Quantifier Elimination for Quartics

Lu Yang[1] and Bican Xia[2,*]

[1] Guangzhou University, Guangzhou 510006, China
[2] School of Mathematical Sciences, Peking University, China
xbc@math.pku.edu.cn

Abstract. Concerning quartics, two particular quantifier elimination (QE) problems of historical interests and practical values are studied. We solve the problems by the theory of *complete discrimination systems* and *negative root discriminant sequences* for polynomials that provide a method for real (positive/negative) and complex root classification for polynomials. The equivalent quantifier-free formulas are obtained mainly be hand and are simpler than those obtained automatically by previous methods or QE tools. Also, applications of the results to program verification and determination of positivity of symmetric polynomials are showed.

1 Introduction

The elementary theory of real closed fields is the first-order theory with atomic formulas of the forms $A = B$ and $A > B$ where A and B are multivariate polynomials with integer coefficients and an axiom system consisting of the real closed fields axioms. The problem of quantifier elimination (QE) for real closed fields can be expressed as: for a given standard prenex formula ϕ find a standard quantifier-free formula ψ such that ψ is equivalent to ϕ. The problem of quantifier elimination for real closed field is an important problem originating from mathematical logic with applications to many significant and difficult mathematical problems with various backgrounds.

Many researchers contribute to QE problem. A. Tarski gave a first quantifier elimination method for real closed fields in 1930s though his result was published almost 20 years later [Ta51]. G. E. Collins introduced a so-called *cylindrical algebraic decomposition* (CAD) algorithm in the early 1970s [Co75] for QE problem. Since then, the algorithm and its improved variations have become one of the major tools for performing quantifier elimination. Through these years, some new algorithms have been proposed and several important improvements on CAD have been made to the original method. See, for example, [ACM84b, ACM88, Br01a, Br01b, BM05, Co98, CH91, DSW98, Hong90, Hong92] and [Hong96, Mc88, Mc98, Re92, Wei94, Wei97, Wei98]. Most of the works including Tarski's algorithm were collected in a book [CJ98].

---

* Corresponding author.

J. Calmet, T. Ida, and D. Wang (Eds.): AISC 2006, LNAI 4120, pp. 131–145, 2006.
© Springer-Verlag Berlin Heidelberg 2006

In this paper, we consider the following two QE problems:

$$(\forall \lambda > 0) \; (\; \lambda^4 + p\,\lambda^3 + q\,\lambda^2 + r\lambda + s > 0 \;) \tag{1}$$

and

$$(\forall \lambda \geq 0) \; (\; \lambda^4 + p\,\lambda^3 + q\,\lambda^2 + r\lambda + s \geq 0 \;), \tag{2}$$

where $s \neq 0$.

Many researchers studied the following problem of quantifier elimination (see, for example, [AM88, CH91, La88, Wu92, Wei94]),

$$(\forall x)(x^4 + px^2 + qx + r \geq 0).$$

Problems (1) and (2) are similar to this famous QE problem but have obviously different points, that is, the variable λ has to be positive or non-negative in our problems. The two problems attract us not only because they are related to the above famous QE problem but also because we encounter them when studying some problems concerning program termination [YZXZ05] and positivity of symmetric polynomials of degree 4.

Let $Q(\lambda) = \lambda^4 + p\,\lambda^3 + q\,\lambda^2 + r\lambda + s$ with $s \neq 0$[1]. Problem (1) is equivalent to finding the necessary and sufficient condition such that $Q(\lambda)$ does not have positive zeros and Problem (2) is equivalent to finding the necessary and sufficient condition such that $Q(\lambda)$ does not have non-negative zeros or the non-negative zeros of $Q(\lambda)$ (if any) are all of even multiplicities. Therefore, if one has an effective tool for root classification or positive-root-classification of polynomials, the problems can be solved in this way which is different from existing algorithms for QE.

There do exist such tools. Actually, one can deduce such a method from the Chapters 10 and 15 of Gantmacher's book [Ga59] in 1959[2]. González-Vega etc. proposed a theory on root classification of polynomials in [GLRR89] which is based on the Sturm-Habicht sequence and the theory of subresultants. For QE problems in the form $(\forall x)(f(x) > 0)$ or $(\forall x)(f(x) \geq 0)$ where the degree of $f(x)$ is a positive even integer, González-Vega proposed a combinatorial algorithm [Gon98] based on the work in [GLRR89]. Other applications of the theory in [GLRR89] to QE problems in the form $(\forall x > 0)(f(x) > 0)$ and other variants in the context of control system design were studied by Anai etc., see [AH00] for example.

The authors also have such kind of tools [YHZ96, Yang99, YX00] at hand. The theory of *complete discrimination systems* for polynomials proposed in [YHZ96] and the *negative root discriminant sequences* for polynomials proposed in [Yang99, YX00] are just appropriate tools for root classification and positive-root-classification of polynomials[3]. With the aid of these tools, determining the

---

[1] If $s = 0$, the problems essentially degenerate to similar problems with polynomials of degree 3 which are much easier.

[2] The Russian version of the book is published in 1953.

[3] The theories in [GLRR89] and [YHZ96] are both essentially based on the relations between subresultant chains and Sturm sequences (or Sturm-Habicht sequences), *i.e.*, based on the *subresultant theorem*. However, the main results in these two theories are expressed in different forms.

number of real (or complex/positive) zeros of a polynomial f is reduced to discussing the number of sign changes in a list of polynomials in the coefficients of f (see also Section 2 of this paper for details). There are many research works making heavy use of complete discrimination systems, see for example [WH99, WH00, WY00].

Solutions to those two problems presented in this paper are obtained mainly by hand with some computation by computer. Our formulas, especially for the semi-definite case (Problem 2), are simpler than those generated automatically by previous methods or QE tools and thus make them possible for AI applications. Hopefully, our "manual" method presented here could be turned into a systematic algorithm later on.

The rest of the paper is organized as follows. Section 2 devotes to some basic concepts and results concerning complete discrimination systems and negative root discriminant sequences for polynomials. Section 3 presents our solutions to Problems (1) and (2). Applications of our results to program termination and determination of positivity of symmetric polynomials are showed in Section 4.

2   Preliminaries

For convenience of readers, in this section we provide preliminary definitions and theorems (without proof) concerning complete discrimination systems and negative root discriminant sequences for polynomials. For details, please be referred to [YHZ96, Yang99, YX00].

Definition 1. *Given a polynomial with general symbolic coefficients* $f(x) = a_0 x^n + a_1 x^{n-1} + \cdots + a_n$, *the following* $(2n+1) \times (2n+1)$ *matrix is called the discrimination matrix of* $f(x)$ *and denoted by* Discr (f).

$$
\begin{bmatrix}
a_0 & a_1 & a_2 & \cdots & a_n & & & & \\
0 & na_0 & (n-1)a_1 & \cdots & a_{n-1} & & & & \\
& a_0 & a_1 & \cdots & a_{n-1} & a_n & & & \\
& 0 & na_0 & \cdots & 2a_{n-2} & a_{n-1} & & & \\
& & \cdots & \cdots & & & & & \\
& & \cdots & \cdots & & & & & \\
& & & a_0 & a_1 & \cdots & \cdots & a_n & \\
& & & 0 & na_0 & \cdots & \cdots & a_{n-1} & \\
& & & & a_0 & a_1 & \cdots & \cdots & a_n
\end{bmatrix}
$$

Denote by d_k $(k = 1, 2, \cdots, 2n+1)$ the determinant of the submatrix of Discr (f) formed by the first k rows and the first k columns.

Definition 2. *Let* $D_k = d_{2k}, k = 1, \cdots, n$. *We call* $[D_1, \cdots, D_n]$ *the discriminant sequence of* $f(x)$ *and denote it by* DiscrList(f, x). *Furthermore, we call* $[d_1 d_2, d_2 d_3, \cdots, d_{2n} d_{2n+1}]$ *the negative root discriminant sequence of* $f(x)$ *and denote it by* n.r.d.(f).

Definition 3. *We call* $[\text{sign}(B_1), \text{sign}(B_2), \cdots, \text{sign}(B_n)]$ *the sign list of a given sequence* $[B_1, B_2, \cdots, B_n]$.

Definition 4. *Given a sign list* $[s_1, s_2, \cdots, s_n]$, *we construct its revised sign list* $[t_1, t_2, \cdots, t_n]$ *as follows:*

- *If* $[s_i, s_{i+1}, \cdots, s_{i+j}]$ *is a section of the given list, where*

$$s_i \neq 0, s_{i+1} = \cdots = s_{i+j-1} = 0, s_{i+j} \neq 0,$$

  then, we replace the subsection $[s_{i+1}, \cdots, s_{i+j-1}]$ *by the first* $j-1$ *terms of* $[-s_i, -s_i, s_i, s_i, -s_i, -s_i, s_i, s_i, \cdots]$, *i.e., let*

$$t_{i+r} = (-1)^{[(r+1)/2]} \cdot s_i, \quad r = 1, 2, \cdots, j-1.$$

- *Otherwise, let* $t_k = s_k$, *i.e., no changes for other terms.*

For example, the revised one of the sign list $[1, 0, 0, 0, 1, -1, 0, 0, 1, 0, 0]$ is $[1, -1, -1, 1, 1, -1, 1, 1, 1, 0, 0]$.

Theorem 1 ([YHZ96, Yang99]). *Given a polynomial* $f(x)$ *with real coefficients,* $f(x) = a_0 x^n + a_1 x^{n-1} + \cdots + a_n$, *if the number of sign changes of the revised sign list of* $[D_1(f), D_2(f), \cdots, D_n(f)]$ *is* v, *then the number of distinct pairs of conjugate imaginary roots of* $f(x)$ *equals* v. *Furthermore, if the number of non-vanishing members of the revised sign list is* l, *then the number of distinct real roots of* $f(x)$ *equals* $l - 2v$.

Definition 5. *Let* $M = \text{Discr}(f)$. *Denote by* M_k *the submatrix formed by the first* $2k$ *rows of* M, *for* $k = 1, \cdots, n$; *and* $M(k, i)$ *denotes the submatrix formed by the first* $2k-1$ *columns and the* $(2k+i)$-*th column of* M_k, *for* $k = 1, \cdots, n, i = 0, \cdots, n-k$. *Then, construct polynomials*

$$\Delta_k(f) = \sum_{i=0}^{k} \det(M(n-k, i)) x^{k-i},$$

for $k = 0, 1, \cdots, n-1$, *where* $\det(M)$ *stands for the determinant of the square matrix* M. *We call the* n-*tuple*

$$\{\Delta_0(f), \Delta_1(f), \cdots, \Delta_{n-1}(f)\}$$

the multiple factor sequence of $f(x)$.

Lemma 1. *If the number of the* 0's *in the revised sign list of the discrimination sequence of* $f(x)$ *is* k, *then* $\Delta_k(f) = \gcd(f(x), f'(x))$, *i.e. the greatest common divisor of* $f(x)$ *and* $f'(x)$.

Definition 6. *By* $\mathcal{U}$ *denote the set of* $\{\gcd^0(f), \gcd^1(f), \cdots, \gcd^k(f)\}$, *where* $\gcd^0(f) = f, \gcd^{i+1}(f) = \gcd(\gcd^i(f), \frac{\partial}{\partial x} \gcd^i(f))$ *and* $\gcd^k(f) = 1$, *i.e., all the greatest common divisors at different levels. Each polynomial in* $\mathcal{U}$ *has a discriminant sequence, and all of the discriminant sequences are called a complete discrimination system (CDS) of* $f(x)$.

Theorem 2 ([Yang99]). *If* $\gcd^j(f)$ *has* k *real roots with multiplicities* $n_1, n_2,$ $..., n_k$ *and* $\gcd^{j-1}(f)$ *has* m *distinct real roots, then* $\gcd^{j-1}(f)$ *has* k *real roots with multiplicities* $n_1 + 1, n_2 + 1, \cdots, n_k + 1$ *and* $m - k$ *simple real roots.*

And the same argument is applicable to the imaginary roots.

Example 1. Let $f(x) = x^{18} - x^{16} + 2x^{15} - x^{14} - x^5 + x^4 + x^3 - 3x^2 + 3x - 1$. The sign list of the discrimination sequence of $f(x)$ is

$$[1, 1, -1, -1, -1, 0, 0, 0, -1, 1, 1, -1, -1, 1, -1, -1, 0, 0].$$

Hence, the revised sign list is

$$[1, 1, -1, -1, -1, 1, 1, -1, -1, 1, 1, -1, -1, 1, -1, -1, 0, 0],$$

of which the number of sign changes is seven, so $f(x)$ has seven pairs of distinct conjugate imaginary roots. Moveover, it has two distinct real roots and two repeated roots. Since $\gcd(f, f') = x^2 - x + 1$, we know that f has two distinct real roots, one pair of conjugate imaginary roots with multiplicity 2 and six pairs of conjugate imaginary roots with multiplicity 1.

Theorem 3 ([Yang99, YX00]). *Let* $[d_1, d_2, \cdots, d_{2n+1}]$ *be the principal minor sequence of the discrimination matrix of the following polynomial*

$$f(x) = a_0 x^n + a_1 x^{n-1} + \cdots + a_n \quad (a_0 \neq 0, \ a_n \neq 0).$$

1. *Denote the number of sign changes and the number of non-vanishing members of the revised sign list of* n.r.d.(f)*,* $[d_1 d_2, d_2 d_3, \cdots, d_{2n} d_{2n+1}]$*, by* v *and* $2l$*, respectively. Then, the number of distinct negative roots of* $f(x)$ *equals* $l - v$*;*
2. *Denote* $[d_2, d_4, ..., d_{2n}]$*,* $[d_1, d_3, ..., d_{2n+1}]$ *and* $[d_1 d_2, d_2 d_3, ..., d_{2n} d_{2n+1}]$ *by* L_1*,* L_2 *and* L_3*, respectively. If we denote the numbers of non-vanishing members and the numbers of sign changes of the revised sign lists of* L_i $(1 \leq i \leq 3)$ *by* l_i *and* v_i*, respectively, then* $l_3 = l_1 + l_2 - 1$*,* $v_3 = v_1 + v_2$*.*
3. *If* $d_{2m-1} = d_{2m+1} = 0$ *for some* m $(1 \leq m \leq n)$*, then* $d_{2m} = 0$*.*

Eliminating the quantifier in the formula

$$(\forall x > 0) \ (f(x) > 0) \tag{3}$$

is equivalent to finding the necessary and sufficient condition for $f(x)$ not having positive zeros. Similarly,

$$(\forall x \geq 0) \ (f(x) \geq 0) \tag{4}$$

is equivalent to the necessary and sufficient condition such that $f(x)$ does not have non-negative zeros or the non-negative zeros of $f(x)$ (if any) are all of even multiplicities. On the other hand, Theorems 1, 2 and 3 imply that, for a given polynomial $f(x)$, those conditions can be obtained by discussing on the signs of elements in the *negative root discriminant sequences* of $f(x)$ and $\gcd^i(f)$. Thus, a sketch of an algorithm for solving (3) can be described as follows which is similar to the combinatorial algorithm in [Gon98].

Algorithm: Def-Con

Input: A polynomial $f(x)$ with degree n and $f(0) \neq 0$

Output: The condition on the coefficients of $f(x)$ such that (3) holds

Step 1. Let $g(x) = f(-x)$ and denote by $[d_1, ..., d_{2n+1}]$ the list of principal minors of Discr (g).

Step 2. Discuss on all the possibilities of the signs of d_{2i}. Output those sign lists such that $l_1 - 2v_1 = 0$ (i.e., $g(x)$ has no real zeros by Theorem 1) where v_1 and l_1 are the numbers of sign changes and non-vanishing members of the revised sign lists.

Step 3. For each list $[d_2, ..., d_{2n}]$ which makes $g(x)$ have real zeros, discuss on all the possibilities of the signs of d_{2i+1}. Output those sign lists of $[d_1, d_2, ..., d_{2n+1}]$ such that $l/2 - v = 0$ (i.e., $g(x)$ has no negative zeros by Theorem 3) where v and l are the numbers of sign changes and non-vanishing members of the revised sign lists of n.r.d.(f).

Analogously, we may have an algorithm, named **Semi-Def-Con**, for solving (4) which is a little bit complicated since we have to use Theorem 2 to discuss on multiple zeros. In order to simplify the description, we suppose the first 3 steps in **Semi-Def-Con** are the same as those in **Def-Con**. So, we only need to consider those sign lists which make $g(x)$ have negative zeros and multiple zeros at the same time. For this case, we replace $f(x)$ by $\gcd^i(f)$ with a suitable i, and run the first 3 steps recursively. By Theorem 2, we can get the condition for the negative zeros of $g(x)$ being all of even multiplicities.

By **Def-Con** and **Semi-Def-Con**, we can solve Problems (1) and (2) automatically. However, the results are much more complicated than those we shall give in the next section.

3   Main Results

Proposition 1. *Given a quartic polynomial of real coefficients,*

$$Q(\lambda) = \lambda^4 + p\lambda^3 + q\lambda^2 + r\lambda + s,$$

with $s \neq 0$, then

$$(\forall \lambda > 0) \; Q(\lambda) > 0$$

is equivalent to

$$
\begin{aligned}
s > 0 \wedge (&(p \geq 0 \wedge q \geq 0 \wedge r \geq 0) \vee \\
&(d_8 > 0 \wedge (d_6 \leq 0 \vee d_4 \leq 0)) \vee \\
&(d_8 < 0 \wedge d_7 \geq 0 \wedge (p \geq 0 \vee d_5 < 0)) \vee \\
&(d_8 < 0 \wedge d_7 < 0 \wedge p > 0 \wedge d_5 > 0) \vee \\
&(d_8 = 0 \wedge d_6 < 0 \wedge d_7 > 0 \wedge (p \geq 0 \vee d_5 < 0)) \vee \\
&(d_8 = 0 \wedge d_6 = 0 \wedge d_4 < 0))
\end{aligned}
\tag{5}
$$

where

$$d_4 = -8\,q + 3\,p^2,$$
$$d_5 = 3\,r\,p + q\,p^2 - 4\,q^2,$$
$$d_6 = 14\,q\,r\,p - 4\,q^3 + 16\,s\,q - 3\,p^3\,r + p^2\,q^2 - 6\,p^2\,s - 18\,r^2,$$
$$d_7 = 7\,r\,p^2\,s - 18\,q\,p\,r^2 - 3\,q\,p^3\,s - q^2\,p^2\,r + 16\,s^2\,p + 4\,r^2\,p^3 + 12\,q^2\,p\,s$$
$$\qquad + 4\,r\,q^3 - 48\,r\,s\,q + 27\,r^3,$$
$$d_8 = p^2\,q^2\,r^2 + 144\,q\,s\,r^2 - 192\,r\,s^2\,p + 144\,q\,s^2\,p^2 - 4\,p^2\,q^3\,s + 18\,q\,r^3\,p$$
$$\qquad - 6\,p^2\,s\,r^2 - 80\,r\,p\,s\,q^2 + 18\,p^3\,r\,s\,q - 4\,q^3\,r^2 + 16\,q^4\,s - 128\,s^2\,q^2$$
$$\qquad - 4\,p^3\,r^3 - 27\,p^4\,s^2 - 27\,r^4 + 256\,s^3.$$

Proof. We need to find the necessary and sufficient condition such that $Q(\lambda)$ does not have positive zeros. First of all, by Cartesian sign rule we have the following results:

1. $s > 0$ must hold. Otherwise, the sequence $[1, p, q, r, s]$ will have an odd number of sign changes which implies $Q(\lambda)$ has at least one positive zero.
2. If the zeros of $Q(\lambda)$ are all real, $Q(\lambda)$ does not have positive zeros if and only if $s > 0$ and p, q, r are all non-negative.

Therefore, in the following we always assume $s > 0$ and do not consider the case when $Q(\lambda)$ has four real zeros (counting multiplicity).

Let $P(\lambda) = Q(-\lambda)$, then we discuss the condition such that $P(\lambda)$ does not have negative zeros. We compute the principal minors d_i $(1 \leq i \leq 9)$ of $\mathrm{Discr}(P)$ and consider the following two lists:

$$L_1 = [1, d_4, d_6, d_8] \quad \text{and} \quad L_2 = [1, d_3, d_5, d_7, d_9]$$

where $d_3 = -p$, $d_9 = s d_8$ and d_i $(4 \leq i \leq 8)$ are showed above in the statement of this proposition. In the following, we denote the numbers of non-vanishing elements and sign changes of the revised sign list of L_i by l_i and v_i $(i = 1, 2)$, respectively.

Case I. $d_8 > 0$.

In this case, by Theorem 1 $P(\lambda)$ has either four imaginary zeros or four real zeros. $P(\lambda)$ has four imaginary zeros if and only if $d_6 \leq 0 \vee d_4 \leq 0$ by Theorem 1. As stated above, we need not to consider the case when $P(\lambda)$ has four real zeros. Thus,

$$d_8 > 0 \ \wedge \ (d_6 \leq 0 \ \vee \ d_4 \leq 0)$$

must be satisfied under Case I.

Case II. $d_8 < 0$.

In this case, L_1 becomes $[1, d_4, d_6, -1]$ with $l_1 = 4, v_1 = 1$ which implies by Theorem 1 that $P(\lambda)$ has two imaginary zeros and two distinct real zeros.

If $d_7 > 0$, L_2 becomes $[1, -p, d_5, 1, -1]$. By Theorem 3, v_2 should be 3 which is equivalent to $p \geq 0 \ \vee \ d_5 \leq 0$.

If $d_7 = 0$, L_2 becomes $[1, -p, d_5, 0, -1]$. By Theorem 3, v_2 should be 3 which is equivalent to $p \geq 0 \ \vee \ d_5 < 0$.

To combine the above two conditions, we perform pseudo-division of d_7 and d_5 with respect to r and obtain that

$$27p^3 d_7 = F d_5 + 12G^2 \tag{6}$$

where F, G are polynomials in p, q, r, s. It's easy to see that p should be non-negative if $d_7 > 0$ and $d_5 = 0$. Thus, we may combine the above two sub-cases into

$$d_7 \geq 0 \ \wedge \ (p \geq 0 \ \vee \ d_5 < 0).$$

If $d_7 < 0$, L_2 becomes $[1, -p, d_5, -1, -1]$. By (6) we know that $p = 0 \wedge d_5 > 0$ and $p > 0 \wedge d_5 = 0$ are both impossible. Thus, v_2 is 3 if and only if $p > 0 \wedge d_5 > 0$.

In Case II, We conclude that

$$d_8 < 0 \wedge \ [(d_7 \geq 0 \wedge (p \geq 0 \vee d_5 < 0)) \vee (d_7 < 0 \wedge p > 0 \wedge d_5 > 0)]$$

must be satisfied.

Case **III.** $d_8 = 0$.

If $d_6 > 0$, $P(\lambda)$ has four real zeros (counting multiplicity) and this is the case having been discussed already.

If $d_6 < 0$, then $l_1 = 3$ and $v_1 = 1$. We need to find the condition for $l_2/2 = v_2$ by Theorem 3. Obviously, l_2 must be an even integer. We consider the sign of d_7. First, $d_7 < 0$ implies $l_2/2 = 2$ and v_2 is an odd integer and thus $l_2/2 = v_2$ can not be satisfied. Second, if $d_7 = 0$, by Theorem 3 $d_5 \neq 0$ since $d_6 < 0$. That means l_2 is odd which is impossible. Finally, if $d_7 > 0$, v_2 must be 2 and this is satisfied by $p \geq 0 \ \vee \ d_5 < 0$.

If $d_6 = 0$, L_1 becomes $[1, d_4, 0, 0]$. And $d_4 \geq 0$ implies $P(\lambda)$ has four real zeros (counting multiplicity) and this is the case having been discussed already. If $d_4 < 0$, $P(\lambda)$ has four imaginary zeros (counting multiplicity) and thus no negative zeros.

In Case III, we conclude that

$$d_8 = 0 \ \wedge \ [(d_6 < 0 \wedge d_7 > 0 \wedge (p \geq 0 \vee d_5 < 0)) \ \vee \ (d_6 = 0 \wedge d_4 < 0)]$$

must be satisfied. That completes the proof.   □

Proposition 2. *Given a quartic polynomial of real coefficients,*

$$Q(\lambda) = \lambda^4 + p\lambda^3 + q\lambda^2 + r\lambda + s,$$

with $s \neq 0$, then

$$(\forall \lambda \geq 0) \quad Q(\lambda) \geq 0$$

is equivalent to

$$s > 0 \wedge ((p \geq 0 \wedge q \geq 0 \wedge r \geq 0) \vee$$
$$(d_8 > 0 \wedge (d_6 \leq 0 \vee d_4 \leq 0)) \vee$$
$$(d_8 < 0 \wedge d_7 \geq 0 \wedge (p \geq 0 \vee d_5 < 0)) \vee \qquad (7)$$
$$(d_8 < 0 \wedge d_7 < 0 \wedge p > 0 \wedge d_5 > 0) \vee$$
$$(d_8 = 0 \wedge d_6 < 0) \vee$$
$$(d_8 = 0 \wedge d_6 > 0 \wedge d_7 > 0 \wedge (p \geq 0 \vee d_5 < 0)) \vee$$
$$(d_8 = 0 \wedge d_6 = 0 \wedge (d_4 \leq 0 \vee E_1 = 0)))$$

where d_i $(4 \leq i \leq 8)$ are defined as in Proposition 1 and

$$E_1 = 8\,r - 4\,p\,q + p^3.$$

Proof. Because $Q(0) = s \neq 0$, it is equivalent to consider

$$(\forall \lambda > 0) \quad Q(\lambda) \geq 0.$$

And the formula holds if and only if $Q(\lambda)$ has no positive zeros or each positive zero (if any) of $Q(\lambda)$ is of even multiplicity.

Since the first case that $Q(\lambda)$ has no positive zeros has been discussed in Proposition 1, we only discuss on the later case. So, we assume that $s > 0$ and $d_8 = 0$. All notations are as in Proposition 1.

Case **I.** $d_6 < 0$.

L_1 becomes $[1, d_4, -1, 0]$ which implies $P(\lambda)$ has a pair of imaginary zeros and one real zero with multiplicity 2. Thus, $P(\lambda)$ is positive semi-definite no matter what value λ is.

Case **II.** $d_6 > 0$.

In this case, L_1 becomes $[1, d_4, 1, 0]$ which implies that $P(\lambda)$ has three distinct real zeros of which one is of multiplicity 2. By Cartesian sign rule, the number of positive real zeros (counting multiplicity) is even. Therefore, we need only to find the condition such that $Q(\lambda)$ has one distinct positive real zero (*i.e.*, $P(\lambda)$ has one distinct negative real zero). Because $l_1 = 3$, $v_1 = 0$, by Theorem 3, it must be $l_2/2 = v_2 = 2$. And this is true if and only if $d_7 > 0 \wedge (p \geq 0 \vee d_5 \leq 0)$. From (6), we know that $d_7 > 0 \wedge p < 0 \wedge d_5 = 0$ is impossible. So we conclude that, in this case, the following formula should be true.

$$d_6 > 0 \wedge d_7 > 0 \wedge (p \geq 0 \vee d_5 < 0).$$

Case **III.** $d_6 = 0$.

Since the case that $d_4 \leq 0$ has been discussed, we assume $d_4 > 0$ which implies that $P(\lambda)$ has two distinct real zeros and no imaginary zeros. Because the case that $P(\lambda)$ has no negative zeros has been discussed as stated above, we must find the condition such that each of the two real zeros of $P(\lambda)$ is of multiplicity 2.

We can obtain the condition by discussing on the root classification of the repeated part of $P(\lambda)$ through Theorem 2. But the condition obtained is a little bit complex than the one obtained in the following way.

Suppose $Q = (\lambda^2 + a\lambda + b)^2$, we get

$$(-p + 2a)\lambda^3 + (2b + a^2 - q)\lambda^2 + (2ab - r)\lambda + b^2 - s = 0.$$

So

$$-p + 2a = 0, 2b + a^2 - q = 0, 2ab - r = 0, b^2 - s = 0, \tag{8}$$

where a, b are indeterminates. Substituting $p/2$ for a in the equations, we get $2b + 1/4p^2 - q = 0, -r + pb = 0, b^2 - s = 0$. Suppose $p \neq 0$ and substituting $b = r/p$ into the equalities, we get $E_1 = 0$ and $E_2 = 0$ where $E_1 = 8r - 4pq + p^3$, $E_2 = r^2 - p^2 s$.

If $p = 0, E_1 = 0$ and $E_2 = 0$, then $r = 0, d_4 = -8q, d_6 = 4q(4s - q^2)$. Under the precondition that $d_6 = 0 \wedge d_4 > 0$, we have $4s - q^2 = 0$ which solves equations (8) together with $p = r = 0$. In a word, the equations (8) has common solutions if and only if $E_1 = 0$ and $E_2 = 0$ under the precondition that $d_6 = 0 \wedge d_4 > 0$.

On the other hand, we have

$$p^2 d_6 = 2d_4 E_2 + (2rq - 3rp^2 + pq^2)E_1.$$

If $d_6 = 0$ and $d4 > 0$, $E_2 = 0$ is implied by $E_1 = 0$. Finally, we conclude in this case that

$$d_6 = 0 \wedge d_4 > 0 \wedge E_1 = 0$$

should be true.

That ends the proof.   □

Remark 1. We have tried the two problems by our Maple program DISCOV-ERER [YHX01, YX05] which includes an implementation of the algorithms in Section 2 and obtained some quantifier-free formulas equivalent to those of (5) and (7). However, the formulas are much more complicated than the ones stated in Propositions 1 and 2. For example, the resulting formula for Problem (1) are as follows.

$$s > 0 \wedge [\; [d_8 < 0, d_7 <= 0, d_6 < 0, 0 < d_5, d_4 <> 0, d_3 < 0] \vee$$
$$[d_8 <= 0, 0 < d_7, d_6 < 0, d_5 < 0] \vee$$
$$[d_8 <= 0, 0 < d_7, d_6 < 0, 0 <= d_5, d_4 <> 0, d_3 < 0] \vee$$
$$[d_8 < 0, d_6 < 0, 0 < d_5, d_4 = 0, d_3 < 0] \vee$$
$$[d_8 < 0, 0 < d_7, d_6 <= 0, d_5 = 0, d_4 = 0, d_3 <= 0] \vee$$
$$[d_8 < 0, d_7 < 0, d_6 = 0, d_5 < 0, d_4 = 0, d_3 <= 0] \vee$$
$$[d_8 < 0, d_7 < 0, d_6 = 0, d_5 = 0, d_4 = 0, d_3 = 0] \vee$$
$$[d_8 < 0, d_7 < 0, d_6 = 0, 0 < d_5, 0 <= d_4, d_3 < 0] \vee$$
$$[d_8 < 0, d_7 = 0, d_6 = 0, d_5 = 0, d_4 = 0] \vee$$
$$[d_8 < 0, 0 <= d_7, d_6 = 0, 0 < d_5, d_4 = 0, d_3 <> 0] \vee$$
$$[d_8 < 0, 0 < d_7, d_6 = 0, d_5 < 0, 0 <= d_4] \vee$$
$$[d_8 < 0, 0 <= d_7, d_6 = 0, 0 <= d_5, 0 < d_4, d_3 < 0] \vee$$
$$[d_8 < 0, d_7 <= 0, 0 < d_6, 0 < d_5, 0 < d_4, d_3 < 0] \vee$$

$$[d_8 < 0, 0 < d_7, 0 < d_6, d_5 < 0, 0 < d_4] \vee$$
$$[d_8 < 0, 0 < d_7, 0 < d_6, 0 <= d_5, 0 < d_4, d_3 < 0] \vee$$
$$[d_8 = 0, 0 < d_7, d_6 < 0, d_5 = 0, d_4 = 0, d_3 <= 0] \vee$$
$$[d_8 = 0, 0 < d_7, d_6 < 0, 0 < d_5, d_4 = 0, d_3 < 0] \vee$$
$$[d_8 = 0, d_6 = 0, d_4 < 0] \vee$$
$$[d_8 = 0, d_7 < 0, d_6 = 0, d_5 <> 0, d_4 = 0, d_3 < 0] \vee$$
$$[d_8 = 0, d_7 = 0, d_6 = 0, d_4 = 0, d_3 < 0] \vee$$
$$[d_8 = 0, d_6 = 0, d_4 = 0, d_3 < 0] \vee$$
$$[d_8 = 0, d_6 = 0, 0 < d_5, 0 < d_4, d_3 < 0] \vee$$
$$[0 <= d_8, d_7 < 0, 0 < d_6, 0 < d_5, 0 < d_4, d_3 < 0] \vee$$
$$[0 < d_8, d_6 <= 0] \vee$$
$$[0 < d_8, 0 < d_6, d_4 <= 0] ~]$$

Here, $d_3 = -p$ and the other d_is are defined as in Proposition 1. The above formula contains much more clauses than formula (5). For Problem (2), the resulting formula created by DISCOVERER is even more complicated because we have to add some more clauses for the cases existing positive zeros with even multiplicities.

Remark 2. We use Cartesian sign rule in the proofs of Propositions 1 and 2. This can be integrated into Def-Con and Semi-Def-Con to produce simpler formulas. In fact, a naive use of Cartesian sign rule may decrease the number of clauses. Some optimal strategy on sign discussion and result simplification can also be implemented. However, some computation like pseudo-division in the proofs depends on each concrete problem and thus is hard to be turned into an algorithm.

4   Two Examples in Application

Our first example comes from determination of termination of linear loop programs. Termination analysis plays a central role in formal verification of programs [Cou00]. An ideal solution to the termination problem for a class of programs is to prove the decidability of its termination problem and to establish calculable conditions so that for any given specific program in the class, we can compute these conditions to conclude whether the given program terminates.

The linear programs [BJT99, CH78, HPR97] are a class of programs that is widely studied. A large number of reactive systems can be modelled precisely or approximately as the linear programs [HH95]. Unfortunately, the termination problem of linear programs is undecidable in general [Tiw04]. However, Tiwari proves [Tiw04] the decidability of a specific class of linear loop programs of the form

$$\mathbf{P_1} : \textbf{ while } Bx > b \; \{x := Ax + c\}$$

where x (b and c) is a vector of N program variables (and real numbers), A and B are $N \times N$ and $N \times M$ real matrices respectively, $Bx > b$ represents a conjunction of M linear inequalities in the program variables and $x := Ax + c$ represents the linear assignments to each of the variables.

Theorem 4 ([Tiw04]). *The termination of nonhomogeneous linear program of* **P_1** *is decidable.*

Denote the homogeneous case of the program $\mathbf{P}_1$ where b and c both are 0 by

$$\mathbf{P}_2: \quad \text{while } (Bx > 0) \ \{x := Ax\}.$$

Theorem 5 ([Tiw04]). *If the program $\mathbf{P}_2$ is nonterminating, then there is a real eigenvector v of A, corresponding to a positive real eigenvalue, such that $Bv \geq 0$.*

Definition 7. *Assignment $x := Ax$ of $\mathbf{P}_2$ is called a terminating assignment, if matrix A has no positive eigenvalue.*

Obviously, if $x := Ax$ of $\mathbf{P}_2$ is a terminating assignment, then $\mathbf{P}_2$ terminates for any matrix B. By the above definition, we have the following theorem as a direct result from Proposition 1. The theorem first appeared in [YZXZ05] without proof due to page limitation.

Theorem 6. *Suppose A is a 4×4 matrix*

$$A = \begin{bmatrix} a_{11} & a_{12} & a_{13} & a_{14} \\ a_{21} & a_{22} & a_{23} & a_{24} \\ a_{31} & a_{32} & a_{33} & a_{34} \\ a_{41} & a_{42} & a_{43} & a_{44} \end{bmatrix},$$

$x := Ax$ is a terminating assignment if and only if the condition (5) is satisfied where

$$p = -a_{11} - a_{22} - a_{33} - a_{44}$$

$$q = a_{33}a_{44} + a_{11}a_{22} - a_{41}a_{14} - a_{31}a_{13} - a_{32}a_{23} - a_{34}a_{43} + a_{22}a_{44} + a_{22}a_{33}$$
$$\quad - a_{21}a_{12} - a_{42}a_{24} + a_{11}a_{44} + a_{11}a_{33}$$

$$r = -a_{32}a_{24}a_{43} + a_{11}a_{34}a_{43} - a_{11}a_{33}a_{44} - a_{21}a_{42}a_{14} + a_{11}a_{32}a_{23} + a_{21}a_{12}a_{33} + $$
$$\quad a_{42}a_{24}a_{33} + a_{11}a_{42}a_{24} - a_{31}a_{12}a_{23} + a_{22}a_{34}a_{43} - a_{11}a_{22}a_{33} + a_{31}a_{13}a_{44} - $$
$$\quad a_{11}a_{22}a_{44} - a_{42}a_{23}a_{34} - a_{22}a_{33}a_{44} - a_{41}a_{12}a_{24} + a_{32}a_{23}a_{44} - a_{41}a_{13}a_{34} + $$
$$\quad a_{41}a_{14}a_{33} + a_{21}a_{12}a_{44} + a_{41}a_{22}a_{14} - a_{31}a_{14}a_{43} + a_{31}a_{22}a_{13} - a_{21}a_{32}a_{13}$$

$$s = -a_{11}a_{22}a_{34}a_{43} - a_{21}a_{32}a_{14}a_{43} - a_{21}a_{42}a_{13}a_{34} + a_{11}a_{32}a_{24}a_{43} + $$
$$\quad a_{21}a_{42}a_{14}a_{33} + a_{41}a_{12}a_{24}a_{33} + a_{31}a_{12}a_{23}a_{44} - a_{31}a_{12}a_{24}a_{43} + $$
$$\quad a_{11}a_{22}a_{33}a_{44} - a_{21}a_{12}a_{33}a_{44} + a_{21}a_{12}a_{34}a_{43} - a_{31}a_{22}a_{13}a_{44} - $$
$$\quad a_{41}a_{12}a_{23}a_{34} + a_{31}a_{22}a_{14}a_{43} - a_{31}a_{42}a_{14}a_{23} - a_{11}a_{32}a_{23}a_{44} + $$
$$\quad a_{41}a_{22}a_{13}a_{34} + a_{11}a_{42}a_{23}a_{34} - a_{11}a_{42}a_{24}a_{33} + a_{41}a_{32}a_{14}a_{23} + $$
$$\quad a_{21}a_{32}a_{13}a_{44} - a_{41}a_{22}a_{14}a_{33} - a_{41}a_{32}a_{13}a_{24} + a_{31}a_{42}a_{13}a_{24}$$

Our second example comes from the determination of positivity of symmetric polynomials with degree 4 and arbitrary number of variables. Let $\mathbb{R}$ be the real numbers, $\mathbb{R}^n_+ = \{(x_1, ..., x_n)|x_i \in \mathbb{R}, x_i \geq 0\}$, $1_k = \overbrace{(1, ..., 1)}^{k}$, $0_k = \overbrace{(0, ..., 0)}^{k}$ and $H^{[n]}_d$ the set of real symmetric d-homogeneous polynomials in n variables. For any $x = (x_1, ..., x_n) \in \mathbb{R}^n$, set

$$v(x) = |\{x_i| \ i = 1, ..., n\}|, \quad v^*(x) = |\{x_i| \ x_i \neq 0, \ i = 1, ..., n\}|.$$

That is to say, $v(x)$ is the number of distinct elements in x and $v^*(x)$ is the number of distinct non-zero elements in x. V. Timofte proves the following result.

Theorem 7 ([Tim03]). *Suppose $f(x) \in H_d^{[n]}$. Then $f \geq 0$ holds on $\mathbb{R}_+^n$ if and only if it holds for $x \in \{x|\ x \in \mathbb{R}_+^n,\ v^*(x) \leq \max([\frac{d}{2}], 1)\}$.*

Set $N_n = \{(r, s)|\ r, s$ are positive integers with $r + s \leq n\}$. If $d = 4$, it's easy to see that

$$
\begin{aligned}
&f(x) \geq 0,\ x \in \mathbb{R}_+^n \\
\Longleftrightarrow\ &f(x) \geq 0,\ x \in \mathbb{R}_+^n,\ v^*(x) \leq 2 \\
\Longleftrightarrow\ &f(t_1 \cdot 1_r, t_2 \cdot 1_s, 0_{n-r-s}) \geq 0,\ \forall t_1, t_2 \geq 0,\ \forall (r, s) \in N_n \\
\Longleftrightarrow\ &t_2^a f(\tfrac{t_1}{t_2} \cdot 1_r, 1_s, 0_{n-r-s}) \geq 0,\ \forall t_1 \geq 0, \forall t_2 > 0,\ \forall (r, s) \in N_n \\
\Longleftrightarrow\ &f(t \cdot 1_r, 1_s, 0_{n-r-s}) \geq 0,\ \forall t \geq 0,\ \forall (r, s) \in N_n
\end{aligned}
$$

Therefore, to determine the positivity of a polynomial in $H_4^{[n]}$ on $\mathbb{R}_+^n$, it's sufficient to determine the positivity of a finite number of polynomials in one non-negative variable with degree 4. The result of Proposition 2 is exactly suitable for the determination and if n is very large, the determination will benifit from such off-line condition as (7).

Acknowledgments

The authors would like to thank the anonymous referees for their valuable comments on a previous version of this paper. The work is supported by NKBRPC-2004CB318003, NKBRPC-2005CB321902 and NSFC-60573007 in China.

References

[AH00]   H. Anai, S. Hara: Fixed-structure robust controller synthesis based on sign definite condition by a special quantifier elimination. In: *Proc. American Control Conference 2000*, pp. 1312–1316, 2000.

[ACM84a]   D. S. Arnon, G. E. Collins and S. McCallum: Cylindrical algebraic decomposition I: The basic algorithm. *SIAM J. Comput.* **13** (1984): 865–877.

[ACM84b]   D. S. Arnon, G. E. Collins and S. McCallum: Cylindrical algebraic decomposition II: An adjacency algorithm for the plane. *SIAM J. Comput.* **13** (1984): 878–889.

[ACM88]   D. S. Arnon, G. E. Collins and S. McCallum: Cylindrical algebraic decomposition III: An adjacency algorithm for three-dimensional space. *J. Symb. Comput.*: 163–187 (1988).

[AM88]   D. S. Arnon and M. Mignotte: On mechanical quantifier elimination for elementary algebra and geometry. *J. Symb. Comput.*, 5:237–260, 1988.

[BJT99]   F. Besson, T. Jensen, and J.-P. Talpin: Polyhedral analysis of synchronous languages. In SAS'99, LNCS 1694, pp. 51–69, Springer-Verlag, 1999.

[Br01a]   C. W. Brown: Simple CAD construction and its applications. *J. Symb. Comput.* **31** (2001): 521–547.

[Br01b]   C. W. Brown: Improved projection for cylindrical algebraic decomposition. *J. Symb. Comput.* **32** (2001): 447–465.

[BM05]   C. W. Brown and S. McCallum: On Using Bi-equational Constraints in CAD Construction. In: *Proc. ISSAC2005* (Kauers, M. ed.), 76–83, ACM Press, New York (2005).

[CJ98]   B. F. Caviness and J. R. Johnson(eds.), *Quantifier Elimination and Cylindrical Algebraic Decomposition*, Springer-Verlag, 1998.

[Co75]   G. E. Collins: Quantifier elimination for real closed fields by cylindrical algebraic decomposition. In: *Lecture Notes in Computer Science* **33**, pp. 134–165. Springer-Verlag, Berlin Heidelberg (1975).

[Co98]   G. E. Collins: Quantifier elimination by cylindrical algebraic decomposition - 20 years of progress. In: *Quantifier Elimination and Cylindrical Algebraic Decomposition* (Caviness, B. and Johnson, J. eds.), 8–23. Springer-Verlag, New York (1998).

[CH91]   G. E. Collins and H. Hong: Partial cylindrical algebraic decomposition for quantifier elimination. *J. Symb. Comput.* **12**: 299–328 (1991).

[Cou00]   P. Cousot: Abstract interpretation based formal methods and future challenges. In *Informatics, 10 Years Back - 10 Years Ahead*, R. Wilhelm (Ed.), LNCS 2000, pp. 138–156, 2001.

[CH78]   P. Cousot and N. Halbwachs: Automatic discovery of linear restraints among the variables of a program. In ACM POPL'78, pp. 84-97, 1978.

[DSW98]   A. Dolzmann, Th. Sturm and V. Weispfenning: Real quantifier elimination in practice, *Algorithmic Algebra and Number Theory*, Matzat, B. H., Greuel, G.-M. and Hiss, G. (eds.), Springer 1998, pp. 221–247.

[Ga59]   F. R. Gantmacher: *The Theory of Matrices*. Chelsea Publishing Company, New York, 1959.

[Gon98]   L. González-Vega: A Combinatorial Algorithm Solving Some Quantifier Elimination Problems. In: *Quantifier Elimination and Cylindrical Algebraic Decomposition* (Caviness, B. and Johnson, J. eds.), 365–375. Springer-Verlag, New York (1998).

[GLRR89]   L. González-Vega, H. Lombardi, T. Recio, M.-F. Roy: Sturm-Habicht sequence. In *Proc. of ISSAC'89*, ACM Press, pp. 136–146(1989).

[HPR97]   N. Halbwachs, Y.E. Proy, and P. Roumanoff: Verification of real-time systems using linear relation analysis. *Formal Methods in System Design*, 11(2):157–185, 1997.

[HH95]   T.A. Henzinger and P.-H. Ho: Algorithmic analysis of nonlinear hybrid systems. In CAV'95, LNCS 939, pp.225–238. 1995.

[Hong90]   H. Hong: An improvement of the projection operator in cylindrical algebraic decomposition. In: *Proceedings of ISSAC '90* (Watanabe, S. and Nagata, M. eds.), pp. 261–264. ACM Press, New York (1990).

[Hong92]   H. Hong: Simple solution formula construction in cylindrical algebraic decomposition based quantifier elimination. In: *Proceedings of ISSAC '92* (Wang, P. S., ed.), pp. 177–188. ACM Press, New York (1992).

[Hong96]   H. Hong: An efficient method for analyzing the topology of plane real algebraic curves. *Math. Comput. Simul.* **42** (1996): 571–582.

[La88]   D. Lazard: Quantifier elimination: optimal solution for two classical examples. *J. Symb. Comput.* 5:261–266, 1988.

[Mc88]   S. McCallum: An improved projection operation for cylindrical algebraic decomposition of three-dimensional space. *J. Symb. Comput.*: 141–161(1988).

[Mc98]   S. McCallum: An improved projection operator for cylindrical algebraic
            decomposition. In: *Quantifier Elimination and Cylindrical Algebraic De-
            composition* (Caviness, B. and Johnson, J. eds.), 242–268. Springer-
            Verlag, New York (1998).

[Re92]   J. Renegar: On the Computational Complexity and Geometry of the
            First-order Theory of the Reals. Parts I, II, and III. *J. Symb. Comput.*,
            13, pp. 255–352, 1992.

[Ta51]   A. Tarski: *A Decision for Elementary Algebra and Geometry.* University
            of California Press, Berkeley, May. 1951.

[Tim03]   V. Timofte: On the positivity of symmetric polynomial functions. Part I:
            General results. *J. Math. Anal. Appl.*, **284**, pp. 174–190, 2003.

[Tiw04]   A. Tiwari: Termination of linear programs. In CAV'04, LNCS 3114, pp.
            70–82, 2004.

[WY00]   L. Wang and W. Yu: Complete characterization of strictly positive real
            regions and robust strictly positive real synthesis method, *Science in
            China*, **(E)43**:97-112, 2000.

[WH99]   Z. H. Wang and H. Y. Hu: Delay-independent stability of retarded dy-
            namic systems of multiple degrees of freedom, *Journal of Sound and Vi-
            bration*, **226**(1), 57-81, 1999.

[WH00]   Z. H. Wang and H. Y. Hu: Stability of time-delayed dynamic systems with
            unknown parameters, *Journal of Sound and Vibration* **233**(2), 215-233,
            2000.

[Wei94]   V. Weispfenning: Quantifier elimination for real algebra - the cubic case.
            In *Proc. ISSAC 94*, Oxford, 1994, ACM Press, pp. 258–263.

[Wei97]   V. Weispfenning: Quantifier elimination for real algebra - the quadratic
            case and beyond. *AAECC 8*, 1997, pp. 85–101.

[Wei98]   V. Weispfenning: A New Approach to Quantifier Elimination for Real Al-
            gebra. In: *Quantifier Elimination and Cylindrical Algebraic Decomposition*
            (Caviness, B. and Johnson, J. eds.), 376–392. Springer-Verlag, New York
            (1998).

[Wu92]   W.T. Wu: On problems involving inequalities. In: *MM-Preprints*, No.7,
            pp. 1–13, 1992.

[Yang99]   L. Yang: Recent advances on determining the number of real roots of
            parametric polynomials. *J. Symbolic Computation*, 28:225–242, 1999.

[YHX01]   L. Yang, X. Hou and B. Xia: A complete algorithm for automated dis-
            covering of a class of inequality-type theorems. *Sci. in China (Ser. F)* 44:
            33–49 (2001).

[YHZ96]   L. Yang, X. Hou and Z. Zeng: A complete discrimination system for
            polynomials. *Science in China (Ser. E)*, 39:628–646, 1996.

[YX00]   L. Yang and B. Xia: An explicit criterion to determine the number of roots
            of a polynomial on an interval. *Progress in Natural Science*, 10(12):897–
            910, 2000.

[YX05]   L. Yang and B. Xia: Real solution classifications of a class of parametric
            semi-algebraic systems. In *Proc. of Int'l Conf. on Algorithmic Algebra
            and Logic*, 281–289, 2005.

[YZXZ05]   L. Yang, N. Zhan, B. Xia and C. Zhou: Program Verification by Us-
            ing DISCOVERER. To appear in Proc. of the IFIP Working Conference
            on Verified Software: Tools, Techniques and Experiments, Zurich, 10-13
            October 2005.

On the Mixed Cayley-Sylvester Resultant Matrix[*]

Weikun Sun[1] and Hongbo Li[2]

[1] Department of Mathematics and Physics
Tianjin University of Technology and Education
Tianjin 300222, P.R. China
swk@amss.ac.cn
[2] Key Laboratory of Mathematics Mechanization
Academy of Mathematics and Systems Science, CAS
Beijing 100080, P.R. China
hli@mmrc.iss.ac.cn

Abstract. For a generic n-degree polynomial system which contains $n + 1$ polynomials in n variables, there are two classical resultant matrices, Sylvester resultant matrix and Cayley resultant matrix, lie at the two ends of a gamut of $n + 1$ resultant matrices. This paper gives the construction of the $n - 1$ resultant matrices which lie between the two pure resultant matrices by the combined method of Sylvester dialytic and Cayley quotient. Since the construction involves two steps, Cayley quotient and Sylvester dialytic, the block structure of these mixed resultant matrices are similar to that of Sylvester resultant matrix in large scale, and the detailed submatrices are similar to Dixon resultant matrix.

Keywords: Mixed Cayley-Sylvester resultant matrix, Cayley quotient, Sylvester dialytic, block structure.

1 Introduction

The resultant is a powerful tool in solving a set of nonlinear polynomial equations or deriving conditions for the existence of their solutions. There are several methods which can be used to compute resultants. Among these methods the resultant matrix method is one of the most efficient. With matrix method most variables in the polynomial system can be eliminated[8].

Bézout, Sylvester, Cayley and Dixon introduced the earliest matrix methods of resultants[2, 3, 7, 13]. In Dixon's paper, he gave three resultant expressions for three bi-degree polynomial equations using the dialytic method of Sylvester, the quotient method of Cayley, and a combined Sylvester-Cayley dialytic-quotient method.

The block structure of the above three resultant matrices and their transformations are studied by Chionh et al[4], and base on those results they gave a

---

[*] Supported partially by NKBRSF 2004CB318001 and NSFC 10471143.

J. Calmet, T. Ida, and D. Wang (Eds.): AISC 2006, LNAI 4120, pp. 146–159, 2006.

fast computation algorithm to construct Dixon resultant matrix for generic bi-degree polynomial systems[5]. This fast computation algorithm was generalized by Zhao and Fu to generic n-degree polynomial systems[14].

For a generic n-degree polynomial system which contains $n+1$ n-variable polynomials, the generalizations of pure Sylvester and Cayley resultant matrices are known as the Macaulay and Dixon resultant matrices[6, 11]. These two pure resultant matrices lie in the two ends of a gamut of $n + 1$ resultant matrices. Those resultant matrices which lie between the two ends are the main topics of this paper.

In our construction of mixed Cayley-Sylvester resultant matrices, a new parameter m is introduced to identify different mixed Cayley-Sylvester resultant matrices. When $m = 1$, the definition of mixed Cayley-Sylvester resultant matrix which is given in this paper is coincide with the definition of the classical Macaulay resultant matrix, which is the generalization of Sylvester resultant matrix for multivariate polynomial system; and when $m = n + 1$ this definition is coincide with the definition of the classical Cayley resultant matrix.

Since the construction of mixed Cayley-Sylvester resultant matrix involves Cayley quotient method and Sylvester dialytic method, the block structure of mixed Cayley-Sylvester resultant matrix is similar to Sylvester resultant matrix in large scale, and the detailed submatrix is similar to Dixon resultant matrix.

For a chosen m, the size of mixed Cayley-Sylvester resultant matrix is $\frac{(n+1)!}{m} \prod_{i=1}^{n} k_i \times \frac{(n+1)!}{m} \prod_{i=1}^{n} k_i$, and each entries is either zero or its degree in coefficients of the original polynomial system is exactly m.

This paper is organized as following: The next section is a review of the classical three resultant matrices presented by Dixon: the definition and comparison of three resultant matrices. Section 3 give the construction of generic mixed Cayley-Sylvester resultant matrix. The block structure of mixed Cayley-Sylvester resultant matrix is studied in section 4. In section 5, we discuss how to apply the mixed resultant matrix for general polynomial systems. we also give some empirical data to compare the time to construct different type of resultant by their definition algorithm.

2   The Three Classical Resultant Matrices

A. Dixon presented three resultant formulations for three bivariate polynomials in 1908. Here we give them again to make the paper self-contained.

Consider three unmixed polynomials in two variables:

$$f = \sum_{i=0}^{m} \sum_{j=0}^{n} a_{ij} x^i y^j, \; g = \sum_{i=0}^{m} \sum_{j=0}^{n} b_{ij} x^i y^j, \; h = \sum_{i=0}^{m} \sum_{j=0}^{n} c_{ij} x^i y^j$$

2.1   The Sylvester Matrix

The Sylvester matrix for multivariate polynomial system is constructed by Sylvester dialytic method, it can be extracted from Macaulay resultant matrix for projective case[6]. Consider the $6mn$ polynomials:

$$x^s y^t \cdot [f \quad g \quad h] \tag{1}$$

where

$$s = 0, 1, \ldots, 2m - 1;$$
$$t = 0, 1, \ldots, n - 1.$$

In these polynomials, the highest degree of x, y is $3m - 1, 2n - 1$ respectively, so we get $6mn$ polynomials in $6mn$ monomials $x^{\epsilon_1} y^{\epsilon_2}$.

In matrix form the above system (1) can be written as

$$\begin{bmatrix} 1 \\ x \\ y \\ \vdots \\ x^{3m-1}y^{2n-1} \end{bmatrix}^T S(f, g, h)$$

where the coefficient matrix $S(f, g, h)$ is the Sylvester matrix.

2.2   The Cayley Matrix

The Cayley matrix is also called as Dixon matrix, which is derived from the Cayley quotient expression

$$\Delta(f, g, h) = \frac{\begin{vmatrix} f(x, y) & g(x, y) & h(x, y) \\ f(\bar{x}, y) & g(\bar{x}, y) & h(\bar{x}, y) \\ f(\bar{x}, \bar{y}) & g(\bar{x}, \bar{y}) & h(\bar{x}, \bar{y}) \end{vmatrix}}{(x - \bar{x})(y - \bar{y})}$$

where $\bar{x}, \bar{y}$ are new variables.

When $x = \bar{x}$ the numerator is equal to zero, $x - \bar{x}$ is actually a factor of the numerator. The same conclusion can be derived for $y - \bar{y}$. So $\Delta(f, g, h)$ is a polynomial in $x, y, \bar{x}, \bar{y}$ with the highest degree of $x, y, \bar{x}, \bar{y}$ is $m - 1, 2n - 1, 2m - 1, n - 1$ respectively, i.e.

$$\Delta(f, g, h) = \begin{bmatrix} 1 \\ x \\ y \\ \vdots \\ x^{m-1}y^{2n-1} \end{bmatrix}^T C \begin{bmatrix} 1 \\ \bar{x} \\ \bar{y} \\ \vdots \\ \bar{x}^{2m-1}\bar{y}^{n-1} \end{bmatrix}$$

where the $2mn \times 2mn$ coefficient matrix C is the Cayley matrix.

2.3   The Mixed Cayley-Sylvester Matrix

The construction of mixed Cayley-Sylvester resultant matrix combines Cayley quotient and Sylvester dialytic method. First we consider the polynomial

$$\phi(f,g) = \begin{vmatrix} f(x,y) & g(x,y) \\ f(x,\bar{y}) & g(x,\bar{y}) \end{vmatrix} / (y - \bar{y})$$

By the property of Cayley quotient, we know $\phi(f,g)$ is a polynomial in $x, y, \bar{y}$ with the highest degree $2m, n-1, n-1$ respectively. Gathering the coefficients of monomials $1, \bar{y}, \ldots, \bar{y}^{n-1}$, we can get n polynomials in x, y.

Do the same thing for $\phi(g,h)$ and $\phi(h,f)$. Altogether we can get $3n$ polynomials in x, y and the highest degree of x, y is $2m, n-1$.

Then we can do the dialytic step. Multiply every polynomial in the obtained $3n$ polynomials with the multiple set $\{1, x, \ldots, x^{m-1}\}$, we can get $3mn$ polynomials in x, y and the highest degree of x, y is $3m-1, n-1$ respectively. Rewrite the system in matrix form we can get

$$\begin{bmatrix} 1 \\ x \\ y \\ \vdots \\ x^{3m-1}y^{n-1} \end{bmatrix}^T M$$

where the coefficient matrix M is the mixed Cayley-Sylvester matrix.

2.4   The Comparison of the Three Resultant Matrices

The size of above three resultant matrices S, M and C is $6mn$, $3mn$ and $2mn$ respectively, but the entries of these three resultant matrices are either zero or its degree in the coefficients of the original polynomial system is 1, 2 and 3 respectively.

3   The Construction of Generic Mixed Cayley-Sylvester Resultant Matrix

In the n-variable case, we consider the following polynomial system F which consists of $n+1$ polynomials,

$$F : \begin{cases} f_0 = \sum_{i_1=0}^{k_1} \cdots \sum_{i_n=0}^{k_n} c_{0,i_1\cdots i_n} x_1^{i_1} \cdots x_n^{k_n}, \\ f_1 = \sum_{i_1=0}^{k_1} \cdots \sum_{i_n=0}^{k_n} c_{1,i_1\cdots i_n} x_1^{i_1} \cdots x_n^{k_n}, \\ \vdots \\ f_n = \sum_{i_1=0}^{k_1} \cdots \sum_{i_n=0}^{k_n} c_{n,i_1\cdots i_n} x_1^{i_1} \cdots x_n^{k_n}, \end{cases}$$

If each coefficient $c_{j,i_1\cdots i_n}$ is a distinct indeterminant, the polynomial system F is called *generic n-degree*.

When the polynomial system have n variables, there are $n-1$ generic mixed Cayley-Sylvester resultant matrices. Hence we need a new variable $m, 1 \le m \le n+1$ to describe different matrices.

3.1   Step 1: Cayley Quotient Construction

First we consider following Cayley expression Φ_m,

$$\frac{\begin{vmatrix} f_{n-m+1}(x_1,\ldots,x_{n-m+1},x_{n-m+2},\ldots,x_n) & \cdots & f_n(x_1,\ldots,x_{n-m+1},x_{n-m+2},\ldots,x_n) \\ f_{n-m+1}(x_1,\ldots,x_{n-m+1},x_{n-m+2},\ldots,\bar{x}_n) & \cdots & f_n(x_1,\ldots,x_{n-m+1},x_{n-m+2},\ldots,\bar{x}_n) \\ \vdots & \vdots & \vdots \\ f_{n-m+1}(x_1,\ldots,x_{n-m+1},\bar{x}_{n-m+2},\ldots,\bar{x}_n) & \cdots & f_n(x_1,\ldots,x_{n-m+1},\bar{x}_{n-m+2},\ldots,\bar{x}_n) \end{vmatrix}}{(x_{n-m+2}-\bar{x}_{n-m+2})\cdots(x_n-\bar{x}_n)}$$

where the numerator is an $m \times m$ determinant, and in the i-th row $i-1$ variables $x_{n-i+2},\ldots,x_n$ are replaced by $i-1$ new variables $\bar{x}_{n-i+2},\ldots,\bar{x}_n$. Since the numerator is always divisible by the denominator, the Cayley expression is a actually polynomial in

$$x_1,\ x_2,\ \ldots,\ x_{n-m+1},\ x_{n-m+2},\ \ldots,\ x_n,\ \bar{x}_{n-m+2},\ \ldots,\ \bar{x}_n$$

and the degree of

$$x_1,\ x_2,\ \ldots,\ x_{n-m+1},\ x_{n-m+2},\ \ldots,\ x_n$$

is

$$mk_1,\ mk_2,\ \ldots,\ mk_{n-m+1},\ (m-1)k_{n-m+2}-1,\ \ldots,\ k_n-1$$

respectively, and the degree of

$$\bar{x}_{n-m+2},\ \ldots,\ \bar{x}_n$$

is

$$k_{n-m+2}-1,\ \ldots,\ (m-1)k_n-1$$

Consider the coefficients of $\bar{x}_{n-m+2}^{\epsilon_{n-m+2}}\cdots\bar{x}_n^{\epsilon_n}$, where

$$\epsilon_{n-m+2}=0,\ldots,k_{n-m+2}-1$$
$$\vdots$$
$$\epsilon_n=0,\ldots,(m-1)k_n-1$$

we can get $(m-1)!\prod_{i=n-m+2}^n k_i$ polynomials ϕ_ϵ such that

$$\Phi_m = \sum_{\epsilon_{n-m+2}=0}^{k_{n-m+2}-1}\cdots\sum_{\epsilon_n=0}^{(m-1)k_n-1}\phi_\epsilon(x_1,x_2,\ldots,x_n)\bar{x}_{n-m+2}^{\epsilon_{n-m+2}}\cdots\bar{x}_n^{\epsilon_n} \qquad (2)$$

3.2   Step 2: Sylvester Dialytic Construction

Multiplying these polynomials $\phi_\epsilon(x_1,x_2,\ldots,x_n)$ by the $(n-m+1)!$ $\prod_{i=1}^{n-m+1} k_i$ monomials

$$1,x_1,x_2,\ldots,x_{n-m+1},\ldots\ldots\ldots,x_1^{k_1-1}\cdots x_{n-m+1}^{(n-m+1)k_{n-m+1}-1}$$

we obtain $(n - m + 1)!(m - 1)! \prod_{i=1}^{n} k_i$ polynomials

$$x_1^{\epsilon_1} \cdots x_{n-m+1}^{\epsilon_{n-m+1}} \phi_\epsilon(x_1, x_2, \ldots, x_n)$$

in $\frac{(n+1)!}{m} \prod_{i=1}^{n} k_i$ monomials $x_1^{i_1} \cdots x_n^{i_n}$, where

$$0 \leq \epsilon_1 \leq k_1 - 1$$
$$\vdots$$
$$0 \leq \epsilon_{n-m+1} \leq (n - m + 1)k_{n-m+1}$$

Because Cayley expression Φ_m is constructed by m polynomials which are selected from the original $n + 1$ polynomials, there are $\binom{n+1}{m}$ ways to choose these polynomials (The chosen is independent with the order of polynomials, see Lemma 1). Hence we can do the previous step $\binom{n+1}{m}$ times. Altogether we can get

$$\binom{n+1}{m}(n - m + 1)!(m - 1)! \prod_{i=1}^{n} k_i$$
$$= \frac{(n + 1)!}{(n - m + 1)!m!}(n - m + 1)!(m - 1)! \prod_{i=1}^{n} k_i$$
$$= \frac{(n + 1)!}{m} \prod_{i=1}^{n} k_i$$

polynomials.

Proposition 1. *The number of monomials $x_1^{i_1} \cdots x_n^{i_n}$ is $\frac{(n+1)!}{m} \prod_{i=1}^{n} k_i$.*

Proof. By step 1, ϕ_ϵ are polynomials in variables $x_1, x_2, \ldots, x_n$. The highest degree of

$$x_1, \ x_2, \ \ldots, \ x_{n-m+1}, \ x_{n-m+2}, \ \ldots, \ x_n$$

is

$$mk_1, \ mk_2, \ \ldots, \ mk_{n-m+1}, \ (m - 1)k_{n-m+2} - 1, \ \ldots, \ k_n - 1$$

By multiplying monomials

$$1, x_1, x_2, \ldots, x_{n-m+1}, \ldots\ldots\ldots, x_1^{k_1 - 1} \cdots x_{n-m+1}^{(n-m+1)k_{n-m+1}-1}$$

we can get the highest degree of

$$x_1, \ x_2, \ \ldots, \ x_{n-m+1}, \ x_{n-m+2}, \ \ldots, \ x_n$$

is

$$(m+1)k_1 - 1, \ (m+2)k_2 - 1, \ \ldots, \ (n+1)k_{n-m+1} - 1, \ (m-1)k_{n-m+2} - 1, \ \ldots, \ k_n - 1$$

So the number of monomial $x_1^{i_1} \cdots x_n^{i_n}$ is $\frac{(n+1)!}{m} \prod_{i=1}^{n} k_i$. $\qquad \square$

With this proposition, we can rewrite these $\frac{(n+1)!}{m} \prod_{i=1}^{n} k_i$ polynomials as matrix form, we have a coefficient matrix which is consist by the coefficient of the original polynomial system. The size of this coefficient matrix is $\frac{(n+1)!}{m} \prod_{i=1}^{n} k_i$. We call this coefficient matrix G generic mixed Cayley-Sylvester resultant matrix.

3.3   Two Pure Resultant Matrices

When $m = 1$, the Cayley quotient expression Φ_m is actually the original polynomial f_i. The construction of mixed Cayley-Sylvester resultant matrix only involves Sylvester dialytic step. Hence it is the pure Sylvester resultant matrix for multivariable system.

When $m = n + 1$, the Sylvester dialytic step is skipped. All we can get is the Cayley resultant matrix(apparently, this matrix is different with the Dixon resultant matrix in [11], but we can make these two matrices coincident by interchanging some columns and rows).

3.4   Matrix Form Description

The previous construction can be formulated by matrix form. In the construction of Φ_m, there are $\binom{n+1}{m}$ ways to choose m polynomials from the original $n + 1$ polynomials, so the number of Φ_m must be $\binom{n+1}{m}$ (The order of f_i in the Φ_m only impacts the final resultant by a sign, see Lemma 1 below). Let $C_{n+1}^m = \binom{n+1}{m}$ and rewrite these Φ_m as $\Phi_{m,1}, \Phi_{m,2}, \ldots, \Phi_{m,C_{n+1}^m}$, let

$$\psi(x_1, \ldots, x_{n-m+1}, \bar{x}_1, \ldots, \bar{x}_{n-m+1})$$
$$= \sum_{u_1=0}^{k_1-1} \cdots \sum_{u_{n-m+1}=0}^{(n-m+1)k_{n-m+1}-1} x_1^{u_1} \cdots x_{n-m+1}^{u_{n-m+1}} \bar{x}_1^{u_1} \cdots \bar{x}_{n-m+1}^{u_{n-m+1}}$$

By equation (2) we know

$$\psi(x_1, \ldots, x_{n-m+1}, \bar{x}_1, \ldots, \bar{x}_{n-m+1}) \cdot [\Phi_{m,1} \quad \Phi_{m,2} \quad \cdots \quad \Phi_{m,C_{n+1}^m}]$$

$$= \psi \cdot \left[\sum_{\epsilon_{n-m+2}=0}^{k_{n-m+2}-1} \cdots \sum_{\epsilon_n=0}^{(m-1)k_n-1} \phi_{\epsilon,1} \bar{x}_{n-m+2}^{\epsilon_{n-m+2}} \cdots \bar{x}_n^{\epsilon_n} \right.$$
$$\sum_{\epsilon_{n-m+2}=0}^{k_{n-m+2}-1} \cdots \sum_{\epsilon_n=0}^{(m-1)k_n-1} \phi_{\epsilon,2} \bar{x}_{n-m+2}^{\epsilon_{n-m+2}} \cdots \bar{x}_n^{\epsilon_n}$$

$$\cdots$$

$$\left. \sum_{\epsilon_{n-m+2}=0}^{k_{n-m+2}-1} \cdots \sum_{\epsilon_n=0}^{(m-1)k_n-1} \phi_{\epsilon,C_{n+1}^m} \bar{x}_{n-m+2}^{\epsilon_{n-m+2}} \cdots \bar{x}_n^{\epsilon_n} \right]$$

$$= \sum_{u_1=0}^{k_1-1} \cdots \sum_{u_{n-m+1}=0}^{(n-m+1)k_{n-m+1}-1} \sum_{\epsilon_{n-m+2}=0}^{k_{n-m+2}-1} \cdots \sum_{\epsilon_n=0}^{(m-1)k_n-1} x_1^{u_1} \cdots x_{n-m+1}^{u_{n-m+1}}$$

$$\left[\phi_{\epsilon,1} \quad \phi_{\epsilon,2} \quad \cdots \quad \phi_{\epsilon,C_{n+1}^m} \right] \bar{x}_1^{u_1} \cdots \bar{x}_{n-m+1}^{u_{n-m+1}} \bar{x}_{n-m+2}^{\epsilon_{n-m+2}} \cdots \bar{x}_n^{\epsilon_n} \quad (3)$$

Notice that $\phi_{\epsilon,i}^m$ are polynomials in variables $x_1, x_2, \ldots, x_n$, we know there exist a matrix G such that

$$
\psi \cdot [\Phi_{m,1} \quad \Phi_{m,2} \quad \cdots \quad \Phi_{m,C_{n+1}^m}]
$$

$$
= \begin{pmatrix} 1 \\ x_n \\ \vdots \\ x_n^{\tau_n} \\ \vdots \\ x_1^{\tau_1} \cdots x_n^{\tau_n} \end{pmatrix}^T \quad G \quad \begin{pmatrix} 1 \\ \bar{x}_n \\ \vdots \\ \bar{x}_n^{\epsilon_n} \\ \vdots \\ \bar{x}_1^{\epsilon_1} \cdots \bar{x}_n^{\epsilon_n} \end{pmatrix} \tag{4}
$$

where

$$
\tau_1, \tau_2, \ldots, \tau_{n-m+1}, \tau_{n-m+2}, \ldots, \tau_n
$$

are

$$
(m+1)k_1-1, \ (m+2)k_2-1, \ \ldots, \ (n+1)k_{n-m+1}-1, \ (m-1)k_{n-m+2}-1, \ \ldots, \ k_n-1
$$

respectively, and

$$
\epsilon_1, \epsilon_2, \ldots, \epsilon_{n-m+1}, \epsilon_{n-m+2}, \ldots, \epsilon_n
$$

are

$$
k_1 - 1, 2k_2 - 1, \ldots, (n - m + 1)k_{n-m+1} - 1, k_{n-m+2} - 1, \ldots, (m - 1)k_n - 1
$$

respectively. The number of rows of matrix G in equation (4) is $\frac{(n+1)!}{m} \prod_{i=1}^n k_i$, and the number of columns of matrix G is $(n - m + 1)!(m - 1)! \prod_{i=1}^n k_i$. Actually, each entry of matrix G is a $1 \times C_{n+1}^m$ submatrix. The true size of G is $\frac{(n+1)!}{m} \prod_{i=1}^n k_i \times \frac{(n+1)!}{m} \prod_{i=1}^n k_i$.

3.5 The Resultant

Here we want to prove that the determinant of the coefficient matrix G is the resultant of original polynomial system.

Lemma 1. *The order of polynomials in Cayley expression Φ_m only influence the resultant by a sign, so the construction of Φ_m is well defined.*

Proof. By the definition of Cayley expression Φ_m, if we choose m polynomials $\{f_1, f_2, \ldots, f_m\}$, the order of them only affect the sign of Φ_m. Suppose we get $-\Phi_m$, the corresponding ϕ_j also change to $-\phi_j$. After multiplying monomials and rewrite as matrix form, the only change in resultant matrix is all entries in a row change to its reverse. When we compute the determinant of resultant matrix, the coefficient -1 can be extracted. So the resultant doesn't change, or is multiplied by -1. $\square$

The following lemma can be derived directly from the construction,

Lemma 2. *If there are* $\mathbf{a} = (a_1, a_2, \ldots, a_n) \in \mathbb{C}^n$ *such that* $f_0(\mathbf{a}) = f_1(\mathbf{a}) = \cdots = f_n(\mathbf{a}) = 0$, *then* $\det G = 0$. $\qquad\square$

Hence, $\det G$ must be the multiplier of resultant $\mathrm{Res}(\mathcal{F})$. On the other hand, every entry of matrix G is a polynomial in coefficients of the original polynomial system $\mathcal{F}$, and the degree of coefficients is m. So $\det G$ is a polynomial in coefficients of the original polynomial system of degree $(n+1)! \prod_{i=1}^{n} k_i$.

Lemma 3. *The degree of polynomial* $\det G$ *in coefficients of* f_i *is* $n! \prod_{i=1}^{n} k_i$.

Proof. The mixed Cayley-Sylvester resultant matrix is constructed with $(n+1)! \prod_{i=1}^{n} k_i$ polynomials:

$$coeff(\Delta_S, \bar{X}^\beta) X^\alpha$$

where

- Δ_S is a Cayley quotient with m polynomials with indices in S, $|S| = m$ and $S \subseteq \{0, \ldots, n\}$.
- $\bar{X}^\beta = \bar{x}_{n-m+2}^{\beta_{n-m+2}} \cdots \bar{x}_n^{\beta_n}$ and

$$\beta = (0, \ldots, 0), \ldots, (k_{n-m+2} - 1, \ldots, (m-1)k_n - 1)$$

- $X^\alpha = x_1^{\alpha_1} \cdots x_{n-m+1}^{\alpha_{n-m+1}}$ and

$$\alpha = (0, \ldots, 0), \ldots, (k_1 - 1, \ldots, (n-m+1)k_{n-m+1} - 1)$$

- The highest degree monomial in these polynomials is

$$x_1^{(m+1)k_1-1} \cdots x_{n-m+1}^{(n+1)k_{n-m+1}-1} x_{n-m}^{(m-1)k_{n-m+2}-1} \cdots x_n^{k_n-1}$$

Thus the relative frequency of f_i, for any $i = 0, \ldots, n$, in these polynomials is $\frac{\binom{n}{m-1}}{\binom{n+1}{m}} = \frac{m}{n+1}$ and thus f_i appears a total of $n! \prod_{i=1}^{n} k_i$ times. $\qquad\square$

Because $\mathcal{F} = \{f_0, f_1, \ldots, f_n\}$ is generic n-degree, the supports of f_i are same. By the definition of Mixed Volume[6, 9], we know

$$MV(\mathcal{A}_0, \ldots, \mathcal{A}_{i-1}, \mathcal{A}_{i+1}, \ldots, \mathcal{A}_n)$$
$$= n!\mathrm{Vol}(\mathcal{A}) = n! \prod_{i=1}^{n} k_i$$

BKK Bound shows that in the resultant the degree of the coefficients of f_i is equal to the number of common roots the rest of polynomials have[1, 12]. This implies that the degree of the coefficients of f_i in the $\det G$ is exactly equal to the degree of the coefficients of f_i in $\mathrm{Res}(\mathcal{F})$. So we can get the following theorem:

Theorem 1. $\det G = \mathrm{Res}(\mathcal{F})$ *(up to a sign).* $\qquad\square$

4   The Block Structure of Mixed Cayley-Sylvester Resultant Matrix

In this section, we will study the block structure of mixed Cayley-Sylvester resultant matrix.

Here only the block structure of $m = n$ mixed Cayley-Sylvester resultant matrix is given. The general case corresponding to arbitrary m can be derived similarly.

Recall when $m = n$ the definition of Φ_m is:

$$
\Phi(\hat{f}_0) = \frac{\begin{vmatrix} f_1(x_1, x_2, \ldots, x_{n-1}, x_n) & \cdots & f_n(x_1, x_2, \ldots, x_{n-1}, x_n) \\ f_1(x_1, x_2, \ldots, x_{n-1}, \bar{x}_n) & \cdots & f_n(x_1, x_2, \ldots, x_{n-1}, \bar{x}_n) \\ f_1(x_1, x_2, \ldots, \bar{x}_{n-1}, \bar{x}_n) & \cdots & f_n(x_1, x_2, \ldots, \bar{x}_{n-1}, \bar{x}_n) \\ \vdots & \vdots & \vdots \\ f_1(x_1, \bar{x}_2, \ldots, \bar{x}_{n-1}, \bar{x}_n) & \cdots & f_n(x_1, \bar{x}_2, \ldots, \bar{x}_{n-1}, \bar{x}_n) \end{vmatrix}_{n \times n}}{(x_2 - \bar{x}_2) \cdots (x_n - \bar{x}_n)}
\tag{5}
$$

where numerator is an $n \times n$ determinant, $\Phi(\hat{f}_i)$ means deleting f_i from the original $n + 1$ polynomials $\{f_0, f_1, \ldots, f_n\}$. $\Phi(\hat{f}_0)$ can be written as

$$
\Phi(\hat{f}_0) = \sum_{j=1}^{(n-1)! \prod_{i=2}^{n} k_i} \phi_j(\hat{f}_0) \bar{x}_2^{\epsilon_2} \cdots \bar{x}_n^{\epsilon_n}
$$

Let

$$
\psi(x_1, \bar{x}_1) = \sum_{u=0}^{k_1-1} x_1^u \bar{x}_1^u
$$

$$
L_k = [\phi_k(\hat{f}_0) \quad \phi_k(\hat{f}_1) \quad \cdots \quad \phi_k(\hat{f}_n)]
$$

where $k = 1, 2, \ldots, (n-1)! \prod_{i=2}^{n} k_i$, we can get

$$
\psi(x_1, \bar{x}_1)[\Phi(\hat{f}_0) \quad \Phi(\hat{f}_1) \quad \cdots \quad \Phi(\hat{f}_n)]
$$

$$
= \psi(x_1, \bar{x}_1)[\sum_{j=1}^{\omega} \phi_j(\hat{f}_0) \bar{x}_2^{\epsilon_2} \cdots \bar{x}_n^{\epsilon_n}
$$

$$
\sum_{j=1}^{\omega} \phi_j(\hat{f}_1) \bar{x}_2^{\epsilon_2} \cdots \bar{x}_n^{\epsilon_n} \quad \cdots \quad \sum_{j=1}^{\omega} \phi_j(\hat{f}_n) \bar{x}_2^{\epsilon_2} \cdots \bar{x}_n^{\epsilon_n}]
$$

$$
= \sum_{u=0}^{k_1-1} \sum_{j=1}^{\omega} [\phi_j(\hat{f}_0) \quad \phi_j(\hat{f}_1) \quad \cdots \quad \phi_j(\hat{f}_n)] x_1^u \bar{x}_1^u \bar{x}_2^{\epsilon_2} \cdots \bar{x}_n^{\epsilon_n}
$$

$$
= [L_0 \quad \cdots L_\omega \quad \cdots \quad x_1^{k_1-1} L_\omega] \begin{pmatrix} 1 \\ \bar{x}_n \\ \vdots \\ \bar{x}_n^{(n-1)k_n-1} \\ \vdots \\ \bar{x}_1^{k_1-1} \cdots \bar{x}_n^{(n-1)k_n-1} \end{pmatrix}
$$

$$= \begin{bmatrix} 1 \\ x_n \\ \vdots \\ x_n^{k_n-1} \\ \vdots \\ x_1^{(n+1)k_1-1} \cdots x_n^{k_n-1} \end{bmatrix}^T G \begin{bmatrix} 1 \\ \bar{x}_n \\ \vdots \\ \bar{x}_n^{(n-1)k_n-1} \\ \vdots \\ \bar{x}_1^{k_1-1} \cdots \bar{x}_n^{(n-1)k_n-1} \end{bmatrix} \qquad (6)$$

where $\omega = (n-1)! \prod_{i=2}^{n} k_i$.

Let G_i be the coefficient matrix of polynomials $L_1, L_2, \ldots, L_\omega$ in monomials

$$x_1^i \cdot [1, \quad x_n, \quad \ldots, \quad x_n^{k_n-1}, \quad \ldots, \quad x_2^{(n-1)k_2-1} \cdots x_n^{k_n-1}]$$

where $i = 0, 1, \ldots, nk_1$. This matrix is of size $(n-1)! \prod_{i=2}^{n} k_i \times (n+1)(n-1)! \prod_{i=2}^{n} k_i$. By the definition of Φ_m, we know when $i > nk_1$, $G_i = 0$. By equation (6) we know,

Proposition 2. *For a generic n-degree $(k_1, k_2, \ldots, k_n)$ polynomial system $\mathcal{F}$ which contains $n+1$ polynomials, its mixed Cayley-Sylvester resultant matrix corresponding to $m = n$ has the following structure,*

$$G = \begin{pmatrix} G_0 & & \\ \vdots & \ddots & \\ G_{nk_1} & & G_0 \\ & \ddots & \vdots \\ & & G_{nk_1} \end{pmatrix}_{(n+1)k_1 \times k_1}$$

where G_i is as above definition.

By this proposition, we can see that the block structure of mixed Cayley-Sylvester resultant matrix corresponding to $m = n$ is like the block structure of Sylvester resultant matrix in large scale. This similarity of block structure is generated by the Sylvester dialytic step in construction. The submatrices G_i in G are generated by the coefficients of $\phi_k(\hat{f}_i)$, and $\phi_k(\hat{f}_i)$ are derived from Cayley expression. Hence the entries of G_i are similar with the entries of Cayley (Dixon) resultant matrix. This similarity is the representation of Cayley step in construction.

By now, we know that the Cayley step in construction is to compress the coefficients of original polynomial system into entries of submatrix G_i, and the Sylvester dialytic step in construction is to shift these submatrices G_i in matrix G. This is the essential effect of the two steps in the construction of Cayley-Sylvester resultant matrix.

5   General Cases and Empirical Results

By now, we only consider the polynomial system which is generic n-degree. Because under this condition, we can ensure the number of ϕ_ϵ is $(m-1)! \prod_{i=n-m+2}^{n} k_i$

and every ϕ_ϵ is not zero. So we can get a square matrix after the Sylvester dialytic step.

For general polynomial system, since computing the Cayley expression Φ_m is like computing a Dixon resultant matrix, we cannot get the $(m-1)! \prod_{i=n-m+2}^{n} k_i$ polynomials ϕ_ϵ. To ensure that the final resultant matrix will be square matrix, we could add some zero polynomials to make the number of polynomials to be $(m-1)! \prod_{i=n-m+2}^{n} k_i$. The cost of this procedure is the final resultant matrix may be degenerate. We must do the RSC (Rank Submatrix Computation) procedure to get the projection operator[11].

The first example below consists of three parts, each part is the generic n-degree polynomial system corresponding $n = 2, 3, 4$. The results are given for comparing the construction time of each resultant matrix by their definition algorithms.

The polynomial system given in the second example is not generic n-degree. This example shows that the mixed Cayley-Sylvester resultant matrix can be constructed in the non-generic n-degree case.

All the results is obtained in Maple 9, based on a notebook computer which CPU is PIII-M/1.2GHz, and memory is 512M. The algorithms to generate the three resultant matrices are their definition algorithms.

Example 1. Consider the generic n-degree polynomial systems. These tests is to show the cost of time in constructing mixed Cayley-Sylvester resultant matrices.

- $n = 2$. Since $1 \leq m \leq 3$ there is only mixed Cayley-Sylvester resultant matrix corresponding to $m = 2$.
- $n = 3$. Since $1 \leq m \leq 4$ there are two mixed Cayley-Sylvester resultant matrix corresponding to $m = 2$ and 3.
- $n = 4$. Since $1 \leq m \leq 5$ there are three mixed Cayley-Sylvester resultant matrix corresponding to $m = 2, 3$ and 4.

Example 2. Implicitization problem of parametric curves from [10]

$$\begin{cases} q_1 = 3t(t-1)^2 + (s-1)^3 + 3s - x \\ q_2 = 3s(s-1)^2 + t^3 + 3t - y \\ q_3 = -3s(s^2 - 5s + 5)t^3 - 3(s^3 + 6s^2 - 9s + 1)t^2 \\ \qquad +t(6s^3 + 9s^2 - 18s + 3) - 3s(s-1) - z \end{cases}$$

This is a mixed polynomial system. The degree of variables s, t is $3, 3$ respectively, and since $1 \leq m \leq 3$ there is only mixed Cayley-Sylvester resultant matrix corresponding to $m = 2$.

Remark 1. In example 1 (n=4), because each time of computing the mixed Cayley-Sylvester resultant matrices corresponding $m = 3$ and $m = 4$ is more than 1 hour, the results are omitted in table.

Comparing with the construction of Cayley resultant matrix which deals with $(n+1) \times (n+1)$ determinant and n divisions of polynomials, the construction of

Table 1. Empirical Data

	Mixed Cayley-Sylvester		Sylvester		Cayley	
Example	Size	Time	Size	Time	Size	Time
1 (n=2, m=2)	12×12	0.010s	24×24	0s	8×8	0.050s
(n=3, m=2)	96×96	1.182s	192×192	0.08s	48×48	536.942s
(n=3, m=3)	64×64	8.107s				
(n=4, m=2)	960×960	208.971s	1920×1920	5.217s	384×384	>1h
2 (n=2, m=2)	27×27	0.010s	54×54	0.01s	18×18	0.40s

mixed Cayley-Sylvester resultant matrix only deals with $m \times m$ determinant and $m - 1$ divisions of polynomials. Although this step will be computed C_{n+1}^m times repeatedly in mixed Cayley-Sylvester resultant matrix, the time of generating mixed Cayley-Sylvester resultant matrix still less than the time of generating Cayley resultant matrix. So smaller the Cayley expression, sooner the resultant matrix generating.

Acknowledgments

We thank Dr. E-W Chionh for his helpful comments and corrections on an earlier draft of this paper.

References

1. D. N. Bernstein. The number of roots of a system of equations, *Func. Anal. and Appl.*, 9(2):183-185, 1975.
2. E. Bézout. *Théorie Générale des Équations Algébriques*, Paris, 1770.
3. A. Cayley. On the theory of elimination, *Dublin Math. J.*, II:116-120, 1848.
4. E. W. Chionh, M. Zhang and R. N. Goldman. Block structure of three Dixon resultants and their accompanying transformation Matrices, Technical Report, TR99-341, Department of Computer Science, Rice University, 1999.
5. E. W. Chionh, M. Zhang and R. N. Goldman. Fast computations of the Bezout and the Dixon resultant matrices, *J. Symb. Comp.*, 33(1):13-29, 2002
6. D. Cox, J. Little and D. O'shea. *Using Algebraic Geometry*, Springer-Verlag, New York, 1998.
7. A. L. Dixon. The elimination of three quantics in two independent variables. *Proc. London Math. Soc.*, 6(5):49-69, 473-492, 1908.
8. I. Emiris and B. Mourrain. Matrices in elimination theory. *J. Symb. Comp.*, 28(1):3-43, 1999.
9. I. M. Gelfand, M. M. Kapranov and A. Z. Zelevinsky. *Discriminants, Resultants and Multidimensional Determinants*, Birkhäuser, Boston, 1994.
10. C. M. Hoffman. Algebraic and numeric techniques for offsets and blends. In W. Dahmen, M. Gasca and C. Micchelli editors, *Computations of Curves and Surfaces*, Kluwer Academic Publishers, 1990.
11. D. Kapur, T. Saxena and L. Yang. Algebraic and geometric reasoning using the Dixon resultants. In *ACM ISSAC 94*, pages 99-107, Oxford, England, 1994.

12. P.Pedersen and B. Sturmfels. Product formulas for resultants and chow forms. *Math. Zeitschrift*, 214:377-396, 1993.

13. J. Sylvester. On a theory of syzygetic relations of two rational integral functions, comprising an application to the theory of sturms functions, and that of the greatest algebraic common measure. *Philosophical Trans.*, 143:407-548, 1853.

14. S. Zhao and H. Fu. An extended fast algorithm for constructing the Dixon resultant matrix. *Science in China Series A: Mathematics*, 48(1):131-143, 2005.

Implicitization of Rational Curves

Yongli Sun[1,*] and Jianping Yu[2,**]

[1] Department of Mathematics and Computer Science, BUCT, Beijing 100029,
KLMM, AMSS, Chinese Academy of Sciences, Beijing 100080, China
[2] Department of Mathematics and Mechanics, University of Science and Technology
Beijing, Beijing 100083, China

Abstract. A new technique for finding the implicit equation of a rational curve is investigated. It is based on efficient computation of the Bézout resultant and Lagrange interpolation. One of the main features of our approach is that it considerably reduces the size of intermediate expressions and results in significant speed-up in the algorithm.

1   Introduction

In computer-aided geometric design and modeling, there are two standard forms used to represent plane curves: the parametric form and the implicit form. The advantage of each type of representation depends on the operations one wants to perform with the curve. Implicit equations are convenient for determining whether a point lies on, inside, or outside a curve, while parametric equations are suitable for generating points along a curve and useful in rendering algorithms on computer. When one curve is given by an implicit equation, and the other is presented by parametric equations, we can easily compute the intersection of the given two curves. For the above reasons, the conversion between the implicit equation and the parametric equations of a curve is very important, and it is an old problem in algebraic geometry.

As for the implicitization problem, i.e., finding an implicit representation from the given parametric equations of the curve, some recent algorithms can be seen in Sederberg and Chen (1995), Busé (2001), Cox et al. (1998), Cox (2001, 2003), Marco and Martínez (2001), Wang (2004), Corless et al. (2000). The effective algebraic methods that have been proposed and studied for the implicitization problem belong to three classes.

The first class of methods relies on classical elimination theory (Wang, 2000). Iterated resultants in one variable or resultants in several variables are used to compute the implicit form. The computation of the resultant is not a trivial task (Gelfand et al., 1994; Cox et al., 1998).

The second class of methods is based on Gröbner bases (Becker and Weispfenning, 1993). In practice, this class of methods appears to be more time and

---

* This research is supported by the Open Project Foundation of Key Laboratory of Mathematics Mechanization (No. KLMM0602).
** USTB foundation (No. 00009016).

J. Calmet, T. Ida, and D. Wang (Eds.): AISC 2006, LNAI 4120, pp. 160–169, 2006.

memory consuming (Kapur and Saxena, 1995; Sturmfels, 1998). However, theoretically they are important methods for solving the implicitization problem, and some algorithms whose main tool is Gröbner bases have appeared (Alonso et al., 1995; Hoffmann, 1990).

The third class of methods is based on the theory of characteristic sets founded by Wu (2000). Some researchers have applied the theory and related algorithms to the implicitization problem (Gao, 1992; Li, 1989) and other problems (Shi and Sun, 2002).

Let $\mathbf{P}(t) = (x(t), y(t))$ be a proper parametrization of a plane algebraic curve C, where

$$x(t) = \frac{u(t)}{w(t)}, \ y(t) = \frac{v(t)}{w(t)},$$

and

$$\gcd(u(t), w(t)) = \gcd(v(t), w(t)) = 1,$$

where $\gcd(u(t), w(t)), \gcd(v(t), w(t))$ are the greatest common divisors of $u(t)$, $w(t)$ and $v(t), w(t)$ respectively. A parametrization $\mathbf{P}(t) = (x(t), y(t))$ of a curve C is said to be *proper* if almost every point on the curve C is generated by exactly one value of the parameter t. It is well known that every rational curve has a proper parametrization (Sederberg, 1986), so we can assume that the parametrization is proper.

In this paper, we present a new technique for implicitizing plane rational curves based on the theories of Bézout resultant and Lagrange interpolation. The technique is efficient and novel for implicitizing rational curves and can be extended to the case of surfaces. In Section 2, we will describe the problem of implicitization. Our algorithm will be presented in detail in Section 3. Some examples will be given in Section 4 to illustrate the efficiency of our method. We will analyze the computational complexity of our algorithm and provide comparisons in Section 5 and summarize some of the noticeable advantages of our approach in Section 6.

2   Description of the Problem

A parametrization of a geometric object in a space of dimension n can be described by the following set of parametric equations:

$$
\begin{aligned}
x_1 &= \frac{f_1(t_1, \ldots, t_k)}{g_1(t_1, \ldots, t_k)}, \\
&\vdots \\
x_n &= \frac{f_n(t_1, \ldots, t_k)}{g_n(t_1, \ldots, t_k)},
\end{aligned}
\tag{1}
$$

where $t_1, \ldots, t_k$ are parameters and f_i, g_i are polynomials in the variables t_j for $i = 1, \ldots, n$ and $j = 1, \ldots, k$. The case $n = 2, k = 1$ corresponds to plane curves, and $n = 3, k = 2$ corresponds to surfaces in the space of dimension 3.

Our aim is to find a system of polynomial equations (in x_i for $i = 1, \ldots, n$)

$$
\begin{aligned}
p_1(x_1, \ldots, x_n) &= 0, \\
&\vdots \\
p_m(x_1, \ldots, x_n) &= 0,
\end{aligned}
\tag{2}
$$

such that

$$
p_1\left(\frac{f_1(t_1, \ldots, t_k)}{g_1(t_1, \ldots, t_k)}, \ldots, \frac{f_n(t_1, \ldots, t_k)}{g_n(t_1, \ldots, t_k)}\right) = 0,
$$

$$
\vdots
$$

$$
p_m\left(\frac{f_1(t_1, \ldots, t_k)}{g_1(t_1, \ldots, t_k)}, \ldots, \frac{f_n(t_1, \ldots, t_k)}{g_n(t_1, \ldots, t_k)}\right) = 0,
$$

and $V(p_1(x_1, \ldots, x_n), \ldots, p_m(x_1, \ldots, x_n))$ is the smallest variety containing the parametric geometric object described by (1). More details can be found in Cox et al. (1996).

In the next section, we will present our approach and algorithm for solving the implicitization problem.

3   Derivation of the Algorithm

Let $(x(t), y(t))$ be a proper parametrization of a plane algebraic curve, where

$$
x(t) = \frac{u(t)}{w(t)}, \quad y(t) = \frac{v(t)}{w(t)},
$$

and

$$
\gcd(u(t), w(t)) = \gcd(v(t), w(t)) = 1,
$$
$$
n = \deg_t(u(t)) = \deg_t(v(t)) = \deg_t(w).
$$

The rational curve has no base points because of the assumption $\gcd(u(t), w(t)) = \gcd(v(t), w(t)) = 1$. The parametric and implicit forms are related by the following well-known lemma (Marco and Martínez, 2001).

Lemma 1. *Let* $(x(t) = u_1(t)/v_1(t), y(t) = u_2(t)/v_2(t)$ *be a proper rational parametrization of the irreducible curve defined by* $f(x, y)$, *and let* $\gcd(u_1(t), v_1(t)) = \gcd(u_2(t), v_2(t)) = 1$. *Then*

$$
\max\{\deg_t(u_1), \deg_t(v_1)\} = \deg_y(f),
$$
$$
\max\{\deg_t(u_2), \deg_t(v_2)\} = \deg_x(f).
$$

Finding the polynomial f is the main task of this paper. The following theorem is important for our technique and aim.

Theorem 1. *Let* $f(x)$ *be a polynomial of degree* n. *Then we have*

$$
f(x) = \sum_{i=0}^{n} \frac{f(a_i)F(x)}{(x - a_i)F'(a_i)},
$$

where

$$F(x) = (x - a_0)(x - a_1)(x - a_2) \cdots (x - a_n),$$

$$F'(a_i) = F'(x)|_{a_i}, \; i = 0, 1, \ldots, n,$$

and $F'(x)$ is the derivative of $F(x)$ with respect to x.

Proof. Construct a new polynomial of degree $\leq n$:

$$G(x) = f(x) - \sum_{i=0}^{n} \frac{f(a_i)F(x)}{(x - a_i)F'(a_i)}.$$

We see that $G(a_i) = 0$ for $i = 0, 1, \ldots, n$, so the polynomial equation $G(x) = 0$ of degree $\leq n$ has $n + 1$ roots. It follows that $G(x) \equiv 0$, which implies that

$$f(x) = \sum_{i=0}^{n} \frac{f(a_i)F(x)}{(x - a_i)F'(a_i)}.$$

More details about Lagrange interpolation can be found in Gathen and Gerhard (1999).

Our aim is to compute the polynomial $f(x, y) \in \prod_{n,n}(x, y)$ by means of Bézout resultant and Lagrange interpolation using Lemma 1, where $\prod_{n,n}(x, y)$ is the space of polynomials of degree less than or equal to n in x and in y, its dimension is $(n + 1)(n + 1)$, and its basis is given by

$$\{x^i y^j \mid i = 0, \ldots, n; \; j = 0, \ldots, n\}.$$

By Theorem 1, any polynomial $f(x, y)$ can be described by

$$f(x, y) = \sum_{i=0}^{n} \frac{b_i F(x)}{(x - a_i)F'(a_i)}, \tag{3}$$

where

$$F(x) = (x - a_0)(x - a_1) \cdots (x - a_n),$$

and $F'(a_i) = F'(x)|_{a_i}, F'(x)$ is the derivative of $F(x)$ with respect to x, and $b_i = f(a_i, y), \; i = 0, 1, \ldots, n$.

The following lemma (Sederberg et al., 1997) is crucial for our algorithm presented below.

Lemma 2. *When there are no base points, $R(x, y) = 0$ is the implicit equation of the rational curve $x = \dfrac{x(t)}{w(t)}, y = \dfrac{y(t)}{w(t)}$, where $R(x, y)$ is the Bézout resultant of x and y, and $x(t), y(t), w(t)$ are polynomials of degree n in t.*

We choose $a_i = i \; (i = 0, 1, \ldots, n)$ for our algorithm, so that we can compute the values b_i by means of the following procedure, which uses the symbolic Bézout matrix computed by Maple 8, evaluates it at each interpolation node $a_i = i$, and then computes the determinant of the corresponding matrix. The following is our algorithm (presented using Maple notations).

Algorithm
Input: $p := xw(t) - u(t); q := yw(t) - v(t)$
Output: $b_i, der_i, i = 0, 1, \ldots, n$ and $F(x)$
$F(x) := 1;$
for i from 0 to n do
$\qquad F(x) := F(x)(x - i)$
end do:
$Be := bezout(p, q, t)$: $dF := diff(F(x), x)$:
for i from 0 to n do
$\qquad b_i := det(subs(x = i, op(Be)))$:
$\qquad der_i := subs(x = i, dF)$:
end do:

Now we have all the required b_i and derivatives der_i; thus the required polynomial $f(x, y)$ can be obtained from the formula (3):

$$f(x, y) = \sum_{i=0}^{n} \frac{b_i F(x)}{(x - a_i) der_i}.$$

Although the computational complexity of evaluating the Sylvester matrix is less than that of evaluating the Bézout matrix, the computation of the determinant using the Sylvester matrix is much more expensive than that using the Bézout matrix. To be precise, let us observe that the order of the Sylvester matrix is $2n$ while the order of the Bézout matrix is n. Therefore, taking into account the general form of the entries of each matrix, the number of arithmetic operations needed for evaluating at (i, j) is bounded by $4n^2$ in the case of Sylvester and by $7n^2$ in the case of Bézout. However, taking into account that the cost of computing the determinant of a matrix of order n is $O(n^3)$, we see that computing each determinant (i.e. each interpolation datum) is eight times more expensive if we use the Sylvester matrix. So we choose the Bézout matrix instead of the Sylvester matrix.

In the following section, some examples are presented to illustrate the efficiency of our method.

4   Examples

Example 1. Consider the curve defined by

$$x = \frac{t^2}{1 + t - t^2}, \; y = \frac{t^2 + 1}{1 + t - t^2}.$$

Let

$$p = x(1 + t - t^2) - t^2, \; q = y(1 + t - t^2) - (t^2 + 1),$$

and $x = 0, 1, 2$. By the Bézout resultant, we get

$$b_0 = (-y + 1)^2, \ b_1 = y^2 - 7y + 10, \ b_2 = y^2 - 12y + 29$$

from (3), and the implicit equation is

$$f = y^2 - 5yx - 2y + 5x^2 + 4x + 1 = 0.$$

Example 2. Let the parametrization of the curve be

$$x = \frac{2t^4 - t^3 + t^2 + 2}{t^4 - t^2 + 1}, \ y = \frac{t^4 + 2t^3 + t}{t^4 - t^2 + 1}.$$

We suppose that

$$p = x(t^4 - t^2 + 1) - (2t^4 - t^3 + t^2 + 2), \ q = y(t^4 - t^2 + 1) - (t^4 + 2t^3 + t),$$

and $x = 0, 1, 2, 3, 4$. By the Bézout resultant, we have

$$b_0 = 544 - 782y + 240y^2 - 3y^3 + 73y^4,$$
$$b_1 = -95y + 17 + 181y^2 - 107y^3 + 73y^4,$$
$$b_2 = 73y^4 - 211y^3 + 138y^2,$$
$$b_3 = 271y - 47 + 111y^2 - 315y^3 + 73y^4,$$
$$b_4 = 584 + 100y^2 - 419y^3 + 1486y + 73y^4.$$

From (3), we get the implicit equation

$$f = 544 + 1239yx + 240y^2 - 3y^3 + 73y^4 - 402x^3 + 52x^4 - 67y^2x - 104y^3x$$
$$+ 1097x^2 - 680yx^2 + 8y^2x^2 + 128yx^3 - 1274x - 782y = 0.$$

Example 3. Suppose that the curve is defined by the parametric equations

$$x = \frac{t^5 + 2t^4 - t^3 + t^2 + 2}{t^5 - t^2 + 1}, \ y = \frac{t^5 + t^4 + 2t^3 + t}{t^5 - t^2 + 1}.$$

Let

$$p = x(t^5 - t^2 + 1) - (t^5 + 2t^4 - t^3 + t^2 + 2), \ q = y(t^5 - t^2 + 1) - (t^5 + t^4 + 2t^3 + t),$$

and $x = 0, 1, 2, 3, 4, 5$. By the Bézout resultant, we have

$$b_0 = -710 + 1304y - 1107y^2 + 69y^3 + 294y^4 + 343y^5,$$
$$b_1 = -33 - 191y^2 + 84y + 189y^3 - 392y^4 + 343y^5,$$
$$b_2 = 895y^3 - 1078y^4 + 343y^5 - 291y^2,$$
$$b_3 = 799 - 1300y - 165y^2 + 2187y^3 - 1764y^4 + 343y^5,$$
$$b_4 = 5886 - 6576y + 1429y^2 + 4065y^3 - 2450y^4 + 343y^5,$$
$$b_5 = 23655 - 18996y + 5733y^2 + 6529y^3 - 3136y^4 + 343y^5.$$

Similarly from (3), we get the required implicit equation

$$f = -710 + 1493x + 1304y - 2470yx - 1209x^2 - 1107y^2 + 1838xy^2 + 1557yx^2$$
$$+ 512x^3 + 69y^3 - 1129x^2y^2 - 290yx^3 - 173xy^3 - 142x^4 + 294y^4 + 343y^5$$
$$+ 23x^5 + 293x^2y^3 + 207y^2x^3 - 686xy^4 - 17yx^4 = 0.$$

5   Computational Complexity and Comparisons

In this section we study the computational complexity of our algorithm in terms of arithmetic operations. In view of the algorithmic steps described in Section 3, we must compute:

(1) $n + 1$ derivatives of $F(x)$ with respect to x at the points $x = i$ for $i = 0, \ldots, n$, which requires $O(n^2)$ operations;

(2) $det(subs(x = i, op(bezout(p, q, t))))$ for $i = 0, \ldots, n$, for which $O(n^6)$ operations are required totally.

It is worth noting that the main cost of the process corresponds to the generation of the interpolation data, and not to the computation of the derivatives. For every entry of the matrix is a polynomial of degree at most 1 in y; then for the matrix $bezout(p, q, t)$, $O(n^6)$ operations are required when $n+1$ interpolation data are computed.

The algorithm we have developed in the preceding sections for finding the implicit equation of a properly parametrized curve is based on the Bézout resultant and Lagrange interpolation. It avoids some difficulties that arise in the computation of resultants and nodes (Marco and Martínez, 2001). Let $\mathbf{P}(t) = (x(t), y(t))$ be a proper parametrization of a plane algebraic curve C, where

$$x(t) = \frac{u(t)}{w(t)}, \ y(t) = \frac{v(t)}{w(t)}.$$

According to Marco and Martínez (2001), one needs to compute $(n + 1)(n + 1)$ data by the Bézout resultant $Be := bezout(p, q, t)$ in Maple 8, where

$$p = xw(t) - u(t), q = yw(t) - v(t).$$

Then one has to solve a linear system of $(n+1)(n+1)$ equations and finally gets all the coefficients of the implicit equation of curve C. So the procedure is much more complicated.

Compared with the algorithm of Marco and Martínez (2001), our algorithm is much simpler and much more efficient from some aspects, because

(1)   we only need to compute $n+1$ nodes, which is obtained by $n+1$ Bézout resultants; but in Marco and Martínez (2001), one has to compute $(n+1)^2$ nodes, which are obtained from $(n+1)^2$ Bézout resultants;

(2)   we do not need to solve a linear system of $(n + 1)(n + 1)$ equations.

Therefore, we can avoid a lot of unnecessary computations and get the required implicit equation easily by using Lagrange interpolation.

In what follows, we give an example to compare the results obtained using our algorithm with those obtained using the Maple command $resultant$.

Consider the S-shaped Bézier curve properly parametrized by

$$(x(t), y(t)) = \frac{\sum_0^8 \omega_i (a_i, b_i) \binom{8}{i} t^i (1 - t)^i}{\sum_0^8 \omega_i \binom{8}{i} t^i (1 - t)^i},$$

where the control polygon is described by the points

$$(a_0, b_0) = (-\sqrt{5}, 1), \quad (a_1, b_1) = (0, \frac{1}{2}), \quad (a_2, b_2) = (\sqrt{2}, 1),$$
$$(a_3, b_3) = (\sqrt{3}, 2), \quad (a_4, b_4) = (0, \sqrt{5}), (a_5, b_5) = (-\sqrt{2}, \sqrt{5}),$$
$$(a_6, b_6) = (-\sqrt{3}, \frac{7}{2}), \quad (a_7, b_7) = (0, 4), \quad (a_8, b_8) = (\sqrt{5}, 4),$$

and where the weights are

$$\omega_0 = 1, \omega_1 = 1, \omega_2 = 1, \omega_3 = 1, \omega_4 = 1,$$
$$\omega_5 = 1, \omega_6 = 1, \omega_7 = 1, \omega_8 = 1.$$

A good introduction to polynomial and rational Bézier curves can be found in Farin (1996) and Hoschek and Lasser (1993).

When implementing our algorithm in Maple 8 on a personal computer (128 M), we observed that (for this example) our algorithm is about 41 seconds faster and requires 22M less space than *resultant*. Consider another parametric plane curve defined by

$$x = \frac{t^{16} + t}{t^{16} + t + 1}, \quad y = \frac{t^{16} + t^2}{t^{16} + t + 1}.$$

In Maple 8 on the above-mentioned PC, our algorithm takes only 0.032 seconds and is 0.234 seconds faster than *resultant*.

The advantages of our method compared with Maple's *resultant* are partly due to the fact that we make use of the special properties of the resultant to be computed. On one hand, we know the exact degree of the resultant in each variable and thus we know precisely the best interpolation space. On the other hand, the entries of the Bézout matrix are quite simple, so the evaluation of the matrix at the nodes $a_i = i$ is not too expensive.

6   Conclusions

We have presented, in the previous sections, a novel algorithm for finding the implicit equations of properly parametrized plane curves. This algorithm is based on the theories of Lagrange interpolation and Bézout matrix. It avoids some of the difficulties that arise in the computation of resultants and thus may perform better than some of the standard algorithms.

Acknowledgements

The authors are grateful to the referees for their valuable comments and suggestions on an earlier version of this paper.

References

1. Alonso, C., Gutiérrez, J., Recio, T.: An implicitization algorithm with fewer variables. *Computer Aided Geometric Design*, 12, 251-258 (1995).
2. Becker, T., Weispfenning, V.: *Gröbner Bases: A Computational Approach to Commutative Algebra*. Springer, Berlin (1993).
3. Busé, L.: Residual resultant over the projective plane and the implicitization problem. In: *Proceedings ISSAC '01*, 48-55. ACM Press, New York (2001).
4. Corless, R.M., Giesbrecht, M.W., Kotsireas, I.S., Watt, S.M.: Numerical implicitization of parametric hypersurfaces with linear algebra. In: *Proceedings AISC 2000*, 174-183. LNAI 1930, Springer, Berlin (2000).
5. Cox, D., Little, J., O'shea, D.: *Ideals, Varieties and Algorithms*. Springer, New York (1996).
6. Cox, D., Little, J., O'shea, D.: *Using Algebraic Geometry*. Springer, New York (1998).
7. Cox, D.: Equations of parametric curves and surfaces via syzygies. In: *Symbolic Computation: Solving Equations in Algebra, Geometry, and Engineering. Contemporary Mathematics*, 286, 1-20 (2001).
8. Cox, D.: Curves, surfaces and syzygies. In: *Topics in Algebraic Geometry and Geometric Modeling. Contemporary Mathematics*, 334, 131-149 (2003).
9. Farin, G.: *Curves and Surfaces for Computer Aided Geometric Design* (4th edn). Academic Press, Boston (1996).
10. Gao, X.S., Chou, S.C.: Implicitization of rational parametric equations. *J. Symbolic Computation*, 14, 459-470 (1992).
11. Gathen, J.v.z., Gerhard, J.: *Modern Computer Algebra*. Cambridge University Press, Cambridge (1999).
12. Gelfand, I.M., Kapranov, M.M., Zelevinsky, A.V.: *Discriminants, Resultants and Multidimensional Determinants*. Birkhäuser, Basel (1994).
13. Hoffmann, C.M.: Algebraic and numerical techniques for offsets and blends. In: Dahmen, W., Gasca, M., Micchelli, C.A. (eds.), *Computation of Curves and Surfaces. NATO ASI Series C: Mathematical and Physical Sciences*, 307, 499-528. Kluwer Academic, Dordrecht (1990).
14. Hoschek, J., Lasser, D.: *Fundamentals of Computer Aided Geometric Design*. A.K. Peters, Wellesley (1993).
15. Kapur, D., Saxena, T.: Comparison of various multivariate resultant formulations. In: *Proceedings ISSAC '95*, 87-194. ACM Press, New York (1995).
16. Li, Z.M.: Automatic implicitization of parametric objects. *MM Research Preprints*, No. 4, 54-62. Institute of Systems Science, Academia Sinica (1989).
17. Marco, A., Martínez, J.J.: Using polynomial interpolation for implicitizing algebraic curves. *Computer Aided Geometric Design*, 18, 309-319 (2001).
18. Sederberg, T.W.: Improperly parametrized rational curves. *Computer Aided Geometric Design*, 3, 67-75 (1986).
19. Sederberg, T., Chen, F.: Implicitization using moving curves and surfaces. In: *Proceedings SIGGRAPH '95*, 301-308. ACM Press, New York (1995).
20. Sederberg, T., Goldman, R., Du, H.: Implicitizing rational curves by the method of moving algebraic curves. *J. Symbolic Computation*, 23, 153-175 (1997).
21. Shi, H., Sun, Y.L.: Blending of triangular algebraic surfaces. *MM Research Preprints*, No. 21, 200-206. Institute of Systems Science, Academia Sinica (2002).
22. Shi, H., Sun, Y.L.: On blending of cylinders. *MM Research Preprints*, No. 21, 207-211. Institute of Systems Science, Academia Sinica (2002).

23. Sturmfels, B.: Introduction to resultants. In: Cox, D., Sturmfels, B. (eds.), *Applications of Computational Geometry, Proceedings of Symposium in Applied Mathematics*, 53, 25-39. American Mathematical Society, Providence (1998).
24. Wang, D.: *Elimination Methods*. Springer, Wien New York (2000).
25. Wang, D.: A simple method for implicitizing rational curves and surfaces. *J. Symbolic Computation*, 38, 899-914 (2004).
26. Wu, W.T.: *Mathematics Mechanization*. Science Press and Kluwer Academic, Beijing and Dordrecht (2000).

Operator Calculus Approach to Solving Analytic Systems

Philip Feinsilver[1] and René Schott[2]

[1] Southern Illinois University, Carbondale, IL. 62901, U.S.A.
pfeinsil@math.siu.edu
[2] Université Henri Poincaré-Nancy 1, BP 239, 54506 Vandoeuvre-lès-Nancy, France
schott@loria.fr

Abstract. Solving analytic systems using inversion can be implemented in a variety of ways. One method is to use Lagrange inversion and variations. Here we present a different approach, based on dual vector fields.

For a function analytic in a neighborhood of the origin in the complex plane, we associate a vector field and its dual, an operator version of Fourier transform. The construction extends naturally to functions of several variables.

We illustrate with various examples and present an efficient algorithm readily implemented as a symbolic procedure in Maple while suitable as well for numerical computations using languages such as C or Java.

1 Introduction

We introduce the operator calculus necessary to present our approach to (local) inversion of analytic functions. It is important to note that this is different from Lagrange inversion and is based on the flow of a vector field associated to a given function. It appears to be theoretically appealing as well as computationally effective.

Acting on polynomials in x, define the operators

$$D = \frac{d}{dx} \text{ and } X = \text{multiplication by } x.$$

They satisfy commutation relations $[D, X] = I$, where I, the identity operator, commutes with both D and X. Abstractly, the Heisenberg-Weyl algebra is the associative algebra generated by operators $\{A, B, C\}$ satisfying $[A, B] = C$, $[A, C] = [B, C] = 0$. The *standard* HW algebra is the one generated by the realization $A = D$, $B = X$, $C = I$. An *Appell system* is a system of polynomials $\{y_n(x)\}_{n \geq 0}$ that is a basis for a representation of the standard HW algebra with the following properties:

1. y_n is of degree n in x;
2. $D y_n = n y_{n-1}$.

In several variables, $\mathbf{x} = (x_1, \ldots, x_N)$, with multi-indices $\mathbf{n} = (n_1, \ldots, n_N)$, the corresponding monomials are

J. Calmet, T. Ida, and D. Wang (Eds.): AISC 2006, LNAI 4120, pp. 170–180, 2006.

$$\mathbf{x^n} = x_1^{n_1} x_2^{n_2} \cdots x_N^{n_N}.$$

Denote the partial derivative operators by $D_i = \dfrac{\partial}{\partial x_i}$ and the corresponding multiplication operators by X_i. Then $[D_j, X_i] = \delta_{ij} I$. An Appell system is a system of polynomials $\{y_\mathbf{n}\}$ in the variables $\mathbf{x}$ such that

1. the top degree term of $y_\mathbf{n}$ is a constant multiple of $\mathbf{x^n}$;
2. $D_i y_\mathbf{n} = n_i y_{\mathbf{n}-\mathbf{e}_i}$, where $\mathbf{e}_i$ has all components zero except for 1 in the ith position.

G.-C. Rota [3] is well-known for his *umbral calculus* development of special polynomial sequences, called *basic sequences*. From our perspective, these are "canonical polynomial systems" in the sense that they provide polynomial representations of the Heisenberg-Weyl algebra, in realizations different from the standard one. Our idea [2, 1] is to illustrate explicitly the rôle of vector fields and their duals, using operator calculus methods for working with the latter (in our volumes — this viewpoint is prefigured in [3]).

The main feature of our approach is that the action of the vector field may be readily calculated while the action of the dual vector field on exponentials is identical to that of the vector field. Then we note that acting iteratively with a vector field on polynomials involves the complexity of the coefficients, while acting iteratively with the dual vector field always produces polynomials from polynomials. So we can switch to the dual vector field for calculations.

Specifically, fix a neighborhood of 0 in $\mathbf{C}$. Take an analytic function $V(z)$ defined there, normalized to $V(0) = 0$, $V'(0) = 1$. Denote $W(z) = 1/V'(z)$ and $U(v)$ the inverse function, i.e., $V(U(v)) = v$, $U(V(z)) = z$. Then $V(D)$ is defined by power series as an operator on polynomials in x and $[V(D), X] = V'(D)$ so that $[V(D), XW(D)] = I$. In other words, $V = V(D)$ and $Y = XW(D)$ generate a representation of the HW algebra on polynomials in x. The basis for the representation is $y_n(x) = Y^n 1$, i.e., Y is a *raising operator*. And $V y_n = n y_{n-1}$ so that V is the corresponding *lowering operator*. The $\{y_n\}_{n \geq 0}$ form a system of *canonical polynomials* or generalized Appell system. The operator of multiplication by x is given by $X = YV'(D) = YU'(V)^{-1}$, which is a *recursion operator* for the system.

We identify vector fields with first-order partial differential operators. Consider a variable A with corresponding partial differential operator ∂_A. Given V as above, let $\tilde{Y}$ be the vector field $\tilde{Y} = W(A)\partial_A$. Then we observe the following identities

$$\tilde{Y} e^{Ax} = xW(A) e^{Ax} = xW(D) e^{Ax}$$

as any operator function of D acts as a multiplication operator on e^{Ax}. The important property of these equalities is that Y and $\tilde{Y}$ commute, as they involve independent variables. So we may iterate to get

$$\exp(t\tilde{Y})e^{Ax} = \exp(tY)e^{Ax}. \tag{1}$$

On the other hand, we can solve for the left-hand side of this equation using the method of characteristics. Namely, if we solve

$$\dot{A} = W(A) \tag{2}$$

with initial condition $A(0) = A$, then for any smooth function f,

$$e^{t\tilde{Y}} f(A) = f(A(t)).$$

Thus

$$\exp(tY)e^{Ax} = e^{xA(t)}.$$

To solve equation (2), multiply both sides by $V'(A)$ and observe that we get

$$V'(A)\,\dot{A} = \frac{d}{dt}\,V(A(t)) = 1.$$

Integrating yields

$$V(A(t)) = t + V(A) \qquad \text{or} \qquad A(t) = U(t + V(A)).$$

Or, writing v for t, we have

$$\exp(vY)e^{Ax} = e^{xU(v+V(A))}. \tag{3}$$

We can set $A = 0$ to get

$$\exp(vY)1 = e^{xU(v)}$$

on the one hand while

$$e^{vY}1 = \sum_{n=0}^{\infty} \frac{v^n}{n!}\, y_n(x).$$

In summary, we have the expansion of the exponential of the inverse function

$$e^{xU(v)} = \sum_{n=0}^{\infty} \frac{v^n}{n!}\, y_n(x)$$

or

$$\sum_{m=0}^{\infty} \frac{x^m}{m!}\, (U(v))^m = \sum_{n=0}^{\infty} \frac{v^n}{n!}\, y_n(x). \tag{4}$$

This yields an alternative approach to inversion of the function $V(z)$ rather than using Lagrange's formula. We see that the coefficient of $x^m/m!$ yields the expansion of $(U(v))^m$. In particular, $U(v)$ itself is given by the coefficient of x on the right-hand side.

Specifically, we have:

Theorem 1. *The coefficient of $x^m/m!$ in $Y^n 1$ is equal to $\tilde{Y}^n A^m\big|_{A=0}$, each giving the coefficient of $v^n/n!$ in the expansion of $U(v)^m$.*

Proof. Expand both sides of equation (1), using v for t, in powers of x and v, and let $A = 0$:

$$\sum_{n=0}^{\infty} \frac{v^n}{n!} \tilde{Y}^n \sum_{m=0}^{\infty} \frac{x^m}{m!} A^m \Big|_{A=0} = \sum_{n=0}^{\infty} \frac{v^n}{n!} Y^n 1$$

and compare with equation (4).

The same idea works in several variables.

We have $\mathbf{V}(\mathbf{z}) = (V_1(z_1, \ldots, z_N), \ldots, V_N(z_1, \ldots, z_N))$ analytic in a neighborhood of 0 in $\mathbf{C}^N$. Denote the Jacobian matrix $\left(\dfrac{\partial V_i}{\partial z_j} \right)$ by V' and its inverse by W. The variables

$$Y_i = \sum_{k=1}^{N} x_k W_{ki}(D)$$

commute and act as raising operators for generating the basis $y_{\mathbf{n}}(\mathbf{x})$. Namely, $Y_i y_{\mathbf{n}} = y_{\mathbf{n}+\mathbf{e}_i}$. And $V_i = V_i(\mathbf{D})$, $\mathbf{D} = (D_1, \ldots, D_N)$ are lowering operators: $V_i y_{\mathbf{n}} = n_i\, y_{\mathbf{n}-\mathbf{e}_i}$.

Denote $\sum_i a_i b_i$ by $a \cdot b$. With variables A_i and corresponding partials ∂_i, define the vector fields

$$\tilde{Y}_i = \sum_k W_{ki}(A)\partial_k.$$

For a vector field $\tilde{Y} = \sum_i W_i(A)\partial_i$, we have the identities

$$\tilde{Y}\, e^{A \cdot x} = x \cdot W(A)\, e^{A \cdot x} = x \cdot W(D)\, e^{A \cdot x}.$$

The method of characteristics applies as in one variable and as in equation (3)

$$\exp(v \cdot Y)e^{A \cdot x} = e^{x \cdot U(v + V(A))}.$$

Thus, we have the expansion

$$\exp(x \cdot U(v)) = \sum_{\mathbf{n}} \frac{\mathbf{v}^{\mathbf{n}}}{\mathbf{n}!} y_{\mathbf{n}}(\mathbf{x}). \tag{5}$$

In particular, the kth component, U_k, of the inverse function is given by the coefficient of x_k in the above expansion.

An important feature of our approach is that to get an expansion to a given order requires knowledge of the expansion of W just to that order. The reason is that when iterating $xW(D)$, at step n it is acting on a polynomial of degree $n-1$, so all terms of the expansion of $W(D)$ of order n or higher would yield zero acting on y_{n-1}. This allows for streamlined computations.

For polynomial systems $\mathbf{V}$, V' will have polynomial entries, and W will be rational in $\mathbf{z}$. Hence raising operators will be rational functions of $\mathbf{D}$, linear in $\mathbf{x}$. Thus the coefficients of the expansion of the entries W_{ij} of W would be computed by finite-step recurrences.

Remark 1. Note that to solve $V(z) = v$ for z near z_0, with $V(z_0) = v_0$, apply the method to $V_1(z) = V(z + z_0) - v_0$, so that $V_1(0) = 0$. The inverse is $U_1(v) = U(v + v_0) - z_0$. Then $U(v) = z_0 + U_1(v - v_0)$.

2   One-Variable Case

In this section we focus on the one-variable case. We illustrate the method with examples, and then present an algorithm suitable for symbolic computation.

Example 1. In one variable, solving a cubic is interesting as the expansion of W can be expressed in terms of Chebyshev polynomials.

Let $V = z^3/3 - \alpha z^2 + z$. Then $V' = z^2 - 2\alpha z + 1$. Thus

$$W = \frac{1}{1 - 2\alpha z + z^2} = \sum_{n=0}^{\infty} z^n U_n(\alpha),$$

where U_n are Chebyshev polynomials of the second kind.

Specializing α provides interesting cases. For example, let $\alpha = \cos(\pi/4)$, or $V = z^3/3 - z^2/\sqrt{2} + z$. Then the coefficients in the expansion of W are periodic with period 8 and, in fact,

$$W = \frac{1 + z^2 + \sqrt{2}\,z}{1 + z^4}.$$

The coefficient of x in the polynomials y_n yield the coefficients in the expansion of the inverse U. Here are some polynomials starting with $y_0 = 1$, $y_1 = x$:

$$y_2 = x^2 + x\sqrt{2}, \quad y_3 = x^3 + 3\,x^2\,\sqrt{2} + 4\,x,$$
$$y_4 = x^4 + 6\,x^3\,\sqrt{2} + 22\,x^2 + 10\,x\,\sqrt{2},$$
$$y_5 = x^5 + 10\,x^4\,\sqrt{2} + 70\,x^3 + 90\,x^2\,\sqrt{2} + 40\,x,$$
$$y_6 = x^6 + 15\,x^5\,\sqrt{2} + 170\,x^4 + 420\,x^3\,\sqrt{2} + 700\,x^2 - 140\,x\,\sqrt{2}.$$

This gives to order 6:

$$U(v) = \left(v + \frac{2}{3}v^3 + \frac{1}{3}v^5 + \cdots \right) + \sqrt{2}\left(\frac{1}{2}v^2 + \frac{5}{12}v^4 - \frac{7}{36}v^6 + \cdots \right).$$

This expansion will give approximate solutions to

$$z^3/3 - z^2/\sqrt{2} + z - v = 0$$

for v near 0.

Example 2. Inversion of the Chebyshev polynomial $T_3(z) = 4z^3 - 3z$ can be used as the basis for solving general cubic equations ([4]).

To get started we have, with $V(z) = 4z^3 - 3z$,

$$W(z) = \frac{-1}{3}\frac{1}{1 - 4z^2} = \frac{-1}{3}\sum_{n=0}^{\infty} 4^n z^{2n}.$$

So $y_1 = (-1/3)x$, $y_2 = (1/9)x^2$, $y_3 = (-1/27)(x^3 + 8x)$, etc. We find

$$U(v) = -\frac{1}{3}v - \frac{4}{81}v^3 - \frac{16}{729}v^5 - \frac{256}{19683}v^7 - \cdots.$$

In this case, we can find the expansion analytically. To solve $T_3(z) = v$, write

$$T_3(\cos\theta) = \cos(3\theta) = v.$$

Invert to get, for integer k, $\theta = (1/3)(2\pi k \pm \arccos v)$, with arccos denoting the principal branch. Then

$$z = \cos((1/3)(2\pi k \pm \arccos v)).$$

We want a branch with $v = 0$ corresponding to $z = 0$. With $\arccos 0 = \pi/2$, we want the argument of the cosine to be $\pi/2 + \pi l$, for some integer l. This yields the condition $\dfrac{1}{3} = \dfrac{2l+1}{4k \pm 1}$. Taking $l = 0$, we get $k = 1$, with the minus sign. Namely,

$$U(v) = \cos((1/3)(2\pi - \arccos v)).$$

Using hypergeometric functions (see next example) and rewriting, we find the form

$$U(v) = -\frac{1}{3} \sum_{n=0}^{\infty} \binom{3n}{n} \left(\frac{4}{27}\right)^n \frac{v^{2n+1}}{2n+1}.$$

If we generate the polynomials y_n, we can find the expansion of $U(v)^m$ to any order.

Example 3. A similar approach is interesting for the Chebyshev polynomial $T_n(z)$.

$F(v) = \cos(\lambda(\mu \pm \arccos v))$ satisfies the hypergeometric differential equation

$$(1 - v^2)\, F'' - v\, F' + \lambda^2\, F = 0$$

which can be written in the form

$$[(vD_v)^2 - D_v^2]F = \lambda^2\, F$$

with here D_v denoting d/dv. For integer λ, this is the differential equation for the corresponding Chebyshev polynomial. In general, these are *Chebyshev functions*. As noted above, for $F(0) = 0$, we take $\mu = 2\pi k$, and, as above, we require

$$\lambda = \frac{2l+1}{4k \pm 1}.$$

With $F'(0) = \pm\lambda$, we have the solution

$$F(v) = \pm\lambda v\, {}_2F_1\left(\begin{array}{c} \dfrac{1+\lambda}{2}, \dfrac{1-\lambda}{2} \\ \dfrac{3}{2} \end{array}\middle|\; v^2\right).$$

2.1 Using Maple

For symbolic computation using Maple, one can use the Ore_Algebra package.

1. First fix the degree of approximation. Expand W as a polynomial to that degree.
2. Declare the Ore algebra with one variable, x, and one derivative, D.
3. Define the operator $xW(D)$ in the algebra.
4. Iterate starting with $y_0 = 1$ using the applyopr command.
5. Extract the coefficient of $x^m/m!$ to get the expansion of $U(v)^m$.

3   Algorithm as a Matrix Computation

Here is a matrix approach that can be implemented numerically.

Fix the order of approximation n. Cut off the expansion

$$W(z) = w_0 + w_1 z + w_2 w^2 + \cdots + w_k z^k + \cdots$$

at $w_n z^n$.

Let the matrix

$$W = \begin{pmatrix} w_1 & w_0 & 0 & \dots & 0 \\ w_2 & w_1 & w_0 & \dots & 0 \\ \vdots & \vdots & \vdots & \ddots & \vdots \\ w_{n-1} & w_{n-2} & w_{n-3} & \dots & w_0 \\ w_n & w_{n-1} & w_{n-2} & \dots & w_1 \end{pmatrix}.$$

Define the auxiliary diagonal matrices

$$P = \begin{pmatrix} 1! & 0 & \dots & 0 \\ 0 & 2! & \dots & 0 \\ \vdots & \vdots & \ddots & \vdots \\ 0 & 0 & \dots & n! \end{pmatrix}, \quad M = \begin{pmatrix} 1 & 0 & \dots & 0 \\ 0 & 2 & \dots & 0 \\ \vdots & \vdots & \ddots & \vdots \\ 0 & 0 & \dots & n \end{pmatrix},$$

$$Q = \begin{pmatrix} 1/\Gamma(1) & 0 & \dots & 0 \\ 0 & 1/\Gamma(2) & \dots & 0 \\ \vdots & \vdots & \ddots & \vdots \\ 0 & 0 & \dots & 1/\Gamma(n) \end{pmatrix}.$$

Note that $QP = M$.

Denoting $y_k(x) = \sum c_j^{(k)} x^j$, we have the recursion

$$[c_1^{(k+1)}, c_2^{(k+1)}, \dots, c_n^{(k+1)}] = [c_1^{(k)}, c_2^{(k)}, \dots, c_n^{(k)}] PWQ.$$

The condition $U(0) = 0$ gives $y_0 = 1$. Then $y_1 = XW(D)y_0$ yields $y_1 = w_0 x$. We see that $c_0^{(k)} = 0$ for $k > 0$. We iterate as follows:

1. Start with w_0 times the unit vector $[1, 0, \ldots, 0]$ of length n.
2. Multiply by W.
3. Iterate, multiplying on the right by MW at each step.
4. Finally, multiply on the right by Q.

The top row will give the coefficients of the expansion of $U(v)$ to order n.

4   Higher-Order Example

Here is a simple 2×2 system for illustration.

$$V_1 = z_1 + z_2^2/2, \quad V_2 = z_2 - z_1 z_2.$$

So

$$V' = \begin{pmatrix} 1 & z_2 \\ -z_2 & 1 - z_1 \end{pmatrix} \quad \text{and} \quad W = \frac{1}{1 - z_1 + z_2^2} \begin{pmatrix} 1 - z_1 & -z_2 \\ z_2 & 1 \end{pmatrix}.$$

The raising operators are

$$Y_1 = \left(x_1(1 - D_1)) + x_2 D_2\right) \left(1 - D_1 + D_2^2\right)^{-1},$$
$$Y_2 = \left(-x_1 D_2 + x_2\right) \left(1 - D_1 + D_2^2\right)^{-1}.$$

Expanding $\left(1 - D_1 + D_2^2\right)^{-1} = \sum\limits_{n=0}^{\infty} (D_1 - D_2^2)^n$ yields, with $y_{00} = 1$,

$$y_{01} = x_2, \qquad y_{10} = x_1,$$
$$y_{02} = x_2^2 - x_1, \quad y_{11} = x_2 + x_1 x_2, \quad y_{20} = x_1^2.$$

Thus

$$\exp\left(\mathbf{x} \cdot \mathbf{U}(\mathbf{v})\right) = 1 + x_1 v_1 + x_2 v_2$$
$$+ (x_2 + x_1 x_2) v_1 v_2 + (x_2^2 - x_1) \frac{v_1^2}{2} + x_1^2 \frac{v_2^2}{2} + \cdots,$$

so

$$U_1(\mathbf{v}) = v_1 - v_1^2/2 + \cdots, \quad U_2(\mathbf{v}) = v_2 + v_1 v_2 + \cdots.$$

5   Another Matrix Approach

For any given order n, the polynomials of degree n are an invariant subspace for the operator Y up until the last step. We can formulate an alternative matrix computation as follows. Let $\bar{D}$ and $\bar{X}$ denote the matrices of the operators of differentiation and multiplication by x respectively on polynomials of degree less than or equal to n. The space is invariant under differentiation, and we cut off multiplication by x to be zero on x^n. We get

$$\bar{D}_{ij} = i \, \delta_{i+1,j} \quad \text{and} \quad \bar{X}_{ij} = \delta_{i-1,j}$$

with the first row of $\bar{X}$ all zeros. We then compute the matrix $\bar{X}$ times $W(\bar{D})$, where $W(\bar{D})$ is computed as a matrix polynomial by substituting in $W(z)$ up to order n. Then Y has a matrix representation, $\bar{Y} = \bar{X}W(\bar{D})$, on the space and we iterate multiplying by $\bar{Y}$ acting on the unit vector $\mathbf{e_1}$. These give the coefficients of the polynomials y_n.

In several variables, one constructs matrices for D_j and X_i using Kronecker products of $\bar{D}$ and $\bar{X}$ with the identity. For example,

$$\bar{D}_j = I \otimes I \otimes \cdots \otimes \bar{D} \otimes I \cdots \otimes I$$

with $\bar{D}$ in the jth spot. Similarly for $\bar{X}_i$. Then one has explicit matrix representations for the dual vector fields and the polynomials can be found accordingly.

This approach is explicit, but seems to much slower than using the built-in Ore_algebra package.

6   Worksheets

```
[> with(linalg):
[> with(Ore_algebra):
[One Variable
[>
[> n:=10;
```
$$n := 10$$
```
[> unassign('z','y','Y','x','V','W'):
[> V:=z-z^2/2;
```
$$V := z - \frac{1}{2}\,z^2$$
```
[> W:=diff(V,z)^(-1);
[> W:=convert(taylor(W,z=0,n),polynom);
```
$$W := \frac{1}{1-z}$$
$$W := 1 + z + z^2 + z^3 + z^4 + z^5 + z^6 + z^7 + z^8 + z^9$$
```
[> ########### ORE ALGEBRA STARTS HERE ###################
[> A:=diff_algebra([z,x]);
```
$$A := Ore\_algebra$$
```
[> YY:=x*W;
```
$$YY := x\,(1 + z + z^2 + z^3 + z^4 + z^5 + z^6 + z^7 + z^8 + z^9)$$
```
[> y[0]:=1:
[> for i from 1 to n-1 do y[i]:=simplify(applyopr(YY,y[i-1],A)) od;
```
$$y_1 := x$$
$$y_2 := x^2 + x$$
$$y_3 := x^3 + 3\,x^2 + 3\,x$$
$$y_4 := x^4 + 6\,x^3 + 15\,x^2 + 15\,x$$
$$y_5 := x^5 + 10\,x^4 + 45\,x^3 + 105\,x^2 + 105\,x$$

Several Variables

Example in two variables from an analytic function

```
> V1:=evalc(Re(expand(subs(z=zc,f))));V2:=evalc(Im(expand(subs(z=zc,f))));
```
$$V1 := z1 - 4\,z1^2 + 4\,z1\,z2 + 4\,z2^2$$
$$V2 := z2 - 2\,z1^2 - 8\,z1\,z2 + 2\,z2^2$$

```
> Jac:=jacobian([V1,V2],[z1,z2]):
> W:=evalm(Jac^(-1)):adj(Jac),factor(det(Jac));
> WMat:=map(mtaylor,W,[z1=0,z2=0],n);
```
$$\begin{bmatrix} 1 - 8\,z1 + 4\,z2 & -4\,z1 - 8\,z2 \\ 4\,z1 + 8\,z2 & 1 - 8\,z1 + 4\,z2 \end{bmatrix}, 1 - 16\,z1 + 8\,z2 + 80\,z1^2 + 80\,z2^2$$

$WMat := [[1 + 8\,z1 - 4\,z2 + 48\,z1^2 - 48\,z2^2 - 128\,z1\,z2 + 128\,z1^3 - 384\,z1\,z2^2$
$\quad - 2112\,z2\,z1^2 + 704\,z2^3, -4\,z1 - 8\,z2 - 64\,z1^2 - 96\,z1\,z2 + 64\,z2^2 - 704\,z1^3$
$\quad - 384\,z2\,z1^2 + 2112\,z1\,z2^2 + 128\,z2^3], [4\,z1 + 8\,z2 + 64\,z1^2 + 96\,z1\,z2 - 64\,z2^2$
$\quad + 704\,z1^3 + 384\,z2\,z1^2 - 2112\,z1\,z2^2 - 128\,z2^3, 1 + 8\,z1 - 4\,z2 + 48\,z1^2 - 48\,z2^2$
$\quad - 128\,z1\,z2 + 128\,z1^3 - 384\,z1\,z2^2 - 2112\,z2\,z1^2 + 704\,z2^3]]$

```
> ########### ORE ALGEBRA STARTS HERE ####################
> A:=diff_algebra([z1,x1],[z2,x2]);
```
$$A := Ore\_algebra$$

```
> for ix to N do YY[ix]:=simplify(add(x||k*WMat[k,ix],k=1..N)) od;
```
$YY_1 := x1 + 8\,x1\,z1 - 4\,x1\,z2 + 48\,x1\,z1^2 - 48\,x1\,z2^2 - 128\,x1\,z1\,z2 + 128\,x1\,z1^3$
$\quad - 384\,x1\,z1\,z2^2 - 2112\,x1\,z2\,z1^2 + 704\,x1\,z2^3 + 4\,x2\,z1 + 8\,x2\,z2 + 64\,x2\,z1^2$
$\quad + 96\,x2\,z1\,z2 - 64\,x2\,z2^2 + 704\,x2\,z1^3 + 384\,x2\,z2\,z1^2 - 2112\,x2\,z1\,z2^2$
$\quad - 128\,x2\,z2^3$

$YY_2 := -4\,x1\,z1 - 8\,x1\,z2 - 64\,x1\,z1^2 - 96\,x1\,z1\,z2 + 64\,x1\,z2^2 - 704\,x1\,z1^3$
$\quad - 384\,x1\,z2\,z1^2 + 2112\,x1\,z1\,z2^2 + 128\,x1\,z2^3 + x2 + 8\,x2\,z1 - 4\,x2\,z2 + 48\,x2\,z1^2$
$\quad - 48\,x2\,z2^2 - 128\,x2\,z1\,z2 + 128\,x2\,z1^3 - 384\,x2\,z1\,z2^2 - 2112\,x2\,z2\,z1^2$
$\quad + 704\,x2\,z2^3$

```
> y[0,0]:=1:
> for i from 0 to n-1 do j:=0; if(i>0) then
  y[i,0]:=simplify(applyopr(YY[1],y[i-1,0],A)) fi; for j from 1 to n-1 do
  y[i,j]:=simplify(applyopr(YY[2],y[i,j-1],A));print("y["||i||","||j||"]=",
  [i,j]) od;  od;
```
$$j := 0$$
$$\text{"y[0,1]="}, x2$$

$$j := 0$$
$$\text{"y[0,1]="}, x2$$
$$\text{"y[0,2]="}, x2^2 - 4\,x2 - 8\,x1$$
$$\text{"y[0,3]="}, x2^3 - 12\,x2^2 - 24\,x2\,x1 + 192\,x1 - 144\,x2$$
$$j := 0$$
$$\text{"y[1,1]="}, x2\,x1 - 4\,x1 + 8\,x2$$
$$\text{"y[1,2]="}, x2^2\,x1 - 12\,x2\,x1 + 16\,x2^2 - 144\,x1 - 192\,x2 - 8\,x1^2$$
$$\text{"y[1,3]="}, 10560\,x1 - 1920\,x2 - 24\,x2^2\,x1 + 24\,x2^3 + x2^3\,x1 - 24\,x2\,x1^2 - 720\,x2\,x1$$
$$- 672\,x2^2 + 288\,x1^2$$

$$j := 0$$
$$\text{"y[2,1]="}, x2\,x1^2 + 24\,x2\,x1 + 4\,x2^2 - 8\,x1^2 - 192\,x1 + 144\,x2$$
$$\text{"y[2,2]="}, -1920\,x1 - 10560\,x2 + 40\,x2^2\,x1 - 8\,x1^3 + 4\,x2^3 + x2^2\,x1^2 - 20\,x2\,x1^2$$
$$- 960\,x2\,x1 + 400\,x2^2 - 320\,x1^2$$
$$\text{"y[2,3]="}, 497664\,x1 + 145152\,x2 + x2^3\,x1^2 - 24\,x2\,x1^3 - 2496\,x2^2\,x1 + 384\,x1^3$$
$$+ 768\,x2^3 - 36\,x2^2\,x1^2 + 56\,x2^3\,x1 - 1392\,x2\,x1^2 + 4\,x2^4 - 13440\,x2\,x1$$
$$- 43200\,x2^2 + 30720\,x1^2$$

$$j := 0$$
$$\text{"y[3,1]="}, x2\,x1^3 + 48\,x2\,x1^2 + 12\,x2^2\,x1 + 720\,x2\,x1 + 288\,x2^2 - 12\,x1^3 - 672\,x1^2$$
$$- 10560\,x1 + 1920\,x2$$
$$\text{"y[3,2]="}, 145152\,x1 - 497664\,x2 - 28\,x2\,x1^3 + 1632\,x2^2\,x1 - 528\,x1^3 + 384\,x2^3$$
$$+ 72\,x2^2\,x1^2 - 8\,x1^4 + 12\,x2^3\,x1 - 2496\,x2\,x1^2 + x2^2\,x1^3 - 73920\,x2\,x1$$
$$+ 7680\,x2^2 - 5760\,x1^2$$
$$\text{"y[3,3]="}, 23553024\,x1 + 21829632\,x2 + 96\,x2^3\,x1^2 - 2160\,x2\,x1^3 - 233280\,x2^2\,x1$$
$$+ 62400\,x1^3 + 19200\,x2^3 - 5760\,x2^2\,x1^2 + 480\,x1^4 - 24\,x2\,x1^4 + 2880\,x2^3\,x1$$
$$- 34560\,x2\,x1^2 - 48\,x2^2\,x1^3 + x2^3\,x1^3 + 480\,x2^4 + 1435392\,x2\,x1 + 12\,x2^4\,x1$$
$$- 2575872\,x2^2 + 2345472\,x1^2$$

>

References

1. P. Feinsilver and R. Schott. *Algebraic structures and operator calculus, Vols I–III.* Kluwer Academic Publishers, 1993, 1994, 1996.
2. P. Feinsilver and R. Schott. Vector fields and their duals. *Adv. in Math.*, 149:182–192, 2000.
3. G.-C. Rota, D. Kahaner, and A. Odlyzko. *Finite operator calculus.* Academic Press, 1975.
4. http://en.wikipedia.org/wiki/Cubic_equation.

Solving Dynamic Geometric Constraints Involving Inequalities

Hoon Hong[1,*], Liyun Li[2,**], Tielin Liang[3], and Dongming Wang[2,4]

[1] Department of Mathematics, North Carolina State University
Box 8205, Raleigh, NC 27695, USA
[2] LMIB – School of Science, Beihang University, Beijing 100083, China
[3] Department of Mathematics, University of Science and Technology of China
Hefei 230026, Anhui, China
[4] Laboratoire d'Informatique de Paris 6, Université Pierre et Marie Curie – CNRS
8 rue du Capitaine Scott, F-75015 Paris, France

Abstract. This paper presents a specialized method for solving dynamic geometric constraints involving equalities and inequalities. The method works by decomposing the system of constraints into finitely many explicit solution representations in terms of parameters with radicals using triangular decomposition and real quantifier elimination. For any given values of the parameters, if they verify some set of computed relations, the values of the dependent variables may be easily computed by direct evaluation of the corresponding explicit expressions. The effectiveness of our method and its experimental implementation is illustrated by some examples of diagram generation.

1 Introduction

Dynamic geometric constraint solving (GCS) aims at dynamically producing diagrams of given geometric objects satisfying given geometric constraint relations. It has been studied extensively in the area of computer aided geometric design and modeling (see the recent survey [6] and references therein), yielding several approaches such as graph-based, algebraic, numerical and logic-based approaches. These approaches are developed mainly for solving geometric constraints that may be expressed algebraically as equalities. In this paper, we provide an approach for solving geometric constraints involving *inequalities* as well.

The main motivation for considering inequalities is that many geometric constraints naturally require the notion of *order* such as "between", "inside," and "outside." For example, we often need to deal with geometric constraints such as external tangency of circles and internal bisection of angles. When those geometric constraints are translated into algebraic ones, they naturally show up as

---

* Research was supported by NSF Grant 0097976 "Solving Quantified Algebraic Constraints."
** Research was supported partially by the National Key Basic Research Projects 2004CB318000 and 2005CB321902 of China.

J. Calmet, T. Ida, and D. Wang (Eds.): AISC 2006, LNAI 4120, pp. 181–195, 2006.

inequalities. One could formulate those GCS problems using equalities only, but then, they become extremely unnatural, involving artificial slack variables and existential quantifiers, also resulting in enormous computational blowup, making them impractical. One could ignore the order notion while formulating geometric constraints, but it often leads to many unexpected extraneous or degenerate solutions, causing confusion, especially when the diagram is moving across some critical points. It is therefore imperative to develop effective methods that directly tackle geometric constraints involving inequalities.

As mentioned above, there is little work on geometric constraints involving inequalities as well as equalities. This may be mainly due to the well known inherent practical difficulties of dealing with general inequality constraints. There are methods (such as cylindrical algebraic decomposition and others) that can handle, in principle, arbitrarily general equality/inequality constraints, but the practical complexity is prohibitive for even moderate size of problems.

One source of computational blowup of the general methods is the need to consider the interaction of arbitrary degree equalities with inequalities. However, it is well known that many interesting and important geometric constraints involve only low degree equalities (usually less than 5). Thus, in this paper, we restrict ourselves and develop a specialized method for constraints involving *equalities of degree less than 5* but *inequalities of arbitrary degree*.

The restriction on degree of equalities allows us to compute *explicit* representations of real solutions of semi-algebraic systems by radicals. The method proceeds by first decomposing the set of equality constraints into finitely many triangular sets. Then with respect to each triangular set together with inequality constraints, the space of parameters is decomposed into finitely many domains by means of real quantifier elimination, such that associated with each domain there is a set of explicit expressions of the dependent variables in terms of the parameters with radicals. For any given values of the parameters, if they verify the relations of some domain, the values of the dependent variables may be easily computed by direct evaluation of the corresponding explicit expressions.

An experimental implementation of our approach has been done in Java with interface to the Epsilon library [12] in Maple and the QEPCAD package [1] in C. We will illustrate the approach and its implementation with a few examples.

2   Representing Real Solutions of Semi-algebraic Systems by Radicals

Consider the following semi-algebraic system of equations and inequalities

$$\begin{cases} F_1(x_1,\ldots,x_n) = 0, \\ \quad\cdots\cdots \\ F_s(x_1,\ldots,x_n) = 0, \\ G_1(x_1,\ldots,x_n) \gtrless 0, \\ \quad\cdots\cdots \\ G_t(x_1,\ldots,x_n) \gtrless 0, \end{cases} \tag{1}$$

where $F_1, \ldots, F_s, G_1, \ldots, G_t$ are polynomials in $x_1, \ldots, x_n$ with rational coefficients and $\gtrless$ may take any of the inequality operators $<, \leqslant, >, \geqslant, \neq$. We wish to represent the real solutions of (1) by means of explicit formulae with radicals. This is not possible in general (as it is well known from Abel/Galois theory that the solutions of polynomial equations of degree greater than 4 in general cannot be expressed in terms of radicals and field operations), so our objective is to compute such representations for polynomials of low degree. Once such representations are available, they can be efficiently instantiated repeatedly for dynamic update of diagrams.

Let $u = (u_1, \ldots, u_d)$ be a subset of the variables $x_1, \ldots, x_n$. Denote by $y_1, \ldots, y_r$ all the other variables x_i not in u. Note that $u_1, \ldots, u_d, y_1, \ldots, y_r$ is a permutation of $x_1, \ldots, x_n$ (so $d + r = n$ and $\{u_1, \ldots, u_d, y_1, \ldots, y_r\} = \{x_1, \ldots, x_n\}$). We call u parameters (or parametric variables) and $y_1, \ldots, y_r$ dependents (or dependent variables).

Definition 1. *Let $\Gamma(u)$ be a quantifier-free formula composed of equality and inequality relations in the parameters u, and h_j a rational expression of u with radicals for $1 \leqslant j \leqslant r$. We call*

$$\Gamma(u), \ y_1 = h_1(u), \ y_2 = h_2(u), \ \ldots, \ y_r = h_r(u) \tag{2}$$

a solution representation by radicals (SRR) *in u.*

We want to decompose the semi-algebraic system (1) into finitely many SRRs of the form (2) such that the set of real solutions of (1) is equal to the union of the sets of real solutions given by the SRRs.

To compute SRRs, we first decompose the set of polynomials $F_1, \ldots, F_s$ into (irreducible) triangular sets by using the method of characteristic sets or other methods [9, 12]. Each triangular set $\mathbb{T}$ may be written in the form

$$\mathbb{T} = [T_1(u, y_1), T_2(u, y_1, y_2), \ldots, T_r(u, y_1, \ldots, y_r)], \tag{3}$$

where $u, y_1, \ldots, y_r$ is a permutation of $x_1, \ldots, x_n$ as above. Then the problem is reduced to considering the following set of constraints for every triangular set $\mathbb{T}$:

$$T_1(u, y_1) = 0, \ \ldots, \ T_r(u, y_1, \ldots, y_r) = 0, \ \Gamma_r(u, y_1, \ldots, y_r), \tag{4}$$

where $\Gamma_r(u, y_1, \ldots, y_r) := G_1 \gtrless 0, \ldots, G_t \gtrless 0, I_1 \neq 0, \ldots, I_r \neq 0$ and where I_i is the initial of T_i.

For any given values $\bar{u}$ of u, one can solve the equations $T_1 = 0, \ldots, T_r = 0$ successively for $y_1, \ldots, y_r$ and then verify which solutions satisfy the formula Γ_r in the *triangular representation* (4). This simple approach works theoretically but has two drawbacks for finding real solutions. First, it is possible that real solutions are found for $y_1, \ldots, y_k$ ($1 \leqslant k < r$) but there is no real solution for y_{k+1}. In this case, the computation of the real solutions for $y_1, \ldots, y_k$ is waste. Second, if a found real solution of $T_1 = 0, \ldots, T_r = 0$ does not satisfy Γ_r, then the computation of this solution is also waste. How to avoid or reduce such waste? In what follows we explain how to do so by eliminating the variables $y_r, \ldots, y_1$

successively from the inequality constraints in Γ_r using $T_r = 0, \ldots, T_1 = 0$ respectively.

Consider first the constraints

$$T_r(u, y_1, \ldots, y_r) = 0, \quad \Gamma_r(u, y_1, \ldots, y_r). \tag{5}$$

In the next section, we will show how to decompose (5) into finitely many explicit root representations of the form

$$y_r = h_r^{(j)}(u, y_1, \ldots, y_{r-1}), \quad \Gamma_{r-1}^{(j)}, \tag{6}$$

such that $\Gamma_{r-1}^{(j)}$ does not contain the variable y_r. Then we can deal with the constraints

$$T_{r-1}(u, y_1, \ldots, y_{r-1}) = 0, \quad \Gamma_{r-1}^{(j)} \tag{7}$$

similarly for each j. Continuing this way, we will be able to decompose (4) into finitely many explicit root representations

$$\Gamma_0^{(i)}(u), \ y_1 = h_1^{(i)}(u), \ y_2 = h_2^{(i)}(u, y_1), \ \ldots, \ y_r = h_r^{(i)}(u, y_1, \ldots, y_{r-1}), \tag{8}$$

with each $\Gamma_0^{(i)}(u)$ a conjunction of disjunctions of equality and inequality relations in u and each $h_j^{(i)}$ a rational expression of $u, y_1, \ldots, y_{j-1}$ with radicals. In other words, the space of parameters u is decomposed into finitely many domains D_i defined by $\Gamma_0^{(i)}(u)$, such that for any given values $\bar{u}$ of u, if $\bar{u} \in D_i$, then the values of the dependent variables $y_1, \ldots, y_r$ are

$$\bar{y}_1 = h_1^{(i)}(\bar{u}), \ \bar{y}_2 = h_2^{(i)}(\bar{u}, \bar{y}_1), \ \ldots, \ \bar{y}_r = h_r^{(i)}(\bar{u}, \bar{y}_1, \ldots, \bar{y}_{r-1}).$$

Note that a domain D_i may be disconnected and two domains may be joined.

We may substitute

$$y_1 = h_1^{(i)}(u), \ \ldots, \ y_j = h_j^{(i)}(u, y_1, \ldots, y_{j-1}) \tag{9}$$

into $h_{j+1}^{(i)}(u, y_1, \ldots, y_j)$, so that (8) become SRRs. In practice, we may wish to keep the form (8) because the substitution of (9) into $h_{j+1}^{(i)}$ may increase the size of the expression considerably. We call the representation (8) a *weak SRR*.

The method of decomposing a triangular representation of the form (4) into SRRs uses root formulae (as long as the degree of each T_i in y_i is less than 5) and real quantifier elimination [2]. From the obtained SRRs, we can easily construct a decomposition of (1) into SRRs as desired.

We say that a polynomial $P(z)$ is *composed* of polynomials $P_1(z), \ldots, P_k(z)$ if $P(z) = P_1(P_2(\cdots P_k(z) \cdots))$. The method sketched above allows us to establish the following main result.

Theorem 1. *For any semi-algebraic system (1) in $x_1, \ldots, x_n$, if the polynomials in the triangular sets obtained from all the irreducible triangular decompositions of the involved sets of polynomial equations are composed of polynomials of degree* *less than 5 with respect to their leading variables, then one can decompose (1) into q SRRs*

$$\Gamma^{(i)}(u^{(i)}),\ y_1^{(i)} = h_1^{(i)}(u^{(i)}),\ y_2^{(i)} = h_2^{(i)}(u^{(i)}),\ \ldots,\ y_{r_i}^{(i)} = h_{r_i}^{(i)}(u^{(i)})$$

($1 \leqslant i \leqslant q$) such that the set of real solutions of (1) is equal to

$$\bigcup_{i=1}^{q}\left\{(\bar{u}^{(i)}, h_1^{(i)}(\bar{u}^{(i)}), \ldots, h_{r_i}^{(i)}(\bar{u}^{(i)}))\ |\ \Gamma^{(i)}(\bar{u}^{(i)})\right\},$$

where $u^{(i)}, y_1^{(i)}, \ldots, y_{r_i}^{(i)}$ is a permutation of $x_1, \ldots, x_n$ for each i and the union of zero sets is carried out after permuting $u^{(i)}, y_1^{(i)}, \ldots, y_{r_i}^{(i)}$ back to $x_1, \ldots, x_n$.

The method explained above and indicated in Theorem 1 has both theoretical and practical interests because the explicit SRRs computed may provide an efficient way for the computation of the real solutions of (1). We will demonstrate how it can be applied effectively to solving dynamic geometric constraints involving inequalities in Section 4. For dynamic animation, the computation of real solutions has to be performed in real time and thus should be kept as inexpensive as possible, while more expensive symbolic precomputation is acceptable.

3   Root Formulae of Univariate Equations with Inequality Constraints

In this section, we explain how to eliminate the variable y_r from the inequality constraints in $\Gamma_r(u, y_1, \ldots, y_{r-1}, y_r)$ as in (4) by using $T_r(u, y_1, \ldots, y_{r-1}, y_r) = 0$ as in (3), when the degree of T_r in y_r is small. For notational convenience, we will write x for y_r, $f(x)$ for $T_r(u, y_1, \ldots, y_{r-1}, y_r)$, and $\Gamma(x)$ for $\Gamma_r(u, y_1, \ldots, y_{r-1}, y_r)$.

Proposition 1 (Linear Case). *Let $f(x) = ax + b$. Then (A) is equivalent to (B):*

(A) $f(x) = 0 \wedge \Gamma(x)$;

(B) $x = -\dfrac{b}{a} \wedge \Gamma(-b/a).$

Proof. Since $a \neq 0$ is contained in $\Gamma(x)$, $f(x) = 0$ is equivalent to $x = -b/a$. The proof immediately follows.   □

Proposition 2 (Quadratic Case). *Let $f(x) = ax^2 + bx + c$. Then (A) is equivalent to (B1) $\vee$ (B2):*

(A) $f(x) = 0 \wedge \Gamma(x)$;

(B1) $x = \dfrac{-b + \sqrt{\Delta}}{2a} \wedge \Gamma^+$;

(B2) $x = \dfrac{-b - \sqrt{\Delta}}{2\,a} \wedge \Gamma^{-},$

where

$$\Gamma^{+} := (\exists x)\,[\Gamma(x) \wedge f(x) = 0 \wedge f'(x) \geqslant 0],$$
$$\Gamma^{-} := (\exists x)\,[\Gamma(x) \wedge f(x) = 0 \wedge f'(x) \leqslant 0];$$
$$\Delta := b^2 - 4\,ac = \text{discriminant}(f).$$

Proof. As $a \neq 0$ is contained in $\Gamma(x)$, $f(x) = 0$ is equivalent to

$$x = \frac{-b + \sqrt{\Delta}}{2\,a} \quad \vee \quad x = \frac{-b - \sqrt{\Delta}}{2\,a}.$$

Now we make the key observation

$$x = \frac{-b + \sqrt{\Delta}}{2\,a} \iff f(x) = 0 \wedge f'(x) \geqslant 0,$$

$$x = \frac{-b - \sqrt{\Delta}}{2\,a} \iff f(x) = 0 \wedge f'(x) \leqslant 0.$$

The proof immediately follows. □

Proposition 3 (Cubic Case). *Let $f(x) = ax^3 + bx^2 + cx + d$ be such that its discriminant $\Delta \neq 0$. Then (A) is equivalent to (B0) $\vee$ (B1) $\vee$ (B2) $\vee$ (B3):*

(A) $f(x) = 0 \wedge \Gamma(x);$

(B0) $x = \dfrac{^{\mathrm{R}}\sqrt[3]{\delta}}{6\,a} - \dfrac{2\,(3\,ac - b^2)}{3\,a\,^{\mathrm{R}}\sqrt[3]{\delta}} - \dfrac{b}{3\,a} \wedge \Gamma^{\mathrm{R}};$

(B1) $x = \dfrac{^{\mathrm{I}}\sqrt[3]{\delta}}{6\,a} - \dfrac{2\,(3\,ac - b^2)}{3\,a\,^{\mathrm{I}}\sqrt[3]{\delta}} - \dfrac{b}{3\,a} \wedge \Gamma^{\mathrm{I}};$

(B2) $x = \dfrac{^{\mathrm{II}}\sqrt[3]{\delta}}{6\,a} - \dfrac{2\,(3\,ac - b^2)}{3\,a\,^{\mathrm{II}}\sqrt[3]{\delta}} - \dfrac{b}{3\,a} \wedge \Gamma^{\mathrm{II}};$

(B3) $x = \dfrac{^{\mathrm{III}}\sqrt[3]{\delta}}{6\,a} - \dfrac{2\,(3\,ac - b^2)}{3\,a\,^{\mathrm{III}}\sqrt[3]{\delta}} - \dfrac{b}{3\,a} \wedge \Gamma^{\mathrm{III}},$

where

$$\Gamma^{\mathrm{R}} := \Delta > 0 \wedge (\exists x)\,[\Gamma(x) \wedge f(x) = 0],$$
$$\Gamma^{\mathrm{I}} := \Delta < 0 \wedge (\exists x)\,[\Gamma(x) \wedge f(x) = 0 \wedge a f'(x) > 0 \wedge f''(x) > 0],$$
$$\Gamma^{\mathrm{II}} := \Delta < 0 \wedge (\exists x)\,[\Gamma(x) \wedge f(x) = 0 \wedge f'(x) > 0 \wedge (f''(x) < 0 \vee a < 0)],$$
$$\Gamma^{\mathrm{III}} := \Delta < 0 \wedge (\exists x)\,[\Gamma(x) \wedge f(x) = 0 \wedge f'(x) < 0 \wedge (f''(x) < 0 \vee a > 0)],$$
$$\delta := 36\,abc - 108\,a^2 d - 8\,b^3 + 12\,\sqrt{3}\sqrt{\Delta}\,a,$$
$$\Delta := 4\,ac^3 - b^2 c^2 - 18\,abcd + 27\,a^2 d^2 + 4\,b^3 d = \text{discriminant}(f).$$

where $\sqrt[R]{\delta}$ denotes the real cubic root of δ and $\sqrt[I3]{\delta}$, $\sqrt[II3]{\delta}$, $\sqrt[III3]{\delta}$ denote the cubic roots of δ respectively in sectors I, II, III (see the proof for the definition of sectors).

Proof. Let us recall the root formula of f for x:

$$x = \frac{\sqrt[3]{\delta}}{6\,a} - \frac{2\,(3\,ac - b^2)}{3\,a\sqrt[3]{\delta}} - \frac{b}{3\,a}, \tag{10}$$

where $\sqrt[3]{\delta}$ stands for the three (complex) cubic roots of δ.

There are two cases according to the sign of the discriminant Δ: $\Delta > 0$ or $\Delta < 0$. First assume that $\Delta > 0$. Then, it is well known that $f = 0$ has one and only one real root for x. Thus, we see immediately that (A) is equivalent to (B0).

From now on, assume that $\Delta < 0$. The three roots of $f = 0$ for x are all real and the key issue is to distinguish them by using inequality relations. For any complex number w, a cubic root $u + iv$ of w is called the *principal cubic root of w* if $u > 0$ and $\arctan(\frac{v}{u}) \in (-\frac{\pi}{3}, \frac{\pi}{3}]$ (or equivalently, $u > 0$ and $-\sqrt{3} < \frac{v}{u} \leqslant \sqrt{3}$). We call the sector in the complex plane in which principal cubic roots reside the *principal sector* or *sector I*. Sectors II and III are obtained from sector I by rotating the region $120°$ and $240°$ anti-clockwise around the origin respectively.

We claim that $\sqrt[3]{\delta}$ lies in sector I if and only if

$$f''(x) > 0 \quad \text{and} \quad af'(x) > 0.$$

In order to prove the claim, we solve the equation (10) for $\sqrt[3]{\delta}$, obtaining

$$\sqrt[3]{\delta} = 3\,ax + b + \sqrt{\Omega}, \tag{11}$$
$$\sqrt[3]{\delta} = 3\,ax + b - \sqrt{\Omega}, \tag{12}$$

where

$$\Omega = 3\,(3\,a^2x^2 + 2\,abx + 4\,ac - b^2).$$

Plunging the expression of δ into the above equalities, taking the third power of both sides, and comparing the resulting two sides, we see that (11) holds if $3\,ax^2 + 2\,bx + c \geqslant 0$, and (12) holds otherwise. As $\Delta < 0$, δ must be complex, and so must $\sqrt[3]{\delta}$. Write (11) and (12) as

$$\sqrt[3]{\delta} = 3\,ax + b + i\sqrt{-\Omega}, \quad \sqrt[3]{\delta} = 3\,ax + b - i\sqrt{-\Omega}, \quad \Omega < 0.$$

Then, for any x satisfying $3\,ax^2 + 2\,bx + c \geqslant 0$, $\sqrt[3]{\delta}$ lies in sector I if and only if

$$3\,ax + b > 0 \quad \text{and} \quad -\sqrt{3} < \frac{\sqrt{-\Omega}}{3\,ax + b} \leqslant \sqrt{3},$$

i.e.,

$$3\,ax + b > 0 \quad \text{and} \quad 3\,a^2x^2 + 2\,abx + ac \geqslant 0.$$

Similarly, for any x satisfying $3\,ax^2 + 2\,bx + c < 0$, $\sqrt[3]{\delta}$ lies in sector I if and only if

$$3\,ax + b > 0, \quad \text{and} \quad -\sqrt{3} < -\frac{\sqrt{-\Omega}}{3\,ax + b} \leqslant \sqrt{3},$$

i.e.,

$$3\,ax + b > 0 \quad \text{and} \quad 3\,a^2x^2 + 2\,abx + ac > 0.$$

Note that $\Delta < 0$ implies that $f = 0$ has no multiple root, so for any root x of $f = 0$, $f'(x) = 3\,ax^2 + 2\,bx + c \neq 0$. As $a \neq 0$, we have $3\,a^2x^2 + 2\,abx + ac \neq 0$. It follows that $\sqrt[3]{\delta}$ lies in sector I if and only if

$$3\,ax + b > 0 \quad \text{and} \quad 3\,a^2x^2 + 2\,abx + ac > 0.$$

Similarly we claim that $\sqrt[3]{\delta}$ lies in sector II if and only if

$$3\,ax + b < 0 \text{ or } a < 0, \quad \text{and} \quad 3\,ax^2 + 2\,bx + c > 0,$$

and $\sqrt[3]{\delta}$ lies in sector III if and only if

$$3\,ax + b < 0 \text{ or } a > 0, \quad \text{and} \quad 3\,ax^2 + 2\,bx + c < 0.$$

The proof is similar to that for sector I, and thus omitted.

From these, it is immediate that when $\Delta < 0$, (A) is equivalent to (B1) $\vee$ (B2) $\vee$ (B3). $\qquad\qquad\Box$

If $\Delta = 0$, then $f = 0$ has a multiple root for x. In this case, squarefree decomposition of f shows that $f = 0$ is equivalent to

$$2\,a\,(3\,ac - b^2)\,x^2 - (9\,a^2d - 7\,abc + 2\,b^3)\,x - 3\,abd + 4\,ac^2 - b^2c = 0$$

when $3\,ac - b^2 \neq 0$, and to $3\,ax + b = 0$ when $3\,ac - b^2 = 0$. So the problem is reduced to the quadratic or linear case.

We can also derive root formulae for quartic equations by variable transformation and from the root formulae of quadratic and cubic equations. Different real roots may be distinguished by forming existentially quantified formulae involving *polynomial* relations only using similar techniques. The details are quite involved and will be described formally elsewhere. Finally, if the degree of f is greater than 4 and f is composed of polynomials of degree less than 5, then we can also find explicit root representations for all the polynomial equations in the composition with inequality constraints. Therefore, the composed case may be reduced to the quadratic, cubic, and quartic cases.

In order to use the above propositions, we need to eliminate the existential quantifier from the formulae. This can be done by using any real quantifier elimination procedure (such as QEPCAD [1,2], REDLOG [13,3], QEQUAD [5], or SturmHabicht [4]).

In summary, the set of constraints $f(x) = 0 \wedge \Gamma(x)$ may be decomposed into explicit root representations (B), (B0), (B1), (B2), and (B3) in Propositions 1–3, or other similar ones.

4 Generation of Dynamic Diagrams

We are given a set $\mathcal{O}$ of geometric objects in a geometric space (e.g., Euclidean plane or space) and a set $\mathcal{C}$ of constraints among the objects in $\mathcal{O}$. The problem of diagram generation is to decide whether the objects in $\mathcal{O}$ can be placed in the space such that the constraints in $\mathcal{C}$ are satisfied; if so, construct, for any given assignment of values to the parameters, a diagram that satisfies the constraints. We may add some inequality constraints to rule out degenerate cases in which diagrams cannot be properly constructed. For dynamic animation, parametric values are given continuously and diagrams are constructed accordingly and their motion may be shown on the screen.

Denote by $x_1, \ldots, x_n$ the coordinates of points and other geometric entities involved in the objects of $\mathcal{O}$; then the set $\mathcal{C}$ of constraints together with the conditions to exclude some degenerate cases may be expressed as a semi-algebraic system of the form (1). The GCS problem is then reduced to solving this semi-algebraic system.

Sometimes, constraints may be specified as quantified formulae. In this case, we may eliminate the quantifiers to obtain quantifier-free formulae using known methods such as PCAD (partial cylindrical algebraic decomposition) [2]. So we can assume that the GCS problem under consideration may be formulated algebraically in the form (1).

By the method presented in the previous two sections, we can decompose the semi-algebraic system (1) into finitely many weak SRRs

$$
\begin{aligned}
\Gamma^{(i)}(u^{(i)}), \quad y_1^{(i)} &= h_1^{(i)}(u^{(i)}), \\
y_2^{(i)} &= h_2^{(i)}(u^{(i)}, y_1^{(i)}), \\
&\cdots \\
y_{r_i}^{(i)} &= h_{r_i}^{(i)}(u^{(i)}, y_1^{(i)}, \ldots, y_{r-1}^{(i)}),
\end{aligned}
\tag{13}
$$

where $u^{(i)}, y_1^{(i)}, \ldots, y_{r_i}^{(i)}$ is a permutation of $x_1, \ldots, x_n$ for each i. System (1) has a real solution if and only if there exist an i and a set $\bar{u}^{(i)}$ of real values of $u^{(i)}$ such that $\Gamma^{(i)}(\bar{u}^{(i)})$ holds true. In case there is no parameter, we will either end up with the conclusion that (1) has no real solution, or find the radical expressions of all the real solutions, yielding finitely many diagrams.

If there are infinitely many real values $\bar{u}$ of u such that $\Gamma_i(\bar{u}_i)$ holds, then the diagrams are dynamic. In this case, from each weak SRR and the identification of parameters and dependents we can determine which points in the geometric objects are free, semi-free, or dependent points. If the GCS problem is well formulated, then all the SRRs should have the same set of parameters. We may assume that this is the case.

To generate a dynamic diagram, we implement the weak SRRs into the drawing program. For initialization, a random set $\bar{u}$ of real values for u is chosen so that some $\Gamma_i(\bar{u}_i)$ holds. Then the real values of u (corresponding to the free or semi-free points) change continuously by the user, for example, using mouse dragging. For any chosen values $\bar{u}$ of u, the program verifies whether some $\Gamma_i(\bar{u}_i)$

holds. If no such Γ_i exists, then the values $\bar{u}$ are not allowed and the diagram remains unmoved. Otherwise, $\Gamma_i(\bar{u}_i)$ holds true for some i. In the case, the corresponding values for the dependent variables

$$\bar{y}_1^{(i)} = h_1^{(i)}(\bar{u}^{(i)}),\ \bar{y}_2^{(i)} = h_2^{(i)}(\bar{u}^{(i)}, \bar{y}_1^{(i)}),\ \ldots,\ \bar{y}_{r_i}^{(i)} = h_{r_i}^{(i)}(\bar{u}^{(i)}, \bar{y}_1^{(i)}, \ldots, \bar{y}_{r-1}^{(i)})$$

are computed. The diagram is then redrawn according to these values that determine the new locations of the geometric objects in the diagram.

Our approach for automated generation of dynamic diagrams (in Euclidean plane) involving inequalities constraints may be sketched as follows.

1. Assign coordinates to the points involved in the geometric objects and introduce other variables if necessary, so that the geometric constraints are expressed as a semi-algebraic system of equalities and inequalities of the form in (1).
2. Decompose the system (1) into finitely many weak SRRs of the form (13).
3. Determine the free and semi-free points according to the identification of the variables $x_1, \ldots, x_n$ into parameters and dependents.
4. Randomly choose a set of real numerical values for the parametric variables satisfying some $\Gamma^{(i)}$ and compute the values of the dependent variables from the corresponding weak SRR.
5. Check whether all the points are within the window range and no two of them are too close. If not, then go back to step 4.
6. Draw the geometric objects and label the points.

The animation of the drawn diagram may be implemented by the following two additional steps.

7. Update the value(s) of the free coordinate(s) of the free or semi-free point being moved with mouse dragging and recompute the values of the dependent variables from the corresponding weak SRR.
8. Redraw the geometric objects and relabel the points.

5   Examples

The approach presented in the preceding section has been implemented in Java with interface to the Epsilon library [12] in Maple and the QEPCAD program [1] in C for our experiments. The implementation is still very primitive, so we do not discuss any detail here.

In what follows we use two examples to illustrate our approach and its preliminary implementation.

5.1   Steiner Problem

Given an arbitrary triangle ABC, draw three equilateral triangles ABC_1, ACB_1, and BCA_1 all outward or all inward. The well-known Steiner theorem claims that the three lines AA_1, BB_1, and CC_1 are concurrent. This theorem may be proved and so may a dynamic diagram be generated automatically on computer, for example, by using GEOTHER [11]. However, for the publicly available version of GEOTHER there is no way to ensure that the triangles drawn in the diagram are outward or inward. Now by using the method described in this paper, we can ensure that the triangles are drawn all outward.

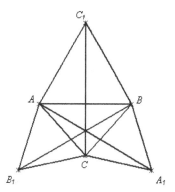

Fig. 1. Steiner problem

Without loss of generality, let the coordinates of the points be assigned as follows:

$$A(0,0),\ B(1,0),\ C(u_1, u_2),\ C_1(x_1, x_2),\ B_1(x_3, x_4),\ A_1(x_5, x_6).$$

Then the geometric constraints may be expressed as the following equalities and inequalities

$$
\begin{cases}
F_1 = 2\,x_1 - 1 = 0, & (|AC_1| = |BC_1|) \\
F_2 = x_2^2 + x_1^2 - 1 = 0, & (|AC_1| = |AB|) \\
F_3 = x_4^2 + x_3^2 - u_2^2 - u_1^2 = 0, & (|AB_1| = |AC|) \\
F_4 = 2\,u_2 x_4 + 2\,u_1 x_3 - u_2^2 - u_1^2 = 0, & (|AB_1| = |CB_1|) \\
F_5 = x_6^2 + x_5^2 - 2\,x_5 - u_2^2 - u_1^2 + 2\,u_1 = 0, & (|BA_1| = |BC|) \\
F_6 = 2\,u_2 x_6 + 2\,u_1 x_5 - 2\,x_5 - u_1^2 - u_2^2 + 1 = 0, & (|BA_1| = |CA_1|) \\
G_1 = u_1 u_2 x_6 - u_2 x_6 - u_2^2 x_5 + u_2^2 < 0, & (BCA_1 \text{ outward}) \\
G_2 = -u_2\, u_1 x_4 + u_2^2 x_3 < 0, & (ACB_1 \text{ outward}) \\
G_3 = u_2 x_2 < 0. & (ABC_1 \text{ outward})
\end{cases}
$$

Assume that C is a free point, so that u_1, u_2 are free parameters. The set of polynomials $F_1, \ldots, F_6$ may be decomposed over $\mathbb{Q}(u_1, u_2)$ into four triangular sets. However, three of them turn out to be inconsistent with the inequality constraints. So we get only one triangular set $[T_1, \ldots, T_6]$, where

$$T_1 = 2\,x_1 - 1,$$
$$T_2 = 4\,x_2^2 - 3,$$
$$T_3 = 2\,x_3 - 2\,u_2 x_2 - u_1,$$

$$T_4 = 2\,x_4 + 2\,u_1\,x_2 - u_2,$$
$$T_5 = 2\,x_5 + 2\,u_2\,x_2 - u_1 - 1,$$
$$T_6 = 2\,x_6 - 2\,u_1\,x_2 + 2\,x_2 - u_2.$$

From the triangular set, we can obtain the following weak SRRs:

$$u_2 < 0, \quad x_1 = \frac{1}{2},\ x_2 = \frac{\sqrt{3}}{2},\ x_3 = \frac{1}{2}\,u_1 + u_2\,x_2,\ x_4 = \frac{1}{2}\,u_2 - u_1\,x_2,$$

$$x_5 = \frac{1}{2}\,u_1 - u_2\,x_2 + \frac{1}{2},\ x_6 = \frac{1}{2}\,u_2 - x_2 + u_1\,x_2;$$

$$u_2 > 0, \quad x_1 = \frac{1}{2},\ x_2 = -\frac{\sqrt{3}}{2},\ x_3 = \frac{1}{2}\,u_1 + u_2\,x_2,\ x_4 = \frac{1}{2}\,u_2 - u_1\,x_2,$$

$$x_5 = \frac{1}{2}\,u_1 - u_2\,x_2 + \frac{1}{2},\ x_6 = \frac{1}{2}\,u_2 - x_2 + u_1\,x_2.$$

With these representations, the dynamic diagram as shown in Fig. 1 can be drawn and animated efficiently.

5.2   Apollonius Circle Problem

The Apollonius circle problem is a classical GCS problem formulated by Apollonius of Perga in the third century B.C. It has been investigated in the recent literature [7, 8]. It comes in three different versions: points, lines, or circles. The most interesting and challenging one is about circles. It asks to construct a circle that touches three given circles (see Fig. 2). It has applications in geometric modeling, biochemistry, and pharmacology. What is of real interest for such applications is the case of *external* contact of the circles.

If the coordinates of the points are chosen as

$$O_1(0,0),\ T_1(1,0),\ O_2(0,1),\ T_2(u_4, x_2),\ O_3(u_6, x_4),\ T_3(u_5, x_3),\ O(x_1, 0),$$

then the constraints for the four circles to be tangent externally may be expressed algebraically as

$$
\begin{cases}
F_1 = x_1 x_2 - x_1 + u_4 = 0, & (O, O_2, T_2 \text{ collinear}) \\
F_2 = -x_2^2 + 2\,u_4 x_1 - 2\,x_1 - u_4^2 + 1 = 0, & (T_1, T_2 \text{ on } O) \\
F_3 = -x_3^2 + 2\,u_5 x_1 - 2\,x_1 - u_5^2 + 1 = 0, & (T_1, T_3 \text{ on } O) \\
F_4 = -x_1 x_4 + u_5 x_4 + x_1 x_3 - u_6 x_3 = 0, & (O, O_3, T_3 \text{ collinear}) \\
G_1 = (\exists \lambda_1)\,[\lambda_1 > 0 \wedge \lambda_1 < 1 \wedge \lambda_1 x_1 = 1], & (O, O_1 \text{ ex-tangent}) \\
G_2 = (\exists \lambda_2)\,[\lambda_2 > 0 \wedge \lambda_2 < 1 \wedge \lambda_2 x_1 = u_4 & (O, O_2 \text{ ex-tangent}) \\
\qquad \wedge -\lambda_2 + 1 = x_2], & \\
G_3 = (\exists \lambda_3)\,[\lambda_3 > 0 \wedge \lambda_3 < 1 \wedge (1 - \lambda_3)\,x_4 = x_3 & (O, O_3 \text{ ex-tangent}) \\
\qquad \wedge \lambda_3 x_1 + (1 - \lambda_3)\,u_6 = u_5].
\end{cases}
$$

We first eliminate the quantified variables $\lambda_1, \lambda_2, \lambda_3$ to get a semi-algebraic system and then compute an irreducible triangular decomposition of $\{F_1, \ldots, F_4\}$

over $\mathbb{Q}(u_4, u_5, u_6)$ with respect to variable ordering $x_1 \prec \cdots \prec x_4$. The decomposition consists of only one triangular set $[T_1, \ldots, T_4]$ with

$$T_1 = -2\,u_4 x_1^3 + 2\,x_1^3 + u_4^2 x_1^2 - 2\,u_4 x_1 + u_4^2,$$
$$T_2 = x_1 x_2 - x_1 + u_4,$$
$$T_3 = x_3^2 - 2\,u_5 x_1 + 2\,x_1 + u_5^2 - 1,$$
$$T_4 = x_1 x_4 - u_5 x_4 - x_1 x_3 + u_6 x_3.$$

From this triangular set and the inequality constraints six weak SRRs were

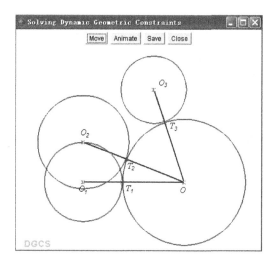

Fig. 2. Dynamic diagram for the Apollonius problem

computed by our program. We list only two of them as follows:

$$\Delta_1(u_4, u_5, u_6), \quad x_1 = X_{11}, \ x_2 = X_2, \ x_3 = X_3, \ x_4 = X_4;$$
$$\Delta_2(u_4, u_5, u_6), \quad x_1 = X_{12}, \ x_2 = X_2, \ x_3 = X_3, \ x_4 = X_4,$$

where

$\Delta_1(u_4, u_5, u_6):\ u_4^3 + 2\,u_4^2 + 11\,u_4 - 16 = 0 \wedge u_5 - 1 \geqslant 0 \wedge [[2\,u_4 u_5^3 - 2\,u_5^3 - u_4^2$
$\quad u_5^2 + 2\,u_4 u_5 - u_4^2 = 0 \wedge 6\,u_4 u_5 - 6\,u_5 - u_4^2 > 0 \wedge u_6 - u_5 = 0] \vee$
$\quad [3\,u_4 u_5^2 - 3\,u_5^2 - 2\,u_4^2 u_5 + 6\,u_4 u_5 - 6\,u_5 - 2\,u_4^2 + 7\,u_4 - 3 \leqslant 0 \wedge$
$\quad 3\,u_4 u_5^2 - 3\,u_5^2 - u_4^2 u_5 + u_4 > 0 \wedge u_6 - u_5 > 0] \vee [3\,u_4 u_5^2 - 3\,u_5^2 -$
$\quad u_4^2 u_5 + u_4 < 0 \wedge u_6 - u_5 < 0] \vee [6\,u_4 u_5 - 6\,u_5 - u_4^2 < 0 \wedge u_6 -$
$\quad u_5 < 0]] \wedge u_4 > 0;$

$\Delta_2(u_4, u_5, u_6):$ $u_4^3 + 2\,u_4^2 + 11\,u_4 - 16 < 0 \wedge u_4 - 1 > 0 \wedge u_5 - 1 \geqslant 0 \wedge [[6\,u_4u_5$

$- 6\,u_5 - u_4^2 > 0 \wedge 3\,u_4u_5^2 - 3\,u_5^2 - u_4^2u_5 + u_4 > 0 \wedge 2\,u_4u_5^3 - 2\,u_5^3$

$- u_4^2u_5^2 + 2\,u_4u_5 - u_4^2 = 0 \wedge u_6 - u_5 = 0] \vee [3\,u_4u_5^2 - 3\,u_5^2 - u_4^2$

$u_5 + u_4 < 0 \wedge u_6 - u_5 < 0] \vee [2\,u_4u_5^3 - 2\,u_5^3 - u_4^2u_5^2 + 2\,u_4u_5 -$

$u_4^2 > 0 \wedge 3\,u_4u_5^2 - 3\,u_5^2 - u_4^2u_5 + u_4 > 0 \wedge 3\,u_4u_5^2 - 3\,u_5^2 - 2\,u_4^2u_5$

$+ 6\,u_4u_5 - 6\,u_5 - 2\,u_4^2 + 7\,u_4 - 3 < 0 \wedge u_6 - u_5 > 0] \vee [2\,u_4u_5^3 -$

$2\,u_5^3 - u_4^2u_5^2 + 2\,u_4u_5 - u_4^2 > 0 \wedge 6\,u_4u_5 - 6\,u_5 - u_4^2 > 0 \wedge u_4u_5^3 -$

$u_5^3 - u_4^2u_5^2 + 3\,u_4u_5^2 - 3\,u_5^2 - 2\,u_4^2u_5 + 7\,u_4u_5 - 3\,u_5 - 5\,u_4^2 + 5\,u_4$

$- 1 \leqslant 0 \wedge u_6 - u_5 > 0] \vee [6\,u_4u_5 - 6\,u_5 - u_4^2 < 0 \wedge u_6 - u_5 < 0]$

$\vee [2u_4u_5^3 - 2u_5^3 - u_4^2u_5^2 + 2u_4u_5 - u_4^2 < 0 \wedge u_6 - u_5 < 0]]$

and

$$X_{11} = \frac{\sqrt[\text{R}\,3]{\delta}}{6\,(u_4 - 1)} + \frac{u_4^4 - 12\,u_4^2 + 12\,u_4}{6\,(u_4 - 1)\,\sqrt[\text{R}\,3]{\delta}} + \frac{u_4^2}{6\,(u_4 - 1)},$$

$$X_{12} = \frac{\sqrt[\text{I}\,3]{\delta}}{6\,(u_4 - 1)} + \frac{u_4^4 - 12\,u_4^2 + 12\,u_4}{6\,(u_4 - 1)\,\sqrt[\text{I}\,3]{\delta}} + \frac{u_4^2}{6\,(u_4 - 1)},$$

$$X_2 = \frac{x_1 - u_4}{x_1},$$

$$X_3 = -\sqrt{-2\,u_5x_1 + 2\,x_1 + u_5^2 - 1},$$

$$X_4 = \frac{x_1x_3 - u_6x_3}{x_1 - u_5};$$

$$\delta = d + 6\,\sqrt{3D}\,u_4(u_4 - 1),$$

$$d = u_4^6 + 36\,u_4^4 - 90\,u_4^3 + 54\,u_4^2,$$

$$D = u_4^6 + 8\,u_4^4 - 36\,u_4^3 + 43\,u_4^2 - 16\,u_4.$$

Figure 2 is a window snapshot that shows a dynamic diagram for the Apollonius problem generated automatically from the above weak SRRs.

6   Summary and Discussion

In this paper we have presented an approach for dynamically solving geometric constraints involving inequalities. It consists of two stages: *preprocessing* and *updating*. During the preprocessing stage (which is carried out only once), we compute, *symbolically*, explicit representations (radical expressions in parameters) of the solutions of the semi-algebraic system representing the geometric constraints. During the updating stage (which is carried out repeatedly), we evaluate, *numerically*, the radical expressions.

Once the preprocessing has been done, the repeated updating can be carried out reliably and efficiently since it only involves evaluating radical expressions, yielding correct and dynamic (real-time) animation.

As expected, the preprocessing, though carried out only once, can be very time-consuming when the involved polynomials are of high degree with many parameters. For example, we have applied our method to the well-known Morley trisector theorem without success. An algebraic formulation of the problem consists of 6 equations and 9 inequalities and the SRRs could not be computed within one hour on a laptop PC. Hence, an important challenge for future work is to improve the speed of the preprocessing stage, in particular, real quantifier elimination. It is well known that general real quantifier elimination is intrinsically difficult. However, it seems that the formulae arising in the context of geometric constraints are not arbitrary, but have certain special structures. Therefore, one should investigate how to utilize those structures in order to develop specialized and thus more efficient real quantifier elimination methods.

References

1. Brown, C. W., Hong, H.: QEPCAD — Quantifier elimination by partial cylindrical algebraic decomposition. http://www.cs.usna.edu/~qepcad/B/QEPCAD.html (2004).
2. Collins, G. E., Hong, H.: Partial cylindrical algebraic decomposition for quantifier elimination. *J. Symb. Comput.* **12**: 299–328 (1991).
3. Dolzmann, A. Sturm, T., Weispfenning, V.: A new approach for automatic theorem proving in real geometry. *J. Automat. Reason.* **21**/3: 357–380 (1998).
4. González-Vega, L.: A combinatorial algorithm solving some quantifier elimination problems. In: *Quantifier Elimination and Cylindrical Algebraic Decomposition* (Caviness, B., Johnson, J., eds.), pp. 300–316. Springer-Verlag, Wien New York (1996).
5. Hong, H.: Quantifier elimination for formulas constrained by quadratic equations via slope resultants. *The Computer J.* **36**/5: 440–449 (1993).
6. Joan-Arinyo, R., Hoffmann, C. M.: A brief on constraint solving. http://www.cs.purdue.edu/homes/cmh/distribution/papers/Constraints/ThailandFull.pdf (2005).
7. Kim, D., Kim, D.-S., Sugihara, K.: Apollonius tenth problem via radius adjustment and Möbius transformations. *Computer-Aided Design* **38**/1: 14–21 (2006).
8. Lewis, R. H., Bridgett, S.: Conic tangency equations and Apollonius problems in biochemistry and pharmacology. *Math. Comput. Simul.* **61**/2: 101–114 (2003).
9. Wang, D.: *Elimination Methods.* Springer-Verlag, Wien New York (2001).
10. Wang, D.: Automated generation of diagrams with Maple and Java. In: *Algebra, Geometry, and Software Systems* (Joswig, M., Takayama, N., eds.), pp. 277–287. Springer-Verlag, Berlin Heidelberg (2003).
11. Wang, D.: GEOTHER 1.1: Handling and proving geometric theorems automatically. In: *Automated Deduction in Geometry* (Winkler, F., ed.), LNAI 2930, pp. 194–215. Springer-Verlag, Berlin Heidelberg (2004).
12. Wang, D.: *Elimination Practice: Software Tools and Applications.* Imperial College Press, London (2004).
13. Weispfenning, V.: Quantifier elimination for real algebra — the cubic case. In: *Proceedings of the 1994 International Symposium on Symbolic and Algebraic Computation* (Oxford, UK, July 20–22, 1994), pp. 258–263. ACM Press, New York (1994).

Constraints for Continuous Reachability in the Verification of Hybrid Systems[*]

Stefan Ratschan[1] and Zhikun She[2]

[1] Institute of Computer Science, Czech Academy of Sciences, Prague, Czech Republic
[2] Max-Planck-Institut für Informatik, Saarbrücken, Germany

Abstract. The method for verification of hybrid systems by constraint propagation based abstraction refinement that we introduced in an earlier paper is based on an over-approximation of continuous reachability information of ordinary differential equations using constraints that do not contain differentiation symbols. The method uses an interval constraint propagation based solver to solve these constraints. This has the advantage that—without complicated algorithmic changes—the method can be improved by just changing these constraints. In this paper, we discuss various possibilities of such changes, we prove some properties about the amount of over-approximations introduced by the new constraints, and provide some timings that document the resulting improvement.

1 Introduction

A hybrid system is a dynamical system that involves both continuous and discrete state and evolution. This can, for example, be used for modeling the behavior of an embedded (digital) computing device influencing its (continuous) environment. An important task is to verify that a given hybrid system is safe, that is, every trajectory starting from an initial state never reaches an unsafe state.

In this paper we study constraints that can be used for modeling the continuous flow in the safety verification of hybrid systems by constraint propagation based abstraction refinement [16, 14]. Especially, we exploit the fact that the underlying solver, which is based on interval constraint propagation as introduced within the field of artificial intelligence [5, 3, 9, 13], allows the use of a rich language of constraints that includes function symbols such as sin, cos, exp. These symbols arise naturally as solutions of linear differential equations.

More specifically, in this paper we study two types of constraints that model the reachability problem: one, for linear differential equations, is related to the explicit solution of such equations; and the second, for general differential constraints, is based on the mean-value theorem. Both constraints are quite simple

---

[*] This work was partly supported by the German Research Council (DFG) as part of the Transregional Collaborative Research Center "Automatic Verification and Analysis of Complex Systems" (SFB/TR 14 AVACS). See www.avacs.org for more information.

J. Calmet, T. Ida, and D. Wang (Eds.): AISC 2006, LNAI 4120, pp. 196–210, 2006.

to derive and similar ones have been used before in the literature. However, here we study in detail, how they behave when using interval based constraint propagation techniques for solving them, and how this behavior reflects in our method of verification of hybrid systems using constraint propagation based abstraction refinement. Our verification software HSOLVER [15], allows the user to experiment with these, and additional, user-defined, constraints.

Regarding additional related work, the approach by Hickey and Wittenberg [7] puts the level of modelling even higher, by employing a constraint logic programming language [8] that directly can deal with differential equations. Internally it solves constraints by transforming them into polynomial constraints using Taylor expansion, and then solves these using a similar solver as ours. The approach does not provide a comparison of different formulations of these constraints, and does not employ abstraction refinement to concentrate on refining the solution which is relevant for a given safety verification problem.

Tiwari [19] derives simple polynomial constraints from linear differential equations by manually doing an over-approximating quantifier elimination on a similar constraint as one of the constraints employed here. In contrast to that, since our solver can handle function symbols such as sin, cos, and exp, we can directly work on the original constraint, and—using an abstraction refinement scheme—approximate it arbitrarily closely.

Anai and Weispfenning [1] provide a classification of the cases when the time variable can be symbolically eliminated from the solution of linear differential equations (which may contain transcendental function symbols).

Similar constraints as employed here, which are based on the mean-value theorem or Taylor expansion, are ubiquitous in the integration of ODE's.

The content of the paper is as follows: In Section 2 we review our method of verification using abstraction refinement; in Section 3 we discuss constraints for modeling reachability of differential equations; in Section 4 we discuss how we solve these constraints; in Section 5 we compare the constraints theoretically; in Section 6 we study empirically, how the constraints behave within our method; and in Section 7 we conclude the paper.

2   Constraint Propagation Based Abstraction Refinement

In this section, we review our previous approach [16, 14] for verifying safety using constraint propagation based abstraction refinement. We fix a variable s ranging over a finite set of discrete modes $\{s_1, \ldots, s_n\}$ and variables $x_1, \ldots, x_k$ ranging over closed real intervals $I_1, \ldots, I_k$. We denote by Φ the resulting state space $\{s_1, \ldots, s_n\} \times I_1 \times \cdots \times I_k$. In addition, for denoting the derivatives of $x_1, \ldots, x_k$ we assume variables $\dot{x}_1, \ldots, \dot{x}_k$, ranging over $\mathbb{R}$ each, and for denoting the targets of jumps, variables $s', x'_1, \ldots, x'_k$ ranging over $\{s_1, \ldots, s_n\}$ and $I_1, \ldots, I_k$, correspondingly. In the following we will use boldface to denote vectors of variables.

For describing hybrid systems, we use a flow constraint $Flow(s, \boldsymbol{x}, \dot{\boldsymbol{x}})$, a jump constraint $Jump(s, \boldsymbol{x}, s', \boldsymbol{x}')$, an initial constraint $Init(s, \boldsymbol{x})$ and an unsafety constraint $UnSafe(s, \boldsymbol{x})$. Now, assuming a given hybrid system $H = (Flow, Jump,$

$Init, Unsafe)$, our safety verification problem is to verify that H is safe, that is, there is no piecewise continuous trajectory of H that starts from an initial state (i.e., a state satisfying the constraint $Init(s, \boldsymbol{x})$ and reaches an unsafe state (i.e., a state satisfying the constraint $UnSafe(s, \boldsymbol{x})$). Here the continuous parts of the trajectory are required to fulfill the flow constraint $Flow(s, \boldsymbol{x}, \dot{\boldsymbol{x}})$, and the discontinuous jumps are required to fulfill $Jump(s, \boldsymbol{x}, s', \boldsymbol{x}')$, where $(s, \boldsymbol{x})$ refers to the state before the jump and $(s', \boldsymbol{x}')$ refers to the state directly after the jump. For formal definitions refer to our previous publication [16].

Our approach decomposes the state space Φ using hyper-rectangles (*boxes*) into finitely many mode/box pairs and then computes a finite over-approximation of the hybrid systems (an *abstraction*). In detail, for each pair of boxes, it sets an abstract transition, only if it cannot prove the absence of trajectories between them; also, it marks boxes as initial (unsafe), if it cannot prove the absence of an initial (unsafe) element in the box. If the resulting finite abstraction is safe, the hybrid system is also safe, since the abstraction over-approximates the hybrid system. If it is not safe, we refine the abstraction by splitting boxes into pieces and recomputing the affected information in the abstraction.

Moreover, we have a mechanism for removing unreachable elements from boxes. For this observe that a point in a box B is reachable only if it is reachable either from the initial set via a flow in B, from a jump via a flow in B, or from a neighboring box via a flow in B. So we formulate constraints corresponding to each of these conditions and then remove points from boxes that do not fulfill at least one of these constraints.

The approach can be used with any constraint describing that $\boldsymbol{y}$ can be reachable from $\boldsymbol{x}$ via a flow in B and mode s, for example, the one introduced in our previous publication and the new ones that will be introduced in the latter sections. We denote the used constraint by $Reachable_B(s, \boldsymbol{x}, \boldsymbol{y})$.

Thus the above three possibilities of reachability allow us to formulate the following theorem:

Theorem 1. *For a set of abstract states $\mathcal{B}$ such that all boxes corresponding to the same mode are non-overlapping, a pair $(s', B') \in \mathcal{B}$ and a point $z \in B'$, if $(s', \boldsymbol{z})$ is reachable, then*

$$initflow_{B'}(s', \boldsymbol{z}) \vee \bigvee_{(s,B) \in \mathcal{B}} jumpflow_{B, B'}(s, s', \boldsymbol{z})$$

$$\vee \bigvee_{(s,B) \in \mathcal{B}, s = s', B \neq B'} boundaryflow_{B, B'}(s', \boldsymbol{z})$$

where $initflow_{B'}(s', \boldsymbol{z})$, $jumpflow_{B, B'}(s, s', \boldsymbol{z})$ and $boundaryflow_{B, B'}(s', \boldsymbol{z})$ denote the following three constraints, respectively:

- $\exists \boldsymbol{y} \in B' \, [Init(s', \boldsymbol{y}) \wedge Reachable_{B'}(s', \boldsymbol{y}, \boldsymbol{z})]$,
- $\exists \boldsymbol{x} \in B \exists \boldsymbol{x}' \in B' \, [Jump(s, \boldsymbol{x}, s', \boldsymbol{x}') \wedge Reachable_{B'}(s', \boldsymbol{x}', \boldsymbol{z})]$,
- $\exists \boldsymbol{x} \in B \cap B' \, [[\forall faces \, F \, of \, B[\boldsymbol{x} \in F \Rightarrow out^F_{s', B, B'}(\boldsymbol{x})]] \wedge Reachable_{B'}(s', \boldsymbol{x}, \boldsymbol{z})]$.

Here, $out^F_{s', B, B'}(\boldsymbol{x})$ is one of the following constraints:

- $\exists \dot{x}_j [Flow(s', \boldsymbol{x}, (\dot{x}_1, \ldots, \dot{x}_k)) \wedge \dot{x}_j \leq 0]$, if F is the j-th *lower face* of B, and
- $\exists \dot{x}_j [Flow(s', \boldsymbol{x}, (\dot{x}_1, \ldots, \dot{x}_k)) \wedge \dot{x}_j \geq 0]$, if F is the j-th *upper face* of B.

We denote the main constraint (i.e., the disjunction) by $reachable_{B'}(s', \boldsymbol{z})$. If we can prove that a certain point does not fulfill this constraint, we know that it is not reachable. In Section 4 we describe pruning algorithm that takes such a constraint and an abstract state (s', B'), and returns a sub-box of B' that still contains all the solutions of the constraint in B'.

Since the constraint $reachable_{B'}(s', \boldsymbol{z})$ depends on all current abstract states, a change of B' might allow further pruning of other abstract states. So we can repeat pruning until a fixpoint is reached. This terminates since we use floating point computation here and there are only finitely many floating point numbers.

We remove the initial mark of an abstract state (s', B') if we can disprove $initflow_{B'}(s', \boldsymbol{z})$ in Theorem 1, and we remove the unsafe mark of an abstract state (s', B') if we can disprove $\exists \boldsymbol{x} \in B' \, UnSafe(s', \boldsymbol{x})$. Moreover, we remove a transition from (s, B) to (s', B') if we can disprove both $boundaryflow_{B,B'}(s', \boldsymbol{z})$ and $jumpflow_{B,B'}(s, s', \boldsymbol{z})$ from Theorem 1. It is easy to show that the resulting system is an abstraction of the original hybrid system.

3   Constraints for Reachability

Assume that the flow constraint contains a differential equation of the form $\dot{\boldsymbol{x}} = A\boldsymbol{x}$, where $A \in \mathbb{Q}^{k \times k}$. Differential equations of the form $\dot{\boldsymbol{x}} = A\boldsymbol{x} + B$ can be reduced to that form by shifting the equilibrium, provided that the equation $A\boldsymbol{x} + B = 0$ has solutions. Given an initial set $Init$, we have the exact solution $\boldsymbol{x}(t) = e^{At}\boldsymbol{x}_0$, where $\boldsymbol{x}_0 \in Init$ and e^{At} is defined by $\sum_{k=0}^{\infty} \frac{t^k}{k!} A^k$. Thus, if $\boldsymbol{x}$ is reachable, then the constraint $\exists t \in \mathbb{R}_{\geq 0} \exists \boldsymbol{x}_0 \in \mathbb{R}^k [\boldsymbol{x}_0 \in Init \wedge \boldsymbol{x} = e^{At}\boldsymbol{x}_0]$ holds.

Since the matrix A appears in an exponent, it is difficult to directly solve this constraint. We use another constraint introduced by Tiwari [19], that over-approximates the reach set, that can be easily computed from the matrix A, and that does not contain matrix exponentiation. For this we re-express the real and complex eigenvalues of A^T (the transpose of A) using the following two sets:

$$\Lambda_1 = \left\{ \lambda \in \mathbb{R} \mid \exists c \in \mathbb{R}^k \left[c \neq 0 \wedge A^T c = \lambda c \right] \right\};$$

$$\Lambda_2 = \left\{ (a, b) \in \mathbb{R} \times \mathbb{R}_{>0} \mid \exists c \in \mathbb{R}^k \left[c \neq 0 \wedge ((A^T)^2 - 2aA^T + (a^2 + b^2)I)c = 0 \right] \right\}.$$

For every $\lambda \in \Lambda_1$, let $c(1, \lambda)$ be an orthonormal basis of $\{c : A^T c = \lambda c\}$; for every $(a, b) \in \Lambda_2$, let $c(2, (a, b))$ be an orthonormal basis of $\{c : ((A^T)^2 - 2aA^T + (a^2 + b^2)I)c = 0\}$. Then we can describe an over-approximation of the set of reachable states as follows.

Lemma 1. *For a differential equation* $\dot{\boldsymbol{x}} = A\boldsymbol{x}$ *and a box* $B \subseteq \mathbb{R}^k$, *if there is a trajectory in* B *from a point* $\boldsymbol{x} = (x_1, \ldots, x_k)^T \in B$ *to a point* $\boldsymbol{y} = (y_1, \ldots, y_k)^T \in B$ *on which* $\dot{\boldsymbol{x}} = A\boldsymbol{x}$ *holds, then*

$$\exists t \in \mathbb{R}_{\geq 0}[eigen^*_{A,B}(t, \boldsymbol{x}, \boldsymbol{y})], \tag{1}$$

where eigen$^*_{A,B}(t, \boldsymbol{x}, \boldsymbol{y})$ *denotes*

$$\left[\bigwedge_{\lambda \in \Lambda_1} \left[\bigwedge_{c \in c(1,\lambda)} c^T \boldsymbol{y} = c^T \boldsymbol{x} e^{\lambda t} \right] \right] \wedge \left[\bigwedge_{(a,b) \in \Lambda_2} \exists \dot{\boldsymbol{x}} \in \mathbb{R}^k \right.$$
$$\left. \left[\dot{\boldsymbol{x}} = A\boldsymbol{x} \wedge \bigwedge_{c \in c(2,(a,b))} c^T \boldsymbol{y} = e^{at} c^T \left(\boldsymbol{x} \cos(bt) + \frac{\dot{\boldsymbol{x}} - \boldsymbol{x}a}{b} \sin(bt) \right) \right] \right]$$

This expression is a formula in the first-order predicate language over the real numbers—it does *not* contain higher-order expressions such as derivatives ($\dot{\boldsymbol{x}}$ does not denote the derivative of $\boldsymbol{x}$ but simply a new variable). However, the corresponding restriction to the first-order theory of the reals is still undecidable, since we have to deal with function symbols like sin. However, there are over-approximating constraint solvers, that can be used (see Section 4).

We denote the above Constraint 1 by $eigen_{A,B}(\boldsymbol{x}, \boldsymbol{y})$. Note that if A has k different real eigenvalues and $B = \mathbb{R}^k$, then this constraint describes the exact solution of the differential equation.

Now we will describe a constraint describing the reachability not only for linear differential equations, but for much more general descriptions of continuous evolution. We assume that the continuous dynamics is defined by a differential constraint $D(\boldsymbol{x}, \dot{\boldsymbol{x}})$ (or short, D) which can be an arbitrary first-order formula in the theory of the reals over the tuples of variables $\boldsymbol{x}$ and $\dot{\boldsymbol{x}}$. This includes explicit and implicit differential equations and inequalities, and even differential-algebraic equations and inequalities.

Earlier [16] we used a constraint $flow_{D,B}(\boldsymbol{x}, \boldsymbol{y})$ describing the reachability in boxes as follows:

Lemma 2. *For a differential constraint $D(\boldsymbol{x}, \dot{\boldsymbol{x}})$ and a box $B \subseteq \mathbb{R}^k$, if there is a trajectory in B from a point $\boldsymbol{x} = (x_1, \ldots, x_k)^T \in B$ to a point $\boldsymbol{y} = (y_1, \ldots, y_k)^T \in B$ such that for every point $\boldsymbol{u}$ on the trajectory and its derivative $\dot{\boldsymbol{u}}$, the pair $(\boldsymbol{u}, \dot{\boldsymbol{u}})$ satisfies $D(\boldsymbol{x}, \dot{\boldsymbol{x}})$, then*

$$\bigwedge_{1 \le m < n \le k} \exists a_1, \ldots, a_k, \dot{a}_1, \ldots, \dot{a}_k [(a_1, \ldots, a_k) \in B \wedge$$
$$D((a_1, \ldots, a_k), (\dot{a}_1, \ldots, \dot{a}_k)) \wedge \dot{a}_n \cdot (y_m - x_m) = \dot{a}_m \cdot (y_n - x_n)]$$

Observe that whenever a given pair of points $(\boldsymbol{x}, \boldsymbol{y})$ fulfills the above constraint $flow_{D,B}(\boldsymbol{x}, \boldsymbol{y})$—indicating that there is a possible flow from $\boldsymbol{x}$ to $\boldsymbol{y}$— then also the flipped pair $(\boldsymbol{y}, \boldsymbol{x})$ fulfills the constraint. That is, the constraint does not distinguish time flowing forward, and time flowing backward. In order to avoid this loss of information we use the mean value theorem to formulate the following constraint:

Lemma 3. *For a differential constraint $D(\boldsymbol{x}, \dot{\boldsymbol{x}})$ and a box $B \subseteq \mathbb{R}^k$, if there is a trajectory in B from a point $\boldsymbol{x} = (x_1, \ldots, x_k)^T \in B$ to a point $\boldsymbol{y} = (y_1, \ldots, y_k)^T \in B$ such that for every point $\boldsymbol{u}$ on the trajectory and its derivative $\dot{\boldsymbol{u}}$, the pair $(\boldsymbol{u}, \dot{\boldsymbol{u}})$ satisfies $D(\boldsymbol{x}, \dot{\boldsymbol{x}})$, then*

$$\exists t \in \mathbb{R}_{\geq 0}[flow^*_{D,B}(t, \boldsymbol{x}, \boldsymbol{y})], \tag{2}$$

where $flow^*_{D,B}(t, \boldsymbol{x}, \boldsymbol{y})$ denotes

$$\bigwedge_{1 \leq i \leq k} \exists a_1, \ldots, a_k, \dot{a}_1, \ldots, \dot{a}_k[(a_1, \ldots, a_k) \in B$$

$$\wedge D((a_1, \ldots, a_k), (\dot{a}_1, \ldots, \dot{a}_k)) \wedge y_i = x_i + \dot{a}_i \cdot t]$$

We denote the above Constraint 2 by $flow'_{D,B}(\boldsymbol{x}, \boldsymbol{y})$. Although both the constraint $flow_{D,B}(\boldsymbol{x}, \boldsymbol{y})$ and the constraint $flow'_{D,B}(\boldsymbol{x}, \boldsymbol{y})$ are quite simple, it is not at all clear, whether $flow'_{D,B}(\boldsymbol{x}, \boldsymbol{y})$ really allows to derive tighter reach set information than $flow_{D,B}(\boldsymbol{x}, \boldsymbol{y})$.

4   Solving the Constraints

We solve the constraints using our constraint solver RSolver [12], which implements interval constraint propagation techniques [5, 3, 9, 13]. These techniques can, for a given constraint and intervals for all its variables, contract these intervals to smaller ones, without losing any solutions. We illustrate the idea on an example: Given the constraint $x^2 - 1 \leq 0$, and the interval $[-2, 2]$ for x, the method first decomposes this constraint into a conjunction of so-called primitive constraints, arriving at $x^2 = t_0 \wedge t_0 - 1 = t_1 \wedge t_1 \leq 0$. Here t_0 and t_1 are new, auxiliary variables. Then it takes the interval $[-\infty, +\infty]$ for all auxiliary variables and tries to contract all intervals wrt. the primitive constraints: using $x^2 = t_0$ we can contract the interval $[-\infty, \infty]$ for t_0 to $[0, 4]$, using $t_0 - 1 = t_1$ contract $[-\infty, \infty]$ for t_1 to $[-1, 3]$, using $t_1 \leq 0$ contract $[-1, 3]$ for t_1 to $[-1, 0]$, using $t_0 - 1 = t_1$ contract $[0, 4]$ for t_0 to $[0, 1]$, and using $x^2 = t_0$, contract $[-2, 2]$ for x to $[-1, 1]$. This process continues until a fixpoint is reached, which will always happen eventually, due to the finiteness of floating point numbers. We call the resulting algorithm a *pruning function* and, given a constraint ϕ and a box B, we denote the result of applying this function to ϕ and B by $Prune(\phi, B)$. In computer implementations the resulting intervals will enclose the solution sets of the primitive constraints up to rounding to the next floating point number. For the theoretical analysis in this paper we will ignore this rounding, and assume that these intervals are the tightest possible enclosures using real-number endpoints.

Definition 1. *Given a constraint ϕ, pruning is optimal for ϕ iff for all boxes B, $Prune(\phi, B)$ is the smallest box containing all solutions of ϕ in B.*

Although the pruning function will contract optimally for primitive constraints, this will in general not be the case for more complex constraints. However, due to a classical result of interval arithmetic, we have (ignoring floating-point rounding):

Property 1. For every constraint ϕ, that contains every variable just once, pruning is optimal.

There are special techniques for handling disjunctions and quantifiers [13]. Moreover, there are various optimizations and extensions of the techniques discussed above. Most of them spend additional time to deal with the problem that pruning is in general not optimal if some variables occur more than once. Since our constraints usually only have few occurrences of the same variables, and since in our abstraction refinement approach (see Section 2) we will do thousands of prunings, it does not seem promising to use such optimizations here.

5   Theoretical Evaluation of the Constraints

In this section we will do a theoretical comparison of the constraints $flow_{D,B}$ $(\boldsymbol{x}, \boldsymbol{y})$, $flow'_{D,B}(\boldsymbol{x}, \boldsymbol{y})$ and $eigen_{A,B}(\boldsymbol{x}, \boldsymbol{y})$. Moreover, based on the gained insight, we will introduce a new constraint that combines their advantages.

We start with comparing $flow'_{D,B}(\boldsymbol{x}, \boldsymbol{y})$ with $flow_{D,B}(\boldsymbol{x}, \boldsymbol{y})$. Let us first discuss the size of the produced constraints. For k dimensions, $flow_{D,B}(\boldsymbol{x}, \boldsymbol{y})$ has $k(k-1)/2$ conjuncts, whereas $flow'_{D,B}(\boldsymbol{x}, \boldsymbol{y})$ has just k. So, for dimensions larger than 2, $flow'_{D,B}(\boldsymbol{x}, \boldsymbol{y})$ is smaller, and its size increases only linearly instead of quadratically. Hence pruning will take less time on $flow'_{D,B}(\boldsymbol{x}, \boldsymbol{y})$, especially for high dimensions.

Let us now compare the effectiveness of the two constraints. For the one-dimensional case, $flow_{D,B}(\boldsymbol{x}, \boldsymbol{y})$ reduces to a conjunction with zero conjuncts, that is, to a constraint that is trivially true. So in that case, $flow'_{D,B}(\boldsymbol{x}, \boldsymbol{y})$ is definitely better. For higher dimensions, the relationship between the two constraints is more complicated. Therefore, we will first study the relationship between $flow_{D,B}(\boldsymbol{x}, \boldsymbol{y})$ and $flow'_{D,B}(\boldsymbol{x}, \boldsymbol{y})$ themselves, and then between the result of applying the pruning function to them.

Here we will use the following notation: Given two constraints $\phi_{S,B}(\boldsymbol{x}, \boldsymbol{y})$ and $\psi_{S,B}(\boldsymbol{x}, \boldsymbol{y})$, where S is a differential constraint D or a matrix A, and B is a box, we will write $\phi_{S,B} \preceq \psi_{S,B}$ ($\phi_{S,B} \equiv \psi_{S,B}$) iff for all S and all B, the solution set of $\phi_{S,B}$ in $B \times B$ is a subset of (equal to) the solution set of $\psi_{S,B}$ in $B \times B$. Analogously, we will write $\phi_{S,B} \preceq_P \psi_{S,B}$ ($\phi_{S,B} \equiv_P \psi_{S,B}$) iff for all S, all B and all B_0, where B_0 is a sub-box of B, $Prune(\phi_{S,B}(\boldsymbol{x}, \boldsymbol{y}), B_0 \times B)$ is a subset of (equal to) $Prune(\psi_{S,B}(\boldsymbol{x}, \boldsymbol{y}), B_0 \times B)$. Note that here we restrict B_0 to be a subset of B because we use the constraints always in such a context.

There is no clearcut relationship between $flow_{D,B}(\boldsymbol{x}, \boldsymbol{y})$ and $flow'_{D,B}(\boldsymbol{x}, \boldsymbol{y})$:

Property 2. Neither $flow_{D,B} \preceq flow'_{D,B}$ nor $flow'_{D,B} \preceq flow_{D,B}$.

Proof. For showing the first part, we use a differential constraint $\dot{x}_1 = 0 \wedge \dot{x}_2 = 0$ and a box $B = [0,2] \times [0,2]$. Obviously, $((1,1),(2,2)) \in \{(\boldsymbol{x}, \boldsymbol{y}) : flow_{D,B}(\boldsymbol{x}, \boldsymbol{y})\}$, but $((1,1),(2,2)) \notin \{(\boldsymbol{x}, \boldsymbol{y}) : flow'_{D,B}(\boldsymbol{x}, \boldsymbol{y})\}$. The reason lies in the fact that the derivatives are zero for this example, and in such a case, the equality in $flow_{D,B}(\boldsymbol{x}, \boldsymbol{y})$ reduces to the trivial equality $0 = 0$ that is true for all $\boldsymbol{x}, \boldsymbol{y}$.

For showing the second part, we use a differential constraint $\dot{x}_1 = x_1 + x_2 + 1 \wedge \dot{x}_2 = x_1 + x_2 + 1$ and a box $B = [0,2] \times [0,2]$. Obviously, $((0,0),(1,\frac{1}{5})) \in \{(\boldsymbol{x}, \boldsymbol{y}) : flow'_{D,B}(\boldsymbol{x}, \boldsymbol{y})\}$, but $((0,0),(1,\frac{1}{5})) \notin \{(\boldsymbol{x}, \boldsymbol{y}) : flow_{D,B}(\boldsymbol{x}, \boldsymbol{y})\}$. This is because we need $\dot{a}_m$ and $\dot{a}_n$ to be derivatives of a_m and a_n at the same point in $flow_{D,B}(\boldsymbol{x}, \boldsymbol{y})$, but in $flow'_{D,B}(\boldsymbol{x}, \boldsymbol{y})$, $\dot{a}_i$, can be the derivative of a different point for every i.   □

However, in our method, instead of computing exact solutions to these constraints, we only use over-approximations computed by the pruning function, and a tighter constraint does not necessarily give rise to a tighter pruning result.

So let us compare these over-approximations. We want to discuss the relation between $\mathrm{Prune}(flow_{D,B}(\boldsymbol{x}, \boldsymbol{y}), B_0 \times B)$ and $\mathrm{Prune}(flow'_{D,B}(\boldsymbol{x}, \boldsymbol{y}), B_0 \times B)$.

Theorem 2. $flow'_{D,B} \preceq_P flow_{D,B}$.

Proof. Given a set of variables V and a box B for which each component corresponds to a certain variable, we denote by $\pi_V(B)$ the projection of B to the components corresponding to the variables in V.

We transform the constraints into a conjunction without existential quantifiers as follows: Rename the variables $a_1, \ldots, a_k$ and $\dot{a}_1, \ldots, \dot{a}_k$ to a different tuple of variables in each branch, and then drop all corresponding existential quantifiers. As a result, in addition to $\boldsymbol{x}$ and $\boldsymbol{y}$, $flow'_{D,B}$ has $2k^2 + 1$ free variables and $flow_{D,B}$ has $k^2(k-1)$ free variables. Obviously, we have to prove

$$\pi_{\boldsymbol{x},\boldsymbol{y}}(\mathrm{Prune}(flow'_{D,B}(\boldsymbol{x}, \boldsymbol{y}), B_0 \times B \times \mathbb{R}_{\geq 0} \times \mathbb{R}^{2k^2})) \subseteq$$

$$\pi_{\boldsymbol{x},\boldsymbol{y}}(\mathrm{Prune}(flow_{D,B}(\boldsymbol{x}, \boldsymbol{y}), B_0 \times B \times \mathbb{R}^{k^2(k-1)})).$$

Let $\phi'_B \doteq \bigwedge_{1 \leq i \leq k} y_i = x_i + \dot{a}_i t$, and $\phi_B \doteq \bigwedge_{1 \leq m < n \leq k} \dot{a}_n(y_m - x_m) = \dot{a}_m(y_n - x_n)$. Let $\dot{A}$ be an arbitrary, but fixed, k-dimensional box. Let P'_s be the exact solution set of ϕ'_B in $B_0 \times B \times \mathbb{R}_{\geq 0} \times \dot{A}$ and $P' = \mathrm{Prune}(\phi'_B, B_0 \times B \times \mathbb{R}_{\geq 0} \times \dot{A})$. Also, let P_s be the exact solution set of ϕ_B in $B_0 \times B \times \dot{A}$ and $P = \mathrm{Prune}(\phi_B, B_0 \times B \times \dot{A})$. Since $(a_1, \ldots, a_k) \in B \wedge D((a_1, \ldots, a_k), (\dot{a}_1, \ldots, \dot{a}_k))$ is shared by both $flow'_{D,B}(\boldsymbol{x}, \boldsymbol{y})$ and $flow_{D,B}(\boldsymbol{x}, \boldsymbol{y})$, it suffices to prove that $\pi_{\boldsymbol{x},\boldsymbol{y}}(P') \subseteq \pi_{\boldsymbol{x},\boldsymbol{y}}(P)$. We will proceed by first proving that $\pi_{\boldsymbol{x},\boldsymbol{y}}(P'_s) \subseteq \pi_{\boldsymbol{x},\boldsymbol{y}}(P_s)$ and then lifting this to $\pi_{\boldsymbol{x},\boldsymbol{y}}(P') \subseteq \pi_{\boldsymbol{x},\boldsymbol{y}}(P)$.

So let $(\boldsymbol{x}, \boldsymbol{y}) \in \pi_{\boldsymbol{x},\boldsymbol{y}}(P'_s)$ be arbitrary, but fixed. We will prove that $(\boldsymbol{x}, \boldsymbol{y})$ is also in $\pi_{\boldsymbol{x},\boldsymbol{y}}(P_s)$. Since $(\boldsymbol{x}, \boldsymbol{y})$ is in the projection of P'_s we know that there are $(t, \dot{a}_1, \ldots, \dot{a}_k)$ such that $(\boldsymbol{x}, \boldsymbol{y}, t, \dot{a}_1, \ldots, \dot{a}_k)$ satisfies the constraint ϕ'_B. Choose a $t^*, \dot{a}_1^*, \ldots, \dot{a}_k^*$ with that property.

For proving that $(\boldsymbol{x}, \boldsymbol{y}) \in \pi_{\boldsymbol{x},\boldsymbol{y}}(P_s)$, it suffices to prove that $(\boldsymbol{x}, \boldsymbol{y}, \dot{a}_1^*, \ldots, \dot{a}_k^*)$ satisfies $\bigwedge_{1 \leq m < n \leq k} \dot{a}_n^*(y_m - x_m) = \dot{a}_m^*(y_n - x_n)$. Letting m, n be arbitrary, but fixed, such that $1 \leq m < n \leq k$, we prove that $(\boldsymbol{x}, \boldsymbol{y}, \dot{a}_1^*, \ldots, \dot{a}_k^*)$ is in the solution set of the corresponding conjunct. Here we have three cases:

- $\dot{a}_m^* \neq 0, \dot{a}_n^* \neq 0$: Then $\frac{y_m - x_m}{\dot{a}_m^*} = t^*$ and $\frac{y_n - x_n}{\dot{a}_n^*} = t^*$, so $\frac{y_m - x_m}{\dot{a}_m^*} = \frac{y_n - x_n}{\dot{a}_n^*}$, and hence $\dot{a}_n^*(y_m - x_m) = \dot{a}_m^*(y_n - x_n)$
- $\dot{a}_m^* = 0$: then $x_m = y_m$, and both sides of the equality $\dot{a}_n^*(y_m - x_m) = \dot{a}_m^*(y_n - x_n)$ are zero,
- $\dot{a}_n^* = 0$, analogous to previous case.

Hence $\pi_{x,y}(P_s') \subseteq \pi_{x,y}(P_s)$. For lifting this to the projected results of pruning, we first observe that each conjunct of ϕ_B' contains each variable just once. So, due to Property 1, pruning is optimal for each conjunct. Due to the fact that the conjuncts only share a single variable t also pruning of ϕ_B' is optimal [3], and since projection commutes with the smallest-box relation, $\pi_{x,y}(P')$ is the smallest box containing $\pi_{x,y}(P_s')$. Moreover, since projection commutes with the subset relation, not only $P_s \subseteq P$, but also $\pi_{x,y}(P_s) \subseteq \pi_{x,y}(P)$, and by transitivity $\pi_{x,y}(P_s') \subseteq \pi_{x,y}(P)$. Now, the following property implies the theorem: If B is the smallest box containing a set S and $S \subseteq S'$, then for every box B' with $S' \subseteq B'$, $B \subseteq B'$. □

So, we can prune at least as tightly using $flow_{D,B}'(x,y)$ as using $flow_{D,B}(x,y)$. In fact, we can prune strictly tighter!

Property 3. Not $flow_{D,B}' \equiv_P flow_{D,B}$.

Proof. Take a differential constraint $\dot{x}_1 = 1 \wedge \dot{x}_2 = 1$ and a box $B = [0,2] \times [0,2]$. Let $B_0 = \{(1,1)\}$, then $\mathrm{Prune}(flow_{D,B}'(x,y), B_0 \times B) = B_0 \times [1,2] \times [1,2]$ and $\mathrm{Prune}(flow_{D,B}(x,y), B_0 \times B) = B_0 \times B$. □

To sum up, the theoretical evaluation shows that although none of the two constraints always has a smaller solution set than the other, $flow_{D,B}'(x,y)$ has a definite advantage in size and in pruning power. Hence we only use the latter from now on.

Next, we will compare $flow_{\dot{x}=Ax,B}'(x,y)$ (or short: $flow_{A,B}'(x,y)$) with the constraint $eigen_{A,B}(x,y)$ for linear differential equations $\dot{x} = Ax$.

If A has k different real eigenvalues and $B = \mathbb{R}^k$, $eigen_{A,B}(x,y)$ describes the exact solutions of the differential equations. But, $flow_{A,B}'(x,y)$ employs the first-order Taylor expansion to over-approximate the exact solutions. Thus, $eigen_{A,B} \preceq flow_{A,B}'$? No! Only in cases where all trajectories leave the box, and do not enter it again. Otherwise, $eigen_{A,B}(x,y)$ also includes the part of the trajectory that enters the box again, but $flow_{A,B}'(x,y)$ does not necessarily.

Now we compare the two constraints wrt. pruning. For linear differential equations, pruning is optimal for $flow_{A,B}'$. This follows from an analysis of the proof of Theorem 2, and the fact that the differential equation constrains each derivative $\dot{x}$ using a constraint $\dot{x} = Ax$, where each equation of this constraint contains only one component of $\dot{x}$, and contains each component of x only once. However, this is in general not the case for $eigen_{A,B}$ due to multiple occurrences of variables.

There is no clearcut relationship between $\mathrm{Prune}(flow_{A,B}'(x,y), B_0 \times B)$ and $\mathrm{Prune}(eigen_{A,B}(x,y), B_0 \times B)$:

Property 4. Neither $flow_{A,B}' \preceq_P eigen_{A,B}$ nor $eigen_{A,B} \preceq_P flow_{A,B}'$.

Proof. This can be directly proven using only one example with a differential equation $(\dot{x}_1, \dot{x}_2) = (-x_1 - x_2, x_1 - x_2)$ and a box $B = [0,4] \times [0,4]$. If we set $B_0 = [2.5,3] \times [0,0]$, then $\mathrm{Prune}(flow_{A,B}', B_0 \times B) = B_0 \times [0,3] \times [0,4]$ and $\mathrm{Prune}(eigen_{A,B}, B_0 \times B) = B_0 \times [0,3.5] \times [0,3]$. □

Moreover, there are even some cases, where pruning $flow'_{A,B}$ returns a strict subset of pruning $eigen_{A,B}$ and vice versa: The former happens for a differential equation $(\dot{x}_1, \dot{x}_2) = (x_1 - x_2, x_1 + x_2)$ and a box $B = [0, 2] \times [0, 4]$. If we set $B_0 = [2, 2] \times [2, 4]$, then $Prune(flow'_{A,B}, B_0 \times B) = B_0 \times [0, 2] \times [2, 4]$ and $Prune(eigen_{A,B}, B_0 \times B) = B_0 \times B$. This is because the left-hand side of $eigen_{A,B}$, $c^T y$, evaluates to zero on some element in B. Hence every solution of $c^T (x \cos(bt) + \frac{\dot{x} - xa}{b} \sin(bt)) = 0$ fulfills the constraint. There is such a solution, and since sin and cos are periodic, the solution set is not bounded for t, and interval $[0, +\infty]$ for t will not be pruned. Since a is positive, the interval derived for the term e^{at} will also stay unbounded, and no intervals will be pruned. But, $t \in [0, 1]$ in $flow'_{A,B}$, which does provide some pruning.

The latter happens for a differential equation $(\dot{x}_1, \dot{x}_2) = (x_1, x_2)$ and a box $B = [0, 2] \times [0, 2]$. If we set $B_0 = \{(0, 0)\}$, $Prune(flow'_{A,B}, B_0 \times B) = B_0 \times B$ and $Prune(eigen_{A,B}, B_0 \times B) = B_0 \times B_0$. This is because $eigen_{A,B}$ here describes the exact solution starting from the initial point. But, since $\dot{x}_1$ and $\dot{x}_2$ can be zero, pruning $flow'_{A,B}$ results in $t \in [0, \infty]$. Thus, also the intervals for y_1 and y_2 cannot be pruned.

Since there is no clearcut relationship between $Prune(flow'_{A,B}, B_0 \times B)$ and $Prune(eigen_{A,B}, B_0 \times B)$, we strengthen both constraints by combining them. Thus, by sharing the same time variable we allow timing information to be propagated between them as follows:

Lemma 4. *For a linear differential equation $\dot{x} = Ax$ and a box B, if there is a trajectory in B from a point $x = (x_1, \ldots, x_k)^T \in B$ to a point $y = (y_1, \ldots, y_k)^T \in B$ on which $\dot{x} = Ax$ holds, then*

$$\exists t \in \mathbb{R}_{\geq 0}[flow^*_{A,B}(t, x, y) \wedge eigen^*_{A,B}(t, x, y)] \tag{3}$$

We denote the above new Constraint 3 by $comb_{A,B}(x, y)$. Clearly, this constraint implies $flow'_{A,B}(x, y)$, and also implies $eigen_{A,B}(x, y)$. That is, $comb_{A,B} \preceq flow'_{A,B}$ and $comb_{A,B} \preceq eigen_{A,B}$. Moreover, we have:

Theorem 3. $comb_{A,B} \preceq_P flow'_{A,B}$ and $comb_{A,B} \preceq_P eigen_{A,B}$.

So, the combination constraint is at least as good as $flow'_{A,B}(x, y)$ and $eigen_{A,B}$ (x, y). But, in fact, it is better!

Property 5. Neither $comb_{A,B} \equiv_P flow'_{A,B}$ nor $comb_{A,B} \equiv_P eigen_{A,B}$.

Proof. This can be seen on an example with a differential equation $(\dot{x}_1, \dot{x}_2) = (-x_1 - x_2, x_1 - x_2)$ and a box $B = [0, 4] \times [0, 4]$. If we set $B_0 = [2.5, 3] \times [0, 0]$, then $Prune(flow'_{A,B}, B_0 \times B) = B_0 \times [0, 3] \times [0, 4]$, $Prune(eigen_{A,B}, B_0 \times B) = B_0 \times [0, 3.5] \times [0, 3]$ and $Prune(comb_{A,B}, B_0 \times B) = B_0 \times [0, 3] \times [0, 3]$. $\qquad \square$

However, the combination constraint is bigger than both $flow'_{A,B}(x, y)$ and $eigen_B(x, y)$. Thus, pruning will take more time on it.

6   Empirical Evaluation

In this section we evaluate the constraints empirically by using them in the verification method introduced in Section 2. That is, we replace all occurrences of $Reachable_B(s, \boldsymbol{x}, \boldsymbol{y})$ introduced in Theorem 1 by $flow_{D,B}$ (or, $flow_{A,B}$), $flow'_{D,B}$ (or, $flow'_{A,B}$), $eigen_{A,B}$ and $comb_{A,B}$, respectively. We illustrate the behavior of our implementation on a few benchmark problems. Note that in the literature on the verification of hybrid systems the habit prevails to test new methods only on 2-3, or even less examples. We do not follow this tradition and do more extensive benchmarking. We will publish the corresponding HSOLVER input files on its web-page [15].

Note that we use the following splitting strategy here: In each mode, we choose the box with the biggest side-length, and then bisect each choice along the variable along which the box has not been split the longest time (i.e., we use a round-robin strategy to choose the variable). The computations were performed on a Pentium IV, 2.60GHz with 1 GB RAM, and they were canceled when computation did not terminate before 5 hours of computation time.

We used the following benchmark problems for comparing $flow_{D,B}$ and $flow'_{D,B}$ with computation results shown in Table 1:

Example 1
Flow: $(\dot{x}_1, \dot{x}_2) = (-x_1 - x_2, x_1 - x_2)$, empty jump relation
Init: $2.5 \leq x_1 \leq 3 \wedge x_2 = 0$, Unsafe: $x_1 > 3 \vee x_2 > 3$
The state space: $[0, 4] \times [0, 4]$

Example 2
Flow: $(\dot{x}_1, \dot{x}_2) = (x_1 - x_2, x_1 + x_2)$, empty jump relation
Init: $2.5 \leq x_1 \leq 3 \wedge x_2 = 0$, Unsafe: $x_1 \leq 2$
The state space: $[0, 4] \times [0, 4]$

Example 3. The flow constraints are constructed by setting all the parameters in the two tanks problem [18] to 1.

Flow: $\left(s = 1 \rightarrow \left(\begin{smallmatrix} \dot{x}_1 \\ \dot{x}_2 \end{smallmatrix}\right) = \left(\begin{smallmatrix} 1 - \sqrt{x_1} \\ \sqrt{x_1} - \sqrt{x_2} \end{smallmatrix}\right)\right) \wedge \left(s = 2 \rightarrow \left(\begin{smallmatrix} \dot{x}_1 \\ \dot{x}_2 \end{smallmatrix}\right) = \left(\begin{smallmatrix} 1 - \sqrt{x_1 - x_2 + 1} \\ \sqrt{x_1 - x_2 + 1} - \sqrt{x_2} \end{smallmatrix}\right)\right)$
Jump: $(s = 1 \wedge 0.99 \leq x_2 \leq 1) \rightarrow (s' = 2 \wedge x'_1 = x_1 \wedge x'_2 = 1)$
Init: $s = 1 \wedge (x_1 - 5.5)^2 + (x_2 - 0.25)^2 \leq 0.0625$
Unsafe: $\left(s = 1 \wedge (x_1 - 4.25)^2 + (x_2 - 0.25)^2 < 0.0625\right)$
The state space: $(1, [4, 6] \times [0, 1]) \cup (2, [4, 6] \times [1, 2])$

Example 4. This is a predator-prey example.
Flow: $\left(s = 1 \rightarrow \left(\begin{smallmatrix} \dot{x}_1 \\ \dot{x}_2 \end{smallmatrix}\right) = \left(\begin{smallmatrix} -x_1 + x_1 x_2 \\ x_2 - x_1 x_2 \end{smallmatrix}\right)\right) \wedge \left(s = 2 \rightarrow \left(\begin{smallmatrix} \dot{x}_1 \\ \dot{x}_2 \end{smallmatrix}\right) = \left(\begin{smallmatrix} -x_1 + x_1 x_2 \\ x_2 - x_1 x_2 \end{smallmatrix}\right)\right)$
Jump: $\big((s = 1 \wedge 0.875 \leq x_2 \leq 0.9) \rightarrow (s' = 2 \wedge (x'_1 - 1.2)^2 + (x'_2 - 1.8)^2 \leq 0.01)$
$\vee \big((s = 2 \wedge 1.1 \leq x_2 \leq 1.125) \rightarrow (s' = 1 \wedge (x'_1 - 0.7)^2 + (x'_2 - 0.7)^2 \leq 0.01)\big)$
Init: $s = 1 \wedge (x_1 - 0.8)^2 + (x_2 - 0.2)^2 \leq 0.01$
Unsafe: $\left(s = 1 \wedge x_1 > 0.8 \wedge x_2 > 0.8 \wedge x_1 <= 0.9 \wedge x_2 \leq 0.9\right)$
State space: $(1, [0.1, 0.9] \times [0.1, 0.9]) \cup (2, [1.1, 1.9] \times [1.1, 1.9])$

Example 5. This is a simple example with a clock variable.
Flow: $(\dot{x}, \dot{y}, \dot{t}) = (-5.5y + y^2, 6x - x^2, 1)$, empty jump relation
Init: $4 \le x \le 4.5 \wedge y = 1 \wedge t = 0$
Unsafe: $(1 \le x < 2 \wedge 2 < y < 3 \wedge 2 \le t \le 4)$
The state space: $[1, 5] \times [1, 5] \times [0, 4]$

Example 6. A three-dimensional and nonlinear example about a simple controller that steers a car along a straight road [4]. The three continuous variables are the position x, the heading angle γ and the internal timer c. Since we cannot prove the safety property described in the original paper, in this paper the unsafe space is reset to be $x \le -4$.

Example 7. A linear collision avoidance example from a part of the car convoi control from a paper by A. Puri and P. Varaiya [11]. Let gap, v_r, v_l and a_r respectively represent the distance between the two cars $(d_{i-1} - d_i$ in the original paper), the velocity of the rear car $(\dot{d}_i)$, the velocity of the leading car $(\dot{d}_{i-1})$ and the acceleration of the rear car $(\ddot{d}_i)$. By using these variables and restricting v_l by $-2 \le \dot{v}_l \le -0.5$ we transformed the original higher-order differential equation into a four-dimensional differential (in)equation of order one.
    We set the state space to $[0, 4] \times [0, 2] \times [0, 2] \times [-2, -0.5]$, and we want to verify that $gap > 0$ when starting from $gap = 1, v_r = 2, v_l = 2$ and $a_r = -0.5$.

Example 8. A four-dimensional and nonlinear example about a mixing-tank-system from a paper by O. Stursberg, S. Kowalewski and S. Engell [17]. In the original paper, the system is simplified to a two-dimensional system. In this paper, we keep the differential equations $(\dot{V}_1, \dot{V}_2) = (0.008, 0.015)$ in the flow constraint, where V_1 and V_2 are two inlet streams. Then, initially, $V_1(0) = 1, V_2(0) = 1$, and $(h(0), c(0)) \in [1.32, 1.5] \times [1.2, 1.32]$, where h is liquid height and c is concentration. We want to verify that the state $\{(V_1, V_2, h, c) : h \in [1.1, 1.3] \wedge c \in [1.68, 1.80]\}$ is unreachable.

Example 9. A two-dimensional and nonlinear example about a tunnel-diode oscillator circuit [6]. It models the voltage drop V and the current I. The original problem was to prove that all trajectories eventually reach a certain set and stay there. We transformed it to a reachability problem, using the state space $[-0.1, 0.6] \times [-0.002, 0.002]$ and the unsafety constraint $V < -0.04 \vee V > 0.54 \vee I < -0.0015 \vee I > 0.00175$.

Example 10. A linear, three-dimensional model of a mutant of V. *fischeri* [2]. Let x_1, x_2 and x_3 respectively represent the protein *LuxI*, the autoinducer *Ai* and the complex *Co* described in the original paper. The model has two modes with dynamics in the form $\dot{x} = Ax + b_i$, $i = 1, 2$, where $x = (x_1, x_2, x_3)^T$ and

$$
A = \begin{pmatrix} -1/3600 & 0 & 0 \\ 7.5e - 5 & -(1/36000 + 7.5e - 9) & 1.5e - 9 \\ 0 & 0.005 & -1/3600 - 0.01 \end{pmatrix}
$$

and $b_1 = (0.00375, 0, 0)^T$ and $b_2 = (3.75375, 0, 0)^T$.

We set the state space to be $[0, 30000] \times [0, 60000] \times [0, 30000]$ and the switches occur when the plane $x_3 = 1000$ is reachable and $x_2 \in [1000, 45000]$. We want to verify that $x_1 \geq 27500 \vee x_2 \geq 50000 \vee x_3 \geq 25000$ cannot be reachable when starting from $[17500, 20000] \times [40000, 45000] \times [5000, 7500]$ in mode 1.

Table 1. Computation results for $flow_{D,B}$ and $flow'_{D,B}$

Example	$flow_{D,B}$			$flow'_{D,B}$		
	CPU time	Splitting steps	Pruning number	CPU time	Splitting steps	Pruning number
1	0.041s	6	79	0.020s	2	30
2	0.34s	71	1572	0.38s	58	1203
3	0.18s	11	397	0.24s	11	397
4	0.57s	43	2250	0.66s	42	1884
5	2.59s	93	3552	1.69s	81	2929
6	0.35s	1	88	0.14s	0	41
7	187.96s	369	69158	105.55s	367	68675
8	7.68s	54	3138	43.95s	294	39281
9	13.209s	165	12215	3.653s	57	4655
10	1876s	1889	366681	686s	1417	270078

The results show that the new constraint improves the number of pruning steps for all examples except for Example 8, which we will discuss below. As expected, this also decreases the run-time of the method except for 2-dimensional examples, where $flow'_{D,B}$ has more conjuncts than $flow_{D,B}$.

We analyzed the anomaly in the behavior on Example 8 in more detail. After applying the pruning algorithm for the first time to $reachable_{B'}(s', z)$, using $flow'_{D,B}$ we can prune the box $[0, 2] \times [0, 2] \times [0.5, 1.5] \times [1.2, 1.8]$ to a new box $[1, 1.53333333333] \times [1, 2] \times [1.22034017148, 1.5] \times [1.2, 1.8]$; but, after we apply the pruning algorithm to $reachable_{B'}(s', z)$ using $flow_{D,B}$, we can only prune the box $[0, 2] \times [0, 2] \times [0.5, 1.5] \times [1.2, 1.8]$ to a new box $[0.466666666667, 1.53333333333] \times [0, 2] \times [1.22034017148, 1.5] \times [1.2, 1.8]$. So, in fact, the new method is better at the beginning! However, it seems that this improvement at the beginning turns out to be bad luck later since our method is very sensitive to splitting heuristics, and the improved pruning results in different choices of boxes for splitting during the algorithm. This suggests that a detailed study of splitting heuristics, will be able to significantly improve the method further.

In addition to the linear examples from above (Examples 1, 2 and 10), we used the following benchmarks for comparing $flow'_{A,B}$, $eigen_{A,B}$ and $comb_{A,B}$ with results shown in Table 2:

Example 11. A linear, three-dimensional example.
Flow: $(\dot{x}_1, \dot{x}_2, \dot{x}_3) = (0.80x_2 + 0.6x_3 - 1.8, 0.8x_1 + 0.7x_3 - 15.2, 0.6x_1 + 0.7x_2 - 1.8)$;
Empty jump relation; Init: $19 \leq x_1 \leq 20 \wedge 19 \leq x_2 \leq 20 \wedge 19 \leq x_3 \leq 20$;
Unsafe: $x_1 \leq 21 \wedge x_2 \leq 20 \wedge x_3 \geq 22.5$;
The state space: $[15, 24] \times [15, 24] \times [15, 24]$.

Table 2. Computation results for $flow'_{A,B}$, $eigen_{A,B}$ and $comb_{A,B}$

Example	$flow'_{A,B}$			$eigen_{A,B}$			$comb_{A,B}$		
	CPU time	Splitting steps	Pruning number	CPU time	Splitting steps	Pruning number	CPU time	Splitting steps	Pruning number
1	0.020s	2	30	unknown			0.038s	0	3
2	0.38s	58	1203	unknown			5.362s	58	1203
10	686s	1417	270078	12534s	3279	560082	122s	235	46540
11	unknown			unknown			0.319s	5	61
12	unknown			0.465s	0	3	0.756s	0	3

Example 12. A linear collision avoidance example similar to Example 7. We restrict v_l by $\dot{v}_l = 0$, and reset the state space to be $[0, 10] \times [0, 30] \times [0, 30] \times [-2, 5]$ and the initial set to be $-0.8522v_r - 0.1478v_l - 0.3177a_r + gap > 10$.

The results show that the combination decreases the size of the abstraction and the number of calls to the constraint solver. However, as expected, this will for some cases increase the run-time of the method, due to the bigger size of this constraint. This phenomenon is reflected by Examples 1, 2 and 12. But, for hard (and thus realistic) problems (e.g., Example 10), the improvement due to the first phenomenon will always dominate: in such cases the time spent on constraint solving will always be dominated by computations on the abstraction, and hence it is essential to keep the abstraction small.

For some cases, the safety property cannot be verified using $eigen_{A,B}$ in our method. For Examples 1 and 2 this can be explained using an observation already discussed in Section 5: the eigenvalues are complex with non-zero imaginary parts, and in such a case, since t occurs several times in the term $c^T(\boldsymbol{x} \cos(bt) + \frac{(\dot{\boldsymbol{x}} - a\boldsymbol{x})}{b} \sin(bt))$, we will get an over-approximating interval for this term.

On Examples 10 and 11 it can be seen nicely that the combined constraint can be stronger than either $flow'_{A,B}$ or $eigen_{A,B}$ in isolation.

Note that we did not use Examples 11 and 12 in Table 1 because their safety properties cannot be verified using either $flow_{A,B}$ or $flow'_{A,B}$.

7 Conclusion

We have provided a detailed study of two types of constraints in the verification of hybrid systems. The overall approach, to formulate reach set computation as a constraint solving problem, and to apply an efficiently over-approximating constraint solver to it, can be extended to various new types of constraints. Specifically we will study the use of higher order Taylor approximations instead of the constraint based on the mean value theorem. Our software is publically available [15], and includes an interface that allows the incorporation of and experimentation with new, user-defined constraints. Based on the gained experience and user feedback, we will optimize the constraint solver especially for the most useful ones.

References

1. H. Anai and V. Weispfenning. Reach set computation using real quantifier elimination. In *Proceedings of 4th International Workshop on Hybrid Systems: Computation and Control (HSCC2001)*, volume 2034 of *LNCS*, pages 63–76. Springer, 2001.

2. C. Belta, J. Schug, T. Dang, V. Kumar, G. Pappas, H. Rubin, and P. Dunlap. Stability and reachability analysis of a hybrid model of luminescence in the marine bacterium vibrio fisheri. In *CDC'01 - Conference on Decision and Control*. Florida, USA, 2001.

3. F. Benhamou and W. J. Older. Applying interval arithmetic to real, integer and Boolean constraints. *Journal of Logic Programming*, 32(1):1–24, 1997.

4. E. Clarke, O. Grumberg, S. Jha, Y. Lu, and H. Veith. Counterexample-guided abstraction refinement for symbolic model checking. *Journal of the ACM*, 50(5):752–794, 2003.

5. E. Davis. Constraint propagation with interval labels. *Artif. Intell.*, 32(3):281–331, 1987.

6. G. Frehse. PHAVer: Algorithmic verification of hybrid systems past HyTech. In Morari and Thiele [10].

7. T. Hickey and D. Wittenberg. Rigorous modeling of hybrid systems using interval arithmetic constraints. In R. Alur and G. J. Pappas, editors, *Hybrid Systems: Computation and Control*, number 2993 in LNCS. Springer, 2004.

8. T. J. Hickey. Analytic constraint solving and interval arithmetic. In *Proceedings of the 27th Annual ACM SIGACT-SIGPLAN Symposium on Principles of Programming Languages*, pages 338–351. ACM Press, 2000.

9. T. J. Hickey. Metalevel interval arithmetic and verifiable constraint solving. *Journal of Functional and Logic Programming*, 2001(7), October 2001.

10. M. Morari and L. Thiele, editors. *Hybrid Systems: Computation and Control*, volume 3414 of *LNCS*. Springer, 2005.

11. A. Puri and P. Varaiya. Driving safely in smart cars. In *Proc. of the 1995 American Control Conference*, pages 3597–3599, 1995.

12. S. Ratschan. RSOLVER. http://rsolver.sourceforge.net, 2004. Software package.

13. S. Ratschan. Efficient solving of quantified inequality constraints over the real numbers. *ACM Transactions on Computational Logic*, 2005. To appear.

14. S. Ratschan and Z. She. Safety verification of hybrid systems by constraint propagation based abstraction refinement. *ACM Journal in Embedded Computing Systems*. to appear.

15. S. Ratschan and Z. She. HSOLVER. http://hsolver.sourceforge.net, 2004. Software package.

16. S. Ratschan and Z. She. Safety verification of hybrid systems by constraint propagation based abstraction refinement. In Morari and Thiele [10].

17. O. Stursberg, S. Kowalewski, and S. Engell. On the generation of timed discrete approximations for continuous systems. *Mathematical and Computer Models of Dynamical Systems*, 6:51–70, 2000.

18. O. Stursberg, S. Kowalewski, I. Hoffmann, and J. Preußig. Comparing timed and hybrid automata as approximations of continuous systems. In P. J. Antsaklis, W. Kohn, A. Nerode, and S. Sastry, editors, *Hybrid Systems*, number 1273 in LNCS, pages 361–377. Springer, 1997.

19. A. Tiwari. Approximate reachability for linear systems. In O. Maler and A. Pnueli, editors, *Hybrid Systems: Computation and Control (HSCC)*, volume 2623 of *LNCS*. Springer, 2003.

Using Hajós' Construction to Generate Hard Graph 3-Colorability Instances*

Sheng Liu[1,2] and Jian Zhang[1]

[1] Laboratory of Computer Science
Institute of Software, Chinese Academy of Sciences
Beijing 100080, China
{lius, zj}@ios.ac.cn
[2] Graduate University, Chinese Academy of Sciences
Beijing 100049, China

Abstract. In this paper we propose a constructive algorithm using constraint propagation to generate 4-critical graph units (4-*CGU*s) which have only one triangle as subgraph. Based on these units we construct 4-critical graphs using Hajós' join construction. By choosing Grotztsch graph as the initial graph and carefully selecting the edge to be joined, we make sure that the generated graphs are 4-critical and triangle-free. Experiments show that these graphs are exceptionally hard for backtracking algorithms adopting Brélaz's heuristics. We also give some preliminary analysis on the source of hardness.

1 Introduction

Given an undirected graph $G = (V, E)$ with V the set of vertices and E the set of edges, let $|V| = m$ and $|E| = n$. A proper coloring of G is an assignment of colors to vertices such that each vertex receives one color and no two vertices connected by an edge receive the same color. A k-coloring of G is a proper coloring that uses k colors. The smallest number of colors needed to color a graph G is its chromatic number, which is denoted by $\chi(G)$.

Graph coloring problem (GCP) is of great importance in both theory and applications and has been studied intensively in computer science and artificial intelligence (AI). It arises in many applications such as scheduling, timetabling, computer register allocation, electronic bandwidth allocation. However, finding the chromatic number of a given graph is very hard, and even determining whether a given graph can be colored with 3 colors (*3-colorability*) is a standard NP-Complete problem [1], thus being not efficiently solvable by current methods. Despite this, the practical importance of the problem makes it necessary to design algorithms with reasonable computational time to solve instances arising in real-world applications. A lot of work has been done [2, 3, 4] and many powerful techniques have been proposed [5, 6, 7, 8]. On the other hand, in order to

* Supported in part by the National Science Foundation of China (grant No. 60125207).

J. Calmet, T. Ida, and D. Wang (Eds.): AISC 2006, LNAI 4120, pp. 211–225, 2006.

compare and evaluate the performance of different graph coloring algorithms, good benchmarks are needed and thus studied by many researchers.

Providing test instances as benchmarks for AI programs never lacks of interest [9]. Good test instances provide a common reference for people involved in developing and testing new search algorithms. What's more, good benchmarks especially those generated systematically make it possible to take a closer look at the instances and scrutinize their structures, and may give hints for the design of more appropriate algorithms.

Of course, for the sake of better discrimination of different algorithms, among all the test instances, hard ones are preferred. Many researchers discuss the source of hardness of some NP-Complete problems [10, 11]. The mechanism that makes *colorability* very hard is also studied. Possible candidates of order parameters proposed include the 3-path [9], the minimum unsolvable subproblems [12], frozen development [13], etc. Some methods that generate hard real instances are also presented. However, some of them are based on generate-and-test approaches [9], while others use handmade graph units [14], so most of them are either non-deterministic or not repeatable. In this paper we first propose a constructive algorithm that generates small 4-critical graph units (4-*CGU*s), then, similar to [14] we use a recursive self-embedding operation on these 4-*CGU*s to generate big instances. However, our small 4-*CGU*s are generated systematically with the guidance of constraint propagation, not by trial-and-error, so it gives chances to investigate the inner structures of the units and have a good understanding of the source of hardness of the big graphs. And absolutely contrary to [9] which favors graphs having as many triangles as possible, the resulting graphs generated by us are completely triangle-free. Experiments show that our generated graphs are very hard to solve for backtracking algorithms adopting Brélaz's heuristics [5] such as Trick[1].

The outline of the paper is as follows. In the next section we give some notations on constraint propagation and propose an algorithm that creates 4-*CGU*s. In section 3, we introduce the Hajós' join construction and generate big 4-critical graphs with it. Experimental details and analysis are listed in section 4. Comparisons with related work are given in section 5 and in section 6 we conclude the paper.

2   Constraint Propagation and 4-*CGU*s Generation

As mentioned above, what we generated are *3-colorability* instances. However, 4-critical graphs have the property that they are 3-colorable if any vertex/edge is removed (we denote them as vertex-critical graphs and edge-critical graphs respectively), so we first generate a 4-critical graph and when a 3-colorable graph is needed we simply remove some vertex/edge. The graphs that we consider and generate in this article are edge-critical.

For the sake of completeness, we recall some basic notations and definitions of the constraint satisfaction problems (CSPs) [15].

---

[1] http://mat.gsia.cmu.edu/COLOR/color.html

A CSP consists of a set of n variables $X_1, X_2, \ldots, X_n$ and a set of n domains $D_1, D_2, \ldots, D_n$ where each D_i defines the set of values that the variable X_i may assume. A solution of a CSP is an assignment of a value to each variable which satisfies a given set of constraints. A binary CSP is one in which all constraints involve only pairs of variables. A binary constraint between X_i, X_j is a subset of the Cartesian product $D_i \times D_j$. A binary CSP can be associated with a constraint graph in which vertices represent variables and edges connect pairs of constrained variables.

Definition 1. *A constraint graph is **Arc Consistent** iff for each of its arcs $< i, j >$ and for any value $a_i \in D_i$, there is a value $a_j \in D_j$ that satisfies the binary constraint between i and j.*

GCP is to assign colors to vertices of a graph with constraints over its edges, so the constraint graph can be obtained directly and easily. In fact, there is a one-to-one mapping between a graph to be colored and its constraint graph, in which vertices correspond to variables and colors to be assigned to a vertex correspond to the domain of the corresponding variable. So many papers on constraint processing take graph coloring problems as examples, and constraint satisfaction thus promotes the research on graph coloring. Our work is carried out with the guidance of constraint satisfaction. Now we propose our algorithm **CGU(4,n)** which constructs 4-CGU with n vertices ($n \geq 9$).

Algorithm 1. CGU(4,n)

Step 1 Let $n = 3 * m + r$ where both m and r are non-negative integers, and $r < 3$.

Step 2 Construct a triangle $\triangle ABC$ and a circle with $3 * (m - 1)$ vertices denoted as $a_1, b_1, c_1, a_2, b_2, c_2, \ldots, a_{m-1}, b_{m-1}, c_{m-1}$ successively.

Step 3 Connect A with all a_i (i=1, 2, ..., m-1);
connect B with all b_i (i=1, 2, ..., m-1);
connect C with all c_i (i=1, 2, ..., m-1).

Step 4 (a) If r=0 then choose two vertices a_k, a_l from a_i (i=1, 2, ..., m-1), connect a_k and a_l;

(b) if r=1 then choose a vertex a_k from a_i (i=1, 2, ..., m-1), a vertex b_l from b_i (i=1, 2, ..., m-1) and a vertex c_m from c_i (i=1, 2, ..., m-1), introduce a new vertex O, connect O with a_k, O with b_l, O with c_m;

(c) if r=2 then choose two vertices a_{k_1}, a_{k_2} from a_i (i=1, 2, ..., m-1), choose two vertices b_{l_1}, b_{l_2} from b_i (i=1, 2, ..., m-1), introduce two new vertices O_1, O_2, connect O_1 with a_{k_1}, O_1 with b_{l_1}, O_2 with a_{k_2}, O_2 with b_{l_2}, O_1 with O_2;

Step 5 Stop.

Note that each graph generated by the algorithm **CGU(4,n)** has a triangle in it. So the chromatic number is at least 3. In fact we have the following theorems:

Theorem 1. *The graphs constructed by **CGU(4,n)** are 4-chromatic.*

Proof. For the sake of convenience, without loss of generality we generate a small graph and take it as an example (Fig. 1).

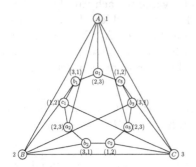

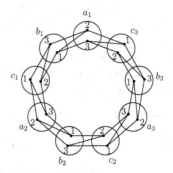

Fig. 1. Part of the graph after the triangle is colored

Fig. 2. Part of the constraint graph on the circle

We prove it by contradiction. Assume that it is 3-colorable. As depicted in Fig. 1, because there is a triangle $\triangle ABC$ in the graph, the three vertices A, B, C of the triangle can be colored by 3 colors denoted as color $1, 2, 3$. Then by constraint propagation we get that the vertices on the circle that are connected with vertex A can only be colored by color 2 or color 3. Similarly, the vertices on the circle that are connected with vertex B (C) can only be colored by color 3 or color 1 (color 1 or color 2). Thus we get a circle with candidate colors in every vertex (Fig. 1). Because the edges on the circle can also denote constraints, we apply arc consistency test on each edge and get part of the constraint graph shown in Fig. 2[2]. Furthermore,we notice that each of the candidate colors of the vertices in Fig. 2 has one edge pointing to one of its two neighbors and two edges pointing to the other neighbor, so once the color of one vertex is fixed, one of its neighbors' color is fixed at the same time. For instance, if we set color 3 to vertex a_1, then vertex b_1 can only be colored with color 1. In the same way vertex c_1 can only be colored with color 2, and the rest may be deduced by analogy. It is the same when vertex a_1 chooses color 2. Thus no matter what color a_1 chooses, once its color is fixed, the colors of all the other vertices are also fixed. Meanwhile it is interesting that all the vertices connected with the same vertex of the triangle have thethe the same color but vertices connected with different vertices of the triangle have different colors. For instance, all a_i $(i=1, 2, \ldots, m\text{-}1)$ have the same color but b_i $(i=1,2, \ldots,m\text{-}1)$ and c_i $(i=1,2, \ldots,m\text{-}1)$ enjoy different colors.

Next we discuss the construction in **Step 4** (Fig. 3, Fig. 4, Fig. 5). (**a**) If $r=0$ then one edge between a_k and a_l is constructed. But from the discussion above we know that a_k and a_l should have the same color, so it makes a contradiction. However, one more color assigned to a_k or a_l (but not both) is enough to solve the contradiction. (**b**) If $r=1$ then because O is connected with a_k, b_l and c_m

---

[2] For the sake of readability, we do not depict the complete constraint network of the vertices on the circle. In fact there are many edges between the vertex a_1 and each of its non-neighbor vertices, but we neglect such edges for simplicity.

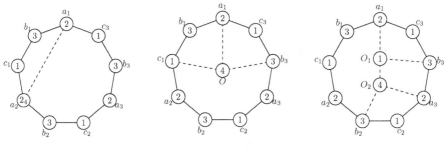

Fig. 3. $r=0$ **Fig. 4.** $r=1$ **Fig. 5.** $r=2$

which have mutually different colors, O can't be colored by any of the three colors. So a fourth color is needed to color O. **(c)** If $r=2$ then because O_1 is connected with a_{k_1} and b_{l_1}, it can't have the same color as a_{k_1} or b_{l_1}, thus it has to choose the color assigned to c_i ($i=1, 2, \ldots, m$-1). It is the same for O_2. But O_1 and O_2 are also connected, so they can't get the same color at the same time. It contradicts again and one more color is needed.

To sum up, the graphs constructed by **CGU(4,n)** are not 3-colorable but 4 colors are adequate to color them, so they are 4-chromatic.   □

Theorem 2. *The graphs constructed by* **CGU(4,n)** *are 4-critical.*

Proof. Now that it has been proved in Theorem 1 that graphs constructed by **CGU(4,n)** are 4-chromatic, we only need to prove that the graphs are 3-colorable if an arbitrary edge is removed, according to the definition of critical graph.

From the analysis above we know that there indeed exists a coloring scheme using 4 colors in which only one vertex (i.e., O in Fig. 4) is colored with the fourth color. Our proof begins with such a scheme. Once an edge is removed from the graph, we prove that by changing the original 4 coloring scheme step by step we can get a new coloring scheme which uses only 3 colors, that is to say, the newly introduced color (the fourth color) can be replaced by one of the original 3 colors because of the removal of one edge.

First we study the edges generated in **Step 4** of the algorithm **CGU(4,n)** (the dashed edges in Fig. 3, Fig. 4 and Fig. 5). Because all these edges are adjacent to the vertex colored by the fourth color[3], once such an edge is removed, the end vertices of that edge can share the same color. So the fourth color is not indispensable any more.

Next we consider the edges forming the circle and the edges between the triangle and the circle. From Fig. 3, Fig. 4 and Fig. 5 we find that the dashed edges divide the region inside the circle into two or more subregions, each of

---

[3] In the case of $r=2$, only one of O_1 and O_2 has to be colored by the fourth color, so if O_1 is colored by the fourth color it seems that the judgment does not hold for the edges that are adjacent to O_2 but not adjacent to O_1. But since the color of O_1 and the color of O_2 can be exchanged, it does not affect the following discussion and conclusions.

which is circumscribed by a closed subcircle (e.g., (a_1, b_1, c_1, O, a_1) of Fig. 4). Each edge of the original circle is on one of these subcircles and each subcircle contains the vertex that is colored by the fourth color[4]. If one of the edges on the circle is removed (e.g., edge (b_2, c_2) of Fig. 4), the constraint over (b_2, c_2) does not exist any more, so vertex b_2 and c_2 can have the same color. Right now b_2 has color 3 and c_2 has color 1, but we can't assign b_2's color 3 to c_2 because c_2 is connected with the triangle vertex C whose color is also 3, so we have no other choice but to assign c_2's color 1 to b_2. Then we propagate the color assignment along the subcircle successively, starting from b_2's color and ending up when we get to O. In this sample, we assign b_2's former color 3 to a_2, assign a_2's former color 2 to c_1, assign c_1's former color 1 to O and at last O's former color 4 is discarded. So far the graph has been colored by 3 colors without any color collisions. In case that the removed edge is one of the edges between the triangle and the circle (i.e., edge (B, b_2)), similarly, we can assign B's color 2 to b_2. But one of b_2's neighbors a_2 on the circle also has color 2 at present, so we first color a_2 with b_2's former color 3 and then propagate the color assignment as we described above, in the direction from b_2 to a_2 along the subcircle. Thus the graph is colored by 3 colors properly.

At last we turn to the edges that form the triangle. If one of the triangle edges (e.g., (A, B)) is removed, the endpoints (A and B) of the edge can share the same color. Since all the a_i-form and b_j-form vertices receive the same color constraint from triangle vertices, it follows that they are equivalent in fact. Thus each a_i-form vertex can change colors with its b_j-form neighbor, vice versa. As for the vertex that is colored by the fourth color (i.e., O'), it has at least one a_i-form or b_j-form neighbor. Next we first discard color 4 and color O' with the color of its a_i-form or b_j-form neighbor (i.e., b_k). Then a conflict arises because the two vertices O' and b_k are connected by an edge but have the same color. In order to overcome the conflict, we first exchange colors between b_k and its a_i-form neighbor on the circle (i.e., a_k). If this leads to a new conflict between b_k (a_k) and B (A), we need only let B (A) have the same color as A (B). Take Fig. 5 for example, first we change the color of O_2 from 4 to 3, after that we exchange colors between a_2 and b_2. This leads to a new conflict between b_2 and B, so we assign A's color 1 to B, then we get a proper 3-coloring of the graph.

After checking all the cases, we reach the conclusion that the generated graph is 3-colorable no matter which edge is removed. Thus Theorem 2 is proved.   □

3   Hajós' Join Construction and Hard Triangle-Free Graph Generation

As mentioned in the introduction, starting with small 4-CGUs, we use Hajós' construction to build big critical graphs. So we introduce Hajós' join construction first [16].

Definition 2 (The Hajós' construction). *Let G and H be two graphs. Let uu' be an edge in G and vv' be an edge in H. The resulting graph $G \triangle H$ is*

---

[4] The case of $r=2$ has been discussed above.

obtained by identifying u and v, deleting the edges uu' and vv', and adding an edge u'v'.

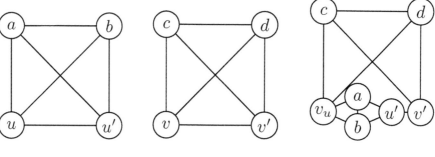

Fig. 6. Graph G   **Fig. 7.** Graph H   **Fig. 8.** Graph $G \triangle H$

Theorem 3 (Hajós). *If both G and H are k-critical graphs (k > 3), then $G \triangle H$ is a k-critical graph.*

Now we can use Hajós' construction iteratively to construct big critical graphs without altering their chromatic number. Details are listed below:

```
procedure HardGraph(k)
begin
    G := G_init;
    for i := 1 to k do
        choose a random number l (l ≥ 9);
        H :=CGU(4,l);
        G := Hajos(G, H);
    od;
end
procedure Hajos(G, H)
begin
    choose an edge uu' from G and remove it;
    choose an edge vv' from H and remove it;
    add edge u'v';
    merge u with v;
end
```

According to Hajós' construction, in order to generate 4-critical graphs the graph G_{init} used in $HardGraph(k)$ must also be 4-critical. [14] finds 7 MUGs (minimal unsolvable graphs, which are also small 4-critical graphs) by trial-and-error and chooses one of them as the initial graph G_{init}. Some MUGs contain more than one triangles, so the resulting instances may contain many triangles. But by choosing a_{k_1} not adjacent to b_{l_1} and a_{k_2} not adjacent to b_{l_2} in **CGU(4,n)**, we easily make sure that each of the 4-CGUs generated by our algorithm has only one triangle. In $Hajos(G, H)$ we choose one of the three edges of the triangle in

H as the joining edge vv', then the remaining part of H is triangle free. Since $Hajos(G, H)$ doesn't introduce any new triangles into the generated graph, if we choose a triangle-free graph as the initial graph G_{init} we can make sure that all our generated graphs are triangle-free. So we use the triangle-free Grotztsch graph as the initial graph (Fig. 9).

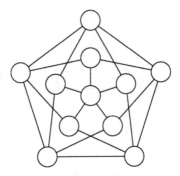

Fig. 9. Grotztsch graph

The size of maximum clique is usually used as a rough lower bound of the chromatic number. Some researchers speculate that the greater the distance between the chromatic number and the lower bound is, the harder the graph is for algorithms that color graphs by detecting lower bound first [17]. Because the maximum clique in triangle-free graphs is of size 2, while the size of that in non-triangle-free graphs is at least 3, we guess that our triangle-free graphs may be harder and our experimental results support our speculation to some extent. We will describe them in the next section in detail.

4   Experiments and Discussion

After the description of the algorithms in section 2 and section 3, we devote this section to some implementation details.

4.1   Generating Better 4-*CGU*s

In **Step 4** of algorithm **CGU(4,n)**, some vertices on the circle such as a_k and b_l have to be chosen. However, there are many choices for these vertices so we have to decide the relative better ones. Of course we prefer choosing vertices that make the generated 4-*CGU*s harder to solve. We compare two versions of implementation. In **CGU₁(4,n)** all the chosen vertices are distributed relatively densely on the circle while in **CGU₂(4,n)** they are distributed as uniformly as possible. We believe that more subproblems lead to more backtracks when the graph is colored by Trick. So we record the number of subproblems when applying Trick to color a graph, which is listed in Table 1. All the experiments were carried out on a P4 2.66GHz Dell computer with 512M memory.

Table 1. Comparison between two kinds of **CGU(4,n)**s on subproblem numbers

n	9	10	11	12	13	14	15	16	17	18	19	20	21	22	23	24	25
$CGU_1(4,n)$	12	13	16	13	16	21	16	19	27	19	22	33	22	25	39	25	28
$CGU_2(4,n)$	12	13	16	16	19	22	19	22	28	25	28	34	28	34	39	34	37

From the table, we notice that there is not much difference between $CGU_1(4,n)$ and $CGU_2(4,n)$ when $r=2$ ($n = 11, 14, 17, 20, 23, \ldots$). However, in other cases, no matter what value n is assigned to, $CGU_2(4,n)$ performs much better than $CGU_1(4,n)$. We also generated some graphs using $CGU_1(4,n)$ and $CGU_2(4,n)$ respectively. Experimental results show that averagely graphs using $CGU_2(4,n)$ are harder to color and all the hardest ones among all the generated graphs are those using $CGU_2(4,n)$. The superiority of $CGU_2(4,n)$ is evident so we adopt the $CGU_2(4,n)$ version in our experiment.

4.2 Making Structure More Regular

From algorithm **CGU(4,n)** we find that all the vertices except the ones on the triangle have lower variance of degrees. However, the degrees of the triangle vertices increase quickly as n increases. In [9] the author finds that more regular instances, with more uniform structures, tend to be much harder. We also generated two kinds of graphs. One kind uses **CGU(4,n)**s with n ranging from 9 to 15 and the other kind uses **CGU(4,n)**s with n ranging from 16 to 22. Experiments show that the latter ones are not so hard as the former ones. As a matter of fact some of the latter graphs are not hard at all although they have more vertices. It seems that regularity[5] is an important factor in the hardness of graphs.

On the other hand, when we use Hajós' construction, the degree of vertex v also increases because of merging u with v. So in order to prevent the degree of v from increasing too much, when choosing the edge vv' we deliberately choose the vertex with degree 3 as v. We also compare such generated graphs with the ones generated by selecting v randomly, denoted by asterisk $*$ and plus $+$ respectively. Figure 10 depicts the average search cost comparison for each n and Fig. 11 depicts the maximal search cost comparison. From the comparison results we find that restricting vertices' degrees to lower variance indeed makes instances harder. It seems that regular instances induce uniformity in the structure of the search space, making the search algorithm confused by the equally promising choices [9]. So in the following experiments we use **CGU(4,n)**s with n ranging from 9 to 15 and deliberately choose a vertex with degree 3 as v, as we did above.

4.3 Generating Hard Triangle-Free Graphs

With **CGU(4,n)**s (n ranges from 9 to 15) and Grotztsch graph we generate triangle-free graphs using $HardGraph(k)$ where k ranges from 5 to 12. For each

---

[5] A graph is regular if all its vertices have the same degree.

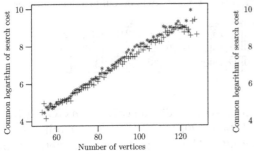

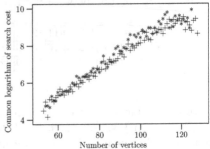

Fig. 10. Comparison on average search cost

Fig. 11. Comparison on maximal search cost

value of k, 200 instances are generated. These instances are tested by Trick and the results are given in Fig. 12 in which search cost is evaluated by the number of subproblems. Figure 12 reveals a linear relationship between the vertical axis and the horizontal axis. However, note that the vertical axis of Fig. 12 represents common logarithm of the number of subproblems, so it is easy to understand that the search cost exhibits exponential growth.

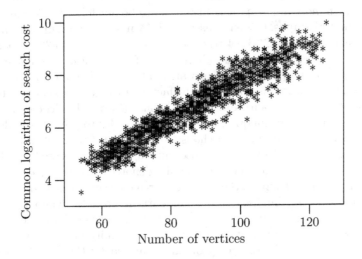

Fig. 12. Experimental results of the triangle-free instances

We also generate instances from [14] and compare them with our triangle-free instances. Figure 13 depicts the average search cost comparison for each n and Fig. 14 depicts the maximal search cost comparison. Here, the asterisk $*$ corresponds to our data while the plus $+$ corresponds to data of [14]. Figure 13 reveals that when n is small, instances of [14] seem superior, but when n grows bigger and bigger our triangle-free instances turn out to be superior. The same

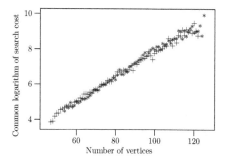

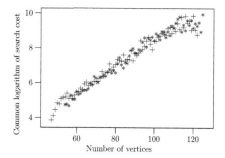

Fig. 13. Comparisons on average search cost

Fig. 14. Comparisons on maximal search cost

result can be obtained from Fig. 14 except that it is not so evident as in Fig. 13. We speculate that the tendency may go on, but it is hard to verify it by further experiment. On one hand, for the sake of statistical accuracy, for each vertex number n we need the sample space as big as possible, on the other hand, when n grows bigger and bigger, each sample needs so much time that the total time cost is too much. However, current results have already shown that our triangle-free instances are at least as hard as and maybe harder than those of [14].

We also compare our instances with those of [14] by running other graph coloring algorithms. Next we give our experimental results with Smallk [13], a sophisticated backtracking coloring program specialized for graphs of small chromatic number.

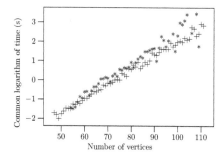

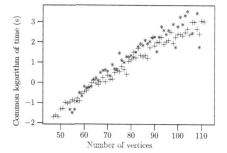

Fig. 15. Comparisons with Smallk on average time

Fig. 16. Comparisons with Smallk on maximal time

From Fig. 15 and Fig. 16 we find that although both kinds of instances exhibit exponential growth, our triangle-free ones seem to be a little harder in general. It is known that Smallk is good at exploiting structural weakness (e.g., frozen pairs in [13]). Maybe our instances just hide such weakness because they are triangle-free.

In summary, our experiments show that our construction is efficient in producing very hard instances whose computational costs seem to be of an exponential order of the vertex number.

4.4   Analysis and Discussion

Decomposition and composition are widely used in graph coloring except that decomposition usually deals with complicated graphs by dividing them into small components and analyzing the small components while composition usually builds complicated graphs from small components. Experiments in [14] and here show that the Hajós' construction is an effective composition method to generate hard *3-colorability* instances. We also notice that, in [14], because the graph components MUGs were found totally by trial-and-error, the authors don't have a good knowledge of the components' structural features, so that they can not explain the reason of the resulting graphs' hardness. However our component 4-*CGU*s are generated systematically with the guidance of constraint propagation, so by scrutinizing their inner structures, we can give some useful information in understanding the reason.

In our construction each 4-*CGU* is a component and it is iteratively embedded into the resulting instance. From the proof of Theorem 1 we know that for the vertices on the circle (Fig. 2) once one of the vertices is colored the other vertices' colors are fixed at the same time. That is to say, only one choice remains for each of the other vertices. This conclusion can be obtained by constraint propagation as described in this paper, but for many algorithms adopting backtracking heuristics they can't foresee the full future, so it is inevitable to make wrong decisions. What's more, wrong decision can occur in every 4-*CGU*. Notice that our instances are generated by iteratively embedding 4-*CGU*s which can also be viewed as the multiplying of 4-*CGU*s. So, even if there is one backtrack in each 4-*CGU*, there will be an exponential number of backtracks altogether in the resulting graphs. For this reason backtracking algorithms will spend exponential time detouring and backtracking before they find the right coloring. So it seems exceptionally hard for them to handle our instances.

5   Related Work

A lot of work has already been done on providing benchmarks for general GCP, but our work focuses on graph *3-colorability*.

Related work on generating hard graph *3-colorability* instances includes [14] and [9]. [14] also uses Hajós' construction. But, in order to generate very hard instances, one must have lots of small 4-*CGU*s at hand first[6]. [14] finds 7 such units by trial-and-error, which shows to some extent the difficulty of finding 4-*CGU*s by hand and the necessity of generating them systematically. So, in

---

[6] What's more, among all 4-*CGU*s the ones including no *near-4-clique* (4-*clique* with an edge removed) as subgraphs are preferred, because such graphs hide a structural weakness that heuristics would be able to exploit (e.g., frozen pairs in [13]).

this paper, we present a constructive algorithm to find 4-CGUs systematically. What's more, because our 4-CGUs are generated automatically with the guidance of constraint propagation, it provides a possibility to investigate the inner structures of the graphs and find some useful rules why they can produce hard graphs. Based on constraint propagation we give some explanations in Section 4, which may help recognize the reason of the hardness and give some hints for researchers working on new coloring heuristics. While in [14] the authors don't have a good knowledge of the reason of the hardness since even their MUGs are found by trial-and-error.

[9] uses a generate-and-test method to produce *3-colorability* instances. The author takes a random graph with fixed vertex number and edge number as an input, then removes an old edge and adds a new edge to the graph iteratively, hoping to minimize the number of 3-paths (denoted by an alternate succession of vertices and edges $x_1e_1x_2e_2x_3e_3x_4, x_1 \neq x_4$). It is easy to find that [9] favors graphs that have as many triangles as possible. Although it avoids 4-clique during construction, it does not avoid *near-4-clique* which appears to be a structural weakness. However, our generated instances do not have such weakness because we make sure that they are all triangle-free by using Grotztsch graph as initial graph and selecting special edges to join. Experiments show that our instances seem to be even harder when tested with sophisticated algorithms such as Smallk.

Although our instances are of small chromatic numbers, they can also be used as general GCP benchmarks. As far as we know, there are already such benchmarks in the DIMACS 2002 Challenge[7] (i.e., mug88_1, mug88_25, mug100_1, mug100_25 are such ones provided by the second author of [14]). We also notice that among the benchmarks in the DIMACS 2002 Challenge, there is a special kind of graphs named Myciel graphs which are based on the Mycielski transformation. Although the chromatic numbers of these graphs range from 4 to 5, 6 and even more, their maximum clique numbers remain 2. Because they are triangle-free, these graphs are difficult to solve. As for chromatic number 3, however, as far as we know, it seems that there are few benchmarks with the same property. But our generated instances (with an edge removed) just have the property. What's more, there are only a fixed number of Myciel graphs for each chromatic number, but many instances can be generated using our method.

6 Conclusions and Future Work

In this paper, a constructive algorithm that generates 4-CGUs systematically is presented. With these 4-CGUs we generate 4-critical and triangle-free graphs using Hajós' construction. Experiments show that our instances are exceptionally hard for backtracking algorithms adopting Brélaz's heuristics. Because our instances are triangle-free which hides some structural weakness, compared with similarly generated instances, they seem to be harder when experimented with sophisticated backtracking algorithms.

---

[7] http://mat.gsia.cmu.edu/COLOR02/

As benchmarks, hard instances are good but we believe that hard instances with known inner structures are better, because they can give some hints for researchers working on new coloring heuristics. Our triangle-free instances just have the property. Since our 4-*CGU*s are generated systematically with the guidance of constraint propagation, we have a good knowledge of their inner structures, which makes it possible for us to give some explanations on the hardness of the resulting 4-critical graphs. We think that one of our contributions is that we present such a constructive algorithm to produce 4-*CGU*s systematically. We plan to find more methods to produce 4-*CGU*s with more sophisticated structures, so as to get more knowledge of the inner structures of the generated hard graphs and develop heuristics to solve them.

Acknowledgments

Thanks to Jin-Kao Hao for offering us the executable that helps to verify our proof. Thanks to the anonymous referees for the valuable and detailed comments.

References

[1] Garey, M.R., Johnson, D.S.: Computers and Intractability - A Guide to the Theory of NP-Completeness. W. H. Freeman, San Francisco (1979)

[2] Kubale, M., Jackowski, B.: A generalized implicit enumeration algorithm for graph coloring. Commun. ACM **28**(4) (1985) 412–418

[3] Mehrotra, A., Trick, M.A.: A column generation approach for graph coloring. INFORMS Journal on Computing **8** (1996) 344–354

[4] Johnson, D.S., Trick, M.A., eds.: Cliques, Coloring, and Satisfiability: Second DIMACS Implementation Challenge, Workshop, October 11-13, 1993. American Mathematical Society, Boston, MA, USA (1996)

[5] Brélaz, D.: New methods to color the vertices of a graph. Commun. ACM **22**(4) (1979) 251–256

[6] Peemöller, J.: A correction to Brélaz's modification of Brown's coloring algorithm. Commun. ACM **26**(8) (1983) 595–597

[7] Hertz, A., de Werra, D.: Using tabu search techniques for graph coloring. Computing **39**(4) (1987) 345–351

[8] Galinier, P., Hao, J.K.: Hybrid evolutionary algorithms for graph coloring. J. Comb. Optim. **3**(4) (1999) 379–397

[9] Vlasie, R.D.: Systematic generation of very hard cases for graph 3-colorability. In: Proceedings of 7th IEEE ICTAI. (1995)

[10] Cheeseman, P., Kanefsky, B., Taylor, W.M.: Where the really hard problems are. In: Proceedings of the 12th IJCAI. (1991) 331–337

[11] Hogg, T., Williams, C.P.: The hardest constraint problems: A double phase transition. Artificial Intelligence **69** (1994) 359–377

[12] Mammen, D.L., Hogg, T.: A new look at the easy-hard-easy pattern of combinatorial search difficulty. Journal of Artificial Intelligence Research **7** (1997) 47–66

[13] Culberson, J., Gent, I.: Frozen development in graph coloring. Theoretical Computer Science **265**(1–2) (2001) 227–264

[14] Nishihara, S., Mizuno, K., Nishihara, K.: A composition algorithm for very hard graph 3-colorability instances. In: Proceedings of the 9th CP. Volume 2833 of Lecture Notes in Computer Science. (2003) 914–919
[15] Dechter, R.: Constraint Processing. Morgan Kaufmann, San Francisco (2003)
[16] Jensen, T.R., Toft, B.: Graph Coloring Problems. Wiley, New York (1995)
[17] Caramia, M., Dell'Olmo, P.: Constraint propagation in graph coloring. Journal of Heuristics 8(1) (2002) 83–107

Finding Relations Among Linear Constraints*

Jun Yan[1,2], Jian Zhang[1], and Zhongxing Xu[1,2]

[1] Laboratory of Computer Science
Institute of Software, Chinese Academy of Sciences
[2] Graduate University, Chinese Academy of Sciences
{yanjun, zj, xzx}@ios.ac.cn

Abstract. In program analysis and verification, there are some constraints that have to be processed repeatedly. A possible way to speed up the processing is to find some relations among these constraints first. This paper studies the problem of finding Boolean relations among a set of linear numerical constraints. The relations can be represented by rules. It is believed that we can not generate all the rules in polynomial-time. A search based algorithm with some heuristics to speed up the search process is proposed. All the techniques are implemented in a tool called MALL which can generate the rules automatically. Experimental results with various examples show that our method can generate enough rules in acceptable time. Our method can also handle other types of constraints if proper numeric solvers are available.

1 Introduction

Constraints play an important role in various applications, and constraint solving has been an important research topic in Artificial Intelligence. A useful technique for constraint solving is to add some redundant constraints so as to improve the algorithms' efficiency [1, 2]. However, there is not much work on the systematic discovery of such constraints.

When we study constraint solving techniques, it is usually helpful if we take the form of constraints into account. This often depends on the application domain. One domain that is quite interesting to us is program analysis and verification. To analyse a program and generate test data for it, we may analyze the program's paths. For each path, we can derive a set of constraints whose solutions represent input data which force the program to be executed along that path [3]. Such a path-oriented method is often used in software testing, and it may also be used in infinite loop detection [4].

Generally speaking, the constraints encountered in program analysis and testing can be represented as Boolean combinations of arithmetic constraints [3, 5]. Here each constraint is a Boolean combination of primitive constraints, and a primitive constraint is a relational expression like $2x + 3y < 4$. In other words, a constraint is a Boolean formula, but each variable in the formula may stand for a relational expression. To solve such constraints, we developed a solver called BoNuS which combines Boolean satisfiability checking with linear programming.

---

* Supported in part by the National Science Foundation of China (grant No. 60125207 and 60421001).

J. Calmet, T. Ida, and D. Wang (Eds.): AISC 2006, LNAI 4120, pp. 226–240, 2006.

In program analysis, we often need to solve many sets of constraints. They may contain some common primitive constraints. So if we can find some logic relationship (i.e. rules) between the primitive constraints (especially those occurring frequently) and add them as lemmas, the search space can be reduced.

Formal verification is also an important way to maintain program quality. Model checking is an effective verification method. But traditionally it is based on the propositional temporal logic. To scale it up to more real programs, one may use abstractio[6, 7]. It is based on the observation that the specifications of systems that include data paths usually involve fairly simple relationships among the data values in the system. A mapping between the actual data values in the system and a small set of abstract data values are given according to these relationships. In fact, since the abstract values are not always independent, these abstractions can be regarded as some rules deduced from the predicates. Therefore we need to find out some logic relations of the predicates before using abstraction.

In this paper, we try to employ a linear programming solver called `lp_solve` [8] to find all the logic relations among a set of linear arithmetic constraints automatically. We implemented a tool and used it to analyze how the attributes of a constraint set affect the number of rules. Since most of our techniques do not rely on any special characteristics of linear constraints, our method can be generalized to other types of constraints such as non-linear constraints if a proper solver is provided.

This paper is organized as follows. The next section will briefly introduce the problem of finding rules from numerical constraints and analyze its complexity. Then section 3 will present the main idea of our algorithm and some improving techniques. Experimental results and some analysis are given in Section 4. Then our approach is compared with some related works in Section 5, and some directions of future research are suggested in the last section.

2   The Problem and Its Complexity

2.1   Linear Arithmetic Constraints

In this paper, a numerical constraint is a *Linear Constraint* in the following form:

$$a_1x_1 + \ldots + a_nx_n \bowtie b$$

where a_i is a coefficient, x_i is a variable, and $\bowtie \in \{=, <, >, \leq, \geq, \neq\}$ is a relational operator. A conjuncion of linear constraints

$$\varphi : \begin{bmatrix} a_{11}x_1 + \cdots + a_{1n}x_n \bowtie b_1 & \wedge \\ \vdots & \cdots & \vdots & \vdots & \vdots \\ a_{m1}x_1 + \cdots + a_{mn}x_n \bowtie b_m & \end{bmatrix}$$

can be written concisely in matrix form as $Ax \bowtie b$ where the bold x and b are n-dimensional and m-dimensional vectors, respectively.

Example 1. Here is an example of a set of numerical constraints.

```
C1 = (2x + y >= 4);
C2 = (x == y);
C3 = (x > 1).
```

We can use a Linear Programming (LP) package like `lp_solve` [8] to process linear constraints. In a linear programming problem, there can be a set of equations and standard inequalities (whose relational operators are $\geq$ and $\leq$). LP has been studied by many people and it can be solved by efficient algorithms [9].

Since most Linear Programming packages do not support the strict inequalities $\bowtie\in\{\neq, <, >\}$ (that are the negations of standard equations and inequalities respectively), we translate them into the standard form in the following way.

1. The expressions $Exp_1 < Exp_2$ and $Exp_1 > Exp_2$ can be replaced by $Exp_1 + \delta \leq Exp_2$ and $Exp_1 \geq \delta + Exp_2$ respectively, where δ is a very small positive number.
2. We use two expressions $A_1 = (Exp_1 < Exp_2)$ and $A_2 = (Exp_1 > Exp_2)$ to translate $A = (Exp_1 \neq Exp_2)$. This strict inequality will be replaced by $A = A_1 \vee A_2$. But LP solvers do not support the disjunction of constraints. So in the worst case, we may have to call `lp_solve` 2^n times if there are n such strict inequalities.

2.2   Constraints-Relation-Finding Problem

For the constraints in Example 1, we can easily find out that the following relation holds:

$$C1 \wedge C2 \rightarrow C3.$$

But if the number of constraints grows, the relations among the constraints can be quite complex and we need a method to obtain these rules automatically.

We can formalize the relations or rules as $A \rightarrow B$ where A and B are numerical constraints or their Boolean combinations. Note that

$$A \rightarrow B \text{ is tautology} \Leftrightarrow A \wedge \neg B \text{ is contradiction}$$

and we can transform $A \wedge \neg B$ into disjunctive normal form (DNF) such that each conjunctive clause is unsatisfiable. (A conjunctive clause is the conjunction of literals. It will be simply called a clause in this paper, unless stated otherwise.) So we only need to find out unsatisfiable clauses in the following form:

$$C_1 \wedge C_2 \ldots \wedge C_m$$

where m is the *length* of this clause. The literals $C_1, C_2, \ldots C_m$ represent the original numeric constraints or their negation.

It is natural to think of a clause as a set of literals. A clause that is a subset of another is called its *subclause*. For any clause ϕ, each constraint C_i may have one of the following 3 statuses: $C_i \in \phi$; $\neg C_i \in \phi$; neither. So the set of conjunctive clauses is the power set 3^C where C is the original constraint set. Our goal is to find out an unsatisfiable subset of this power set.

We define the Constraints-Relation-Finding (CRF) problem as follows:

Definition 1 (CRF). *Given a set of constraints* $\mathcal{C} = \{C_i\}$, *enumerate all unsatisfiable clauses* $C_{j_1} \wedge C_{j_2} \ldots \wedge C_{j_m}$ *where for all* k, $1 \le k \le m$, $C_{j_k} \in \mathcal{C}$ *or* $\neg C_{j_k} \in \mathcal{C}$.

It is well known that integer linear programming (recognizing the satisfiable instances) is NP-complete, so it is trivial that the converse, recognizing the set of unsatisfiable instances, is coNP-complete. So if we restrict the constraints to integer linear constraints, the subproblem is coNP-complete.

The minimal unsatisfiable sub-formula (MUS) is the following problem: Given a Boolean formula in conjunctive normal form with at most three occurrences of each variable, is it true that it is unsatisfiable, yet removing any clause renders it satisfiable? MUS is known to be DP-complete [10]. The subproblem of deciding if a sub-formula $C_{j_1} \wedge C_{j_2} \ldots \wedge C_{j_m}$ is unsatisfiable is also coNP-complete since it's the complementary problem of 3-SAT. If we solve the CRF problem for $\{C_i\}$, we solve the MUS problem for it: we only need to pick up the shortest sub-formula $C_{j_1} \wedge C_{j_2} \ldots \wedge C_{j_m}$ that is unsatisfiable. So the complexity of the CRF problem is at least the same as a DP-complete problem.

If we can solve the CRF problem by calling the constraint solver polynomial times, we can reduce the CRF problem to a coNP-complete problem in polynomial-time. But "DP contains a host of natural problems, whose membership in $NP \cup coNP$ is in serious doubt" [11]. So we believe that we can not solve the CRF problem by calling the constraint solver polynomial times. Otherwise, DP-complete problems would be as easy as coNP-complete problems.

From the above analysis, we believe that we can not solve the CRF problem by calling the constraint solver polynomial times. Therefore it is unlikely for us to have an efficient algorithm which can always generate all the logic relations for a constraint set.

3   The Rule Finding Algorithm

We can make use of linear programming solvers to decide if a clause is unsatisfiable. Due to the exponential size of the power set, we need to use some heuristics to speed up the whole process. The purpose of these heuristics is to reduce the number of times the linear programming solver is invoked.

3.1   The Basic Algorithm

First we give some notations. We use C to denote the set of numeric constraints and the size of C is denoted by NC. Let $C = \{c[1], \ldots, c[NC]\}$. We use two sets S and U to represent the sets of clauses that are satisfiable and unsatisfiable, respectively. S_i and U_i are the sets of satisfiable and unsatisfiable clauses found in loop i.

We use a dynamic programming method to enumerate all the formulae and construct the sets U and S. At first the set S has only one item $TRUE$ that can be regarded as a clause of length 0, and U is empty. For each element of C, add it (and its negation, respectively) to S and call lp_solve to solve the new set of constraints. Then we add the unsatisfiable formulae to U and the others to S. The algorithm is complete and will terminate after we have processed all the elements of C. The algorithm can be represented by a function in Fig 1.

```
void FindRule(){
    for(i = 1;  i ≤ NC;  i++){
        for each clause s ∈ S {
            w₁ = s ∧ c[i], w₂ = s ∧ ¬c[i];
            if (w₁ is unsatisfiable) add w₁ to Uᵢ and w₂ to Sᵢ;
            else {
                add w₁ to Sᵢ;
                if (w₂ is satisfiable) add it to Sᵢ;
                else add w₂ to Uᵢ.
            }
        }
        S = S ∪ Sᵢ,  U = U ∪ Uᵢ;
    }
}
```

Fig. 1. The main algorithm

3.2   Correctness of the Algorithm

First we prove

Lemma 1. *The set S collects all the satisfiable clauses.*

Proof. Assume we have a satisfiable conjunctive clause

$$conj = C_{j_1} \wedge C_{j_2} \ldots \wedge C_{j_m} \text{ where } j_1 < j_2 < \ldots < j_m$$

and $conj \notin S$, we can easily know $C_{j_1}, C_{j_2}, \ldots C_{j_m}$ are all satisfiable.

Since $TRUE \in S$, then the formula $C_{j_1} = (TRUE \wedge C_{j_1}) \in S$. So $conj$ has at least one subclause in S.

Suppose we have two satisfiable subclauses of $conj$:

$$conj_k = C_{j_1} \wedge C_{j_2} \ldots \wedge C_{j_k} \text{ and } conj_{k+1} = C_{j_1} \wedge C_{j_2} \ldots \wedge C_{j_k} \wedge C_{j_{k+1}}$$

where $1 \leq k < m$, $conj_k \in S$ and $conj_{k+1} \notin S$. But according to our algorithm, the satisfiable $conj_{k+1}$ will be judged at the loop $i = j_{k+1}$ and must be added to S. So any satisfiable clause $conj$ will be added to set S.   □

We say a rule is redundant if it has an unsatisfiable subclause. Now we prove our algorithm collects all the rules except some redundant ones.

Theorem 1. *Any unsatisfiable clauses will be included in the set U or it has an unsatisfiable subclause in the set U.*

Proof. For each unsatisfiable clause

$$conj = C_{j_1} \wedge C_{j_2} \ldots \wedge C_{j_m} \text{ where } j_1 < j_2 < \ldots < j_m,$$

if it has no unsatisfiable sub-formulae, its subclause $conj_{m-1}$ must belong to S and $conj$ should be added to U at the loop $i = j_m$. So the set U will collect all the unsatisfiable clauses except for some redundant ones.   □

3.3 A Processing Example

We use Example 1 to show the search process. The clauses added to the two sets U or S at each step of the main loop are given in Table 1. We apply our search algorithm to Example 1 and quickly find that $C1 \wedge C2 \rightarrow C3$ is the only rule.

Table 1. The clauses added to U or S in each loop

The i'th loop	Clauses added to U	Clauses added to S
1	-	$C1, \neg C1$
2	-	$C2, \neg C2, C1 \wedge C2, C1 \wedge \neg C2, \neg C1 \wedge C2, \neg C1 \wedge \neg C2$
3	$C1 \wedge C2 \wedge \neg C3$	$C3, \neg C3, C1 \wedge C3, C1 \wedge \neg C3, \neg C1 \wedge C3, \neg C1 \wedge \neg C3,$ $C2 \wedge C3, C2 \wedge \neg C3, \neg C2 \wedge C3, \neg C2 \wedge \neg C3,$ $C1 \wedge C2 \wedge C3, C1 \wedge \neg C2 \wedge C3, C1 \wedge \neg C2 \wedge \neg C3$ $\neg C1 \wedge \neg C2 \wedge C3, \neg C1 \wedge \neg C2 \wedge \neg C3,$ $\neg C1 \wedge \neg C2 \wedge C3, \neg C1 \wedge \neg C2 \wedge \neg C3$

In the worst case, if the constraint set has no rules (or we can say this instance is very difficult), we have to check all the elements of the power set 3^C. That is to say, we have to solve 3^{NC} conjunctive clauses. Therefore the time complexity of this algorithm is exponential. To make the time cost of this algorithm acceptable, we developed some heuristics listed below to speed it up. The main idea of these techniques is to use other simple preprocessing methods instead of solving the clause to decide its satisfiability.

3.4 Subclause Strategy

Our algorithm has ruled out some redundant formulae which have subclauses in set U, but not all the redundant formulae. For example, if we find $C1 \wedge C3$ is unsatisfiable, then the formula $C1 \wedge C2 \wedge C3$ is obviously unsatisfiable. But it is still possible for the basic algorithm to check the formula. We should devise some technique to avoid checking all the redundant formulae during the search.

Our Subclause Strategy is based on the following observation: If A is a subclause of B and A is unsatisfiable, then B is unsatisfiable.

So our algorithm can be improved in the following way. We first sort the sets S and U by ascending order of clause length, then check if one of the clauses in U is a subclause of the current clause. For the clauses w_1 and w_2, since s has no subclauses in U, we only need to check whether w_1 and w_2 have subclauses in U_i.

3.5 Resolution Principle

Suppose we have two propositional logic expressions: $C_1 \vee p$ and $C_2 \vee \neg p$, where p is a propositional variable. Then we can obtain a new expression $C_1 \vee C_2$ that does not involve the variable p, while preserving satisfiability. That's the Resolution Principle of the propositional logic.

We will modify this deduction rule to form a pruning strategy.

Theorem 2. *Let A and B be two clauses. If we can express A and B as* $A = A' \wedge p$ *and* $B = B' \wedge \neg p$ *where p is a logic variable, let* $D = A' \wedge B'$. *Assume A is unsatisfiable, then we have the following conclusions:*

I *B is unsatisfiable implies D is unsatisfiable;*
II *D is satisfiable implies B is satisfiable.*

This heursitic can be used to decide some clauses' satisfiablity. It can also be used to construct some satisfiable or unsatisfiable clauses. And if a rule is found, a series of rules will be soon generated and that will significantly prune the rest of the search space and speed up the whole process.

3.6   Related Numerical Constraints

The previous two strategies make use of the processed unsatisfiable formulae. In addition, some characteristics of arithmetic expressions can help us to check a clause's satisfiability quickly.

We say two clauses A and B are *related* if they have common numeric variables. For example, the three constraints of Example 1 are all related, while the following two constraints:

```
Cxy = (x == y);   Cz = (z > 0);
```

are not related. Their numeric variable sets $\{x, y\}$ and $\{z\}$ do not have any common element. The constraint $Cxy \wedge Cz$ is obviously satisfiable since Cxy and Cz are both satisfiable.

We summarize this simple principle here:

Theorem 3. *If a clause A can be divided into the conjunction of some satisfiable subclauses* $A = A_1 \wedge \ldots \wedge A_k$, *and any* A_i *and* A_j $(i \neq j)$ *are not related, then A is satisfiable.*

The premises of this strategy can be easily checked in our processing. Firstly, if the subclause stractegy is applied, then obviously all the subclauses are satisfiable. Secondly, a conjunction can be represented as a graph $\langle V, E \rangle$, where the V represents the set of numerical constraints. Two vertices have an edge if they are related. So the problem deciding if a conjunction can be divided is transformed to check the connectivity of an undirected graph, which can be done in polynomial time.

This heuristic is effective when dealing with the sets that each numeric constraints of them have few variables such that two numerical constraints have a considerable probability to be not related.

3.7   Linear Independency of Coefficient Vectors

In our search, we need to process some clauses with the same numeric constraints. For example, in Table 1, in the second loop, $C1 \wedge C2$, $C1 \wedge \neg C2$, $\neg C1 \wedge C2$, $\neg C1 \wedge \neg C2$ have the same constraints $C1$ and $C2$. These clauses have the same coefficient matrix A. For the clause $Ax \bowtie b$, if there exists $b' \bowtie b$ such that $Ax = b'$ is satisfiable, then the clause is also satisfiable. A set of vectors a_j $(1 \leq j \leq m)$ is *linear dependent* if there exist m factors λ_j $(1 \leq j \leq m)$, not all of which are zero, such that $\sum_{1 \leq j \leq m} \lambda_j a_j = 0$.

We have the following lemma:

Lemma 2. *For a clause $Ax \bowtie b$, if the m coefficient vectors a_i $(1 \leq i \leq m)$ are linear independent, then the clause is satisfiable.*

This lemma only need to consider the coefficient matrix, so all the clauses involving the same constraints are satisfiable if the coefficient vectors are linear independent.

However, the constraints of linear dependency can not easily be solved since the solver lp_solve does not work well at the strict inequality $\lambda_j \neq 0$, so we strengthen our previous lemma to the following strategy:

Theorem 4. *For w_1 and w_2 of the main algorithm in Figure 1, if the coefficient vector a_c of $c[i]$ can not be linearly represented by the coefficient vectors of s, then w_1 and w_2 are satisfiable.*

Proof. Let a_j $(1 \leq j \leq m)$ denote the coefficient vectors of s, $A_i = [a_1, \ldots, a_m]^T$ be the coefficient matrix of s and b_s be the constant vector. If a_j $(1 \leq j \leq m)$ are linear independent, then a_c and a_j $(1 \leq j \leq m)$ are linear independent, thus w_1 and w_2 are satisfiable according to Lemma 2.

If a_j $(1 \leq j \leq m)$ are linear dependent, since $c[i]$ and s are satisfiable, so we have a constant vector $b'_s \bowtie b_s$, $r(A_i) = r([A_i, b'_s])$. Here $r(A)$ means the *rank* of matrix A. Consider the constraints

$$\begin{bmatrix} a_c \\ A_i \end{bmatrix} x = \begin{bmatrix} b'_c \\ b'_s \end{bmatrix}$$

where b'_c is an arbitrary numeric value. Since a_c cannot be linearly represented by a_j $(1 \leq j \leq m)$, we have

$$r\left(\begin{bmatrix} a_c \\ A_i \end{bmatrix} \right) = r(A_i) + 1 = r([A_i, b'_s]) + 1 = r\left(\begin{bmatrix} a_c\ b'_c \\ A_i\ b'_s \end{bmatrix} \right)$$

that implies the constraint $a_c x = b'_c \wedge A_i x = b'_s$ is satisfiable and thus $w_1 = s \wedge c[i]$, $w_2 = s \wedge \neg c[i]$ are all satisfiable.

The strategy needs to call the solver once to judge the linear representation. This strategy aims at finding the satisfiable clauses and therefore it is efficient for the constraint sets for which only a few rules can be found. It will save at most $2^m - 1$ callings of the solver where m is the length of clauses. Please refer to Table 3 for the computational results of this strategy.

3.8   Clause Length Restriction

The heuristics introduced before are used for complete search. That is to say, they do not remove the useful rules. These techniques do not decrease the time complexity of the whole search. Here we introduce an approximate technique that can reduce the processing time efficiently.

Theoretically, the maximum length of rules is NC, which is the size of the numerical constraint set. In practice, we do not need to combine arbitrary number of numeric constraints. Firstly, most solvers are more efficient in dealing with short constraints. Secondly, we can see from the experiments 4.3 that the lengths of most clauses are in the domain $[1, NV + 1]$, where NV is the number of numerical variables. So we need not waste time in finding long formulas. We define UL as the upper bound on the formula's length. The clauses with length greater than UL will be abandoned. And the time complexity will be reduced to $\binom{NC}{UL} = O(NC^{UL})$. In practice, we can get a quicker process by choosing a small value for the parameter UL. For the empirical values of UL, please refer to Section 4.3.

4   Experimental Results

We developed a tool called MALL (MAchine Learning for Linear constraints) in the C programming language with all the presented techniques implemented. It invokes lp_solve to decide the satisfiability of clauses. Our tool can generate all the rules of a small constraint set without redundant rules quickly.

We have studied many examples including some random constraint sets. We mainly care about the number of solutions and the solving times. Some experimental results are listed here. We use a PC P4 3.2GHz CPU, 1 GB memory with Gentoo Linux, and the timings are measured in seconds.

4.1   A Real Instance

This strong correlative example comes from [5]. These constraints are collected from a real Middle routine to find the middle value of 3 values. For simplicity, we do not restrict the variables to be integers.

```
bool ba  = (b < a);          bool ac  = (a < c);
bool ca  = (c < a);          bool ab  = (a < b);
bool bc  = (b < c);          bool cb  = (c < b);
bool a_b = (a == b);         bool b_c = (b == c);
bool c_a = (c == a);         bool m_a = (m == a);
bool m_b = (m == b);         bool m_c = (m == c);
bool n_a = (n == a);         bool n_b = (n == b);
bool n_c = (n == c);         bool n_m = (n == m);
```

There are 16 constraints with 5 numerical variables. We tried our tool on this instance with different UL (upper bound of the formula length. See section 3.8). The results are summarized in Table 2. The result of $UL = 16$ reports all the useful logic relations among these constraints.

To check whether an implementation of Middle routine violates the specification, we can employ BoNuS to solve the following Boolean combinations of the numerical constraints:

Table 2. Different UL

UL	2	3	4	5	6	7	16
Number of solutions	9	92	260	557	869	887	887
Time used	0.01	0.11	0.43	2.11	11.55	37.95	184.71
Memory used(M)	1.11	1.56	3.60	12.47	38.11	91.36	444.23
Times to call `lp_solve`	135	548	892	1174	1244	1279	1282

```
% Specification
imp(and(ba, ac), m_a);   imp(and(ca, ab), m_a);
imp(and(ab, bc), m_b);   imp(and(cb, ba), m_b);
imp(and(ac, cb), m_c);   imp(and(bc, ca), m_c);
imp(b_c, m_a);   imp(c_a, m_b);   imp(a_b, m_c);

% Implementation
imp(or(and(ab, bc), and(cb, ba)), n_b);
imp(or(and(ac, cb), and(bc, ca)), n_c);
imp(and(not(or(and(ab, bc), and(cb, ba))), not(or(and(ac, cb),
and(bc, ca)))), n_a);

% Violation of the specification
not(n_m);
```

Here `imp(x,y)` denotes that x implies y, and a line starting with "`%`" is a comment. We apply BoNuS to this problem and BoNuS reports a solution in 0.084 second. If we add the rules with $UL = 3$ to the input of BoNuS, BoNuS will report a solution in 0.008 second. But if we add the rules with $UL = 5$ to BoNuS input, the solving time will be 0.012 second. In fact, for this type of mixed constraint satisfaction problem, some very short lemmas (rules) are enough. Too many long rules will decrease the efficiency because the solver has to spend more time on the logic constraints. If these 16 numerical constraints are checked repeatedly in program analysis, the 0.11 second of preprocessing is worthwhile.

To access the clauses (mainly the elements of S, U and some other clauses) quickly during the search, we use an index tree to record the processed clauses. This tree occupies too much memory if the clause is too long (please refer to Table 2 for memory used). That limits the scalability of the tool.

Our tool reports the rules quickly for some small UL value. But the space and time cost increases notably if UL is increased. On the other hand, the times to call `lp_solve` do not increase so remarkably as the memory cost increases. This is not possible if no heuristic is used.

Also we can see from this instance that $UL = 6$ is enough. In fact, we have experimented with some other small instances derived from real programs. The results are satisfactory. We can get more than 95% rules with proper UL in a few minutes if the number of constraints is not more than 16.

4.2   The Efficiency of Strategies

We also use the example of Section 4.1 with different UL to test the efficiency of four strategies: SC (Subclause Strategy), RP (Resolution Principle), RN (Related Numerical

Constraints) and LC (Linear Independency of Coefficient Vectors). Each time we remove a strategy and at last we remove all the strategies. Since we should check SC
before RN, RN should be removed if SC is removed. We list the processing time and
the number of rules found in Table 3. Note that some redundant rules are found if the
SC strategy is removed.

Table 3. Efficiency of Strategies

Strategies Removed	None	SC+RN	RN	RP	LC	ALL
$UL = 3$	0.11 / 92	0.18 / 172	0.14 / 92	0.19 / 92	0.22 / 92	0.85 / 172
$UL = 4$	0.43 / 260	1.23 / 1571	0.49 / 260	1.35 / 260	0.61 / 260	5.97 / 1571
$UL = 5$	2.11 / 557	16.78 / 11107	2.16 / 557	8.91 / 557	2.20 / 557	34.48 / 11107

4.3   Effects of Various Parameters

In general, the smaller the rule set is, the more difficult the CRF problem is. Finding
rules from a difficult set of constraints may waste much time and the rules may provide
little useful information to the following constraint processing.

To study the characteristics of the CRF problem, we use some random instances.
Here are some parameters to describe a set of linear constraints:

NC The number of numerical constraints.

NV The number of numerical variables.

ANV The ratio of average number of variables of each constraint to NV. It can be
defined as $ANV = \frac{\sum NV_i}{NV * NC}$ where NV_i is the number of numerical variables of
the i'th constraint.

NE The ratio of the number of equations and strict inequalities (whose relational operator is "$\neq$") to NC.

Obviously, with the increasing of NC, we will get more rules, but at the same time,
the space and time cost will increase remarkably. In the following part we generate
some random instances with different parameters to study how the latter 3 parameters
influence the difficulty of the problem. Each random constraint is non-trivial (i.e. each
numerical constraint and its negation are satisfiable). For each set of parameter values,
the result is the average value of 100 runs.

Number of Numerical Variables. We use some 10-sized random sets (i.e., $NC = 10$)
to study the influence of NV on the distribution of the conjunctive clause length. The
results are given in Table 4.

From the results, we can conclude that, if $NV < NC$, most formulae have length
$NV + 1$. Too small or too large NV will not cause many rules. We get the maximum
number of rules near the point $NV = 4$. Our other experiments of small sized constraint
set also indicate that we will get the maximum number of rules near the median of NC.
These phenomena are mainly caused by two reasons:

Table 4. The Effect of the Number of Variables

Length of Clause	2	3	4	5	6	7	8	9	10	Total
$NV = 2$	10.75	53.46			0.00					64.21
$NV = 4$	1.86	3.76	8.36	117.79		0.00				131.77
$NV = 6$	0.62	0.70	1.15	1.07	4.29	70.5		0.00		78.33
$NV = 8$	0.26	0.20	0.18	0.29	0.49	0.29	0.62	6.49	0.00	8.82
$NV = 10$	0.12	0.07	0.06	0.11	0.06	0.11	0.06	0.00	0.00	0.59

1. The NV defines the dimensions of the variables' domain, which is a Euclidean space. According to linear programming, a set of standard inequalities have solution if we can get a solution along the edges of boundary. That is to say, the solvers just check the solutions of boundary constraints (which are a set of equations derived from the original inequalities just by replacing the relational operators with "="). According to linear algebra, $NV + 1$ linear equations of NV dimensions are linear dependent. That is why so many rules have length $NV + 1$.
2. Generally speaking, a small NV implies that these constraints are located in a "small" (low dimension) space and have more opportunities to be related. But on the other hand, because the number of numerical variables is small, many of these relations are redundant. Therefore, we may get a maximum number of rules in the middle of $[1, NC]$.

So from the results, we can get the experimental value of $UL = NV + 1$, and we have no need to find rules whose length is more than that.

Number of Equations. The constraints of equations usually come from the assignments of program. Here we use some instances with 10 constraints to test its effects on the difficulty of the problem. In these tests, the UL value is set as $NV + 2$ to get the most rules, and the results are listed in Table 5.

Table 5. The Effect of Equations

NE	0.2	0.4	0.6	0.8
Number of Rules	58.83	58.23	55.97	58.26
Times calling `lp_solve`	4303.88	4471.27	4911.94	5573.89
Time used	1.26	1.33	1.45	1.65

From the result, we see that the parameter NE has no significant effect on the problem's difficulty. In practice, a strict inequality is translated into two inequalities, and it will cost more time.

Average Number of Numerical Variables. The parameter ANV affects the size of rules. Here we also use some random constraint set sized 10 to test it. The value NE is set as 0.33. The results are summarized in Table 6.

Table 6. The Effect of ANV

ANV	0.2	0.4	0.6	0.8
Number of rules	14.73	44.12	80.09	98.31
Times calling `lp_solve`	698.20	4453.92	6200.64	6481.27
Time used	0.22	1.25	1.83	1.99

From the result, we get that the number of rules increases as ANV increases. Meanwhile, the time cost is also increased. The main reason is that the constraints with high ANV tend to be related.

5   Related Work

There are many research work about analysis of numerical constraints. Constraint Logic Programming systems (e.g. CLP($\mathcal{R}$)[12]) can find solutions if constraints are satisfiable or detect unsatisfiability of constraints. This type of systems focus on checking the satisfiability of mixed types of constraints. Different from our constraint solving method, these works treat equations and inequalities differently. Also on the analysis of linear constraints from programs, the paper [13] presents a method for generating linear invariants for transition systems. Compared with our work, this type of works focus on inducing invariants (which are some numerical expressions) from all the constraints, while we try to find all the logic rules that each may fit for only a small subset of the constraints. Our work, if properly modified, can be used as preprocessing in these systems.

The Inductive Logic Programming (ILP) [14] is a research area formed at the intersection of Machine Learning and Logic Programming. ILP systems have been applied to various learning problem domains. ILP systems develop predicate descriptions from examples and background knowledge. The examples, background knowledge and final descriptions are all described as logic programs. The theory of ILP is based on proof theory and model theory for the first order predicate calculus. In many occasions, the ILP background knowledge can be described as a series of numerical constraints. If we can find the relations between these constraints, we can remove much redundancy and reduce the time of induction or searching.

Conflict-driven Lemma Learning [15] has been proved quite successful in improving SAT algorithms. This type of techniques recognize and record the causes of conflicts, preempt the occurrence of similar conflicts later on the search. When solving complex constraints, recognizing the cause of conflicts is difficult due to the numeric constraints. So we use a pre-learning approach as described in this article instead of dynamic analysis. Also in solving a constraint problem, sometimes we need to add redundant constraints which can lead to new simplifications. For example, the paper [2] examined the impact of redundant domain constraints on the effectiveness of a real-time scheduling algorithm. But the forms of redundant constraints are restricted.

Similar to our CRF problem, the problem MUS (finding minimal unsatisfiable subformula) is used in the solution of SAT and other problems. Some approximate search algorithms for MUS have been developed. For example, the paper [16] discussed an adaptive search method to find the MUS of an unsatisfiable CNF instance.

6   Conclusion

In this paper we study the problem of finding rules from a set of linear constraints and give its theoretic complexity. We present a search method to solve this problem efficiently and developed a tool to generate the rules automatically. Since we can not get all the rules in polynomial-time, we propose a number of strategies to speed it up. Most of these strategies are independent of the constraint type, therefore, our method is general and can be used on other kind of constraints if there are proper solvers.

We also study the difficulty of the problem using some randomly generated instances. We find that some parameters of the instances affect the number of rules notably. We obtain some empirical values to increase the performance of our tool.

Future works are needed to improve the proposed method. Firstly, the algorithm introduced in this paper still has some trouble in processing large scale constraint sets. This deficiency may be resolved by some approximate techniques such as dividing the original constraint set into several small subsets:

$$S = S_1 \cup S_2 \ldots \cup S_k$$

and processing each subset separately. Secondly, we can try other type of constraints. Each constraint can be non-linear, a Boolean combination of numerical constraints or some other kind of mixed constraint. At last, we still need some efficient heuristics to improve the processing speed.

Acknowledgements

The authors are very grateful to the anonymous reviewers for their detailed comments and suggestions.

References

1. Cheng, B.M.W., Lee, J.H.M., Wu, J.C.K.: Speeding up constraint propagation by redundant modeling. In: Proceedings of CP-96, LNCS 1118. (1996) 91–103
2. Getoor, L., Ottosson, G., Fromherz, M., Carlson, B.: Effective redundant constraints for online scheduling. In: Proceedings of the Fourteenth National Conference on Artificial Intelligence(AAAI'97). (1997) 302–307
3. Zhang, J., Wang, X.: A constraint solver and its application to path feasibility analysis. International Journal of Software Engineering & Knowledge Engineering 11 (2001) 139–156
4. Zhang, J.: A path-based approach to the detection of infinite looping. In: Second Asia-Pacific Conference on Quality Software (APAQS'01). (2001) 88–94
5. Zhang, J.: Specification analysis and test data generation by solving Boolean combinations of numeric constraints. In: Proceedings of the First Asia-Pacific Conference on Quality Software, Hong Kong. (2000) 267–274
6. Clarke, E.M., Grumberg, O., Long, D.E.: Model checking and abstraction. ACM Transactions on Programming Languages and Systems 16 (1994) 1512–1542
7. Ball, T., Rajamani, S.K.: Bebop: A symbolic model checker for Boolean programs. In: Proc. of the 7th International SPIN Workshop, LNCS 1885. (2000) 113–130

8. Berkelaar, M.: LP_solve (May 2003) A public domain Mixed Integer Linear Program solver, availiable at http://groups.yahoo.com/group/lp_solve/.
9. Karmarkar, N.: A new polynomial-time algorithm for linear programming. Combinatorica **4** (1984) 373–395
10. Papadimitriou, C.H., Wolfe, D.: The complexity of facets resolved. Journal of Computer and System Sciences **37** (1988) 2–13
11. Papadimitriou, C.H., Yannakakis, M.: The complexity of facets (and some facets of complexity). In: Proceedings of the fourteenth annual ACM symposium on Theory of computing (STOC '82). (1982) 255–260
12. Jaffar, J., Michaylov, S., Stuckey, P.J., Yap, R.H.C.: The CLP($\mathcal{R}$) language and system. ACM Transactions on Programming Languages and Systems (July 1992) 339–395
13. Sankaranarayanan, S., Sipma, H.B., Manna, Z.: Scalable analysis of linear systems using mathematical programming. In: Proceedings of VMCAI 2005, LNCS 3385. (2005)
14. Muggleton, S., Raedt, L.D.: Inductive logic programming: Theory and methods. Journal of Logic Programming **19/20** (1994) 629–679
15. Silva, J., Sakallah, K.: Conflict analysis in search algorithms for propositional satisfiability. In: Proceedings of the 8th International Conference on Tools with Artificial Intelligence (ICTAI '96). (1996) 467
16. Bruni, R., Sassano, A.: Restoring satisfiability or maintaining unsatisfiability by finding small unsatisfiable subformulae. In: Proceedings of the Workshop on Theory and Applications of Satisfiability Testing. (2001)

A Search Engine for Mathematical Formulae

Michael Kohlhase and Ioan Sucan

Computer Science, International University Bremen
m.kohlhase@iu-bremen.de, i.sucan@iu-bremen.de

Abstract. We present a search engine for mathematical formulae. The
MATHWEBSEARCH system harvests the web for content representations
(currently MATHML and OPENMATH) of formulae and indexes them with
substitution tree indexing, a technique originally developed for access-
ing intermediate results in automated theorem provers. For querying, we
present a generic language extension approach that allows constructing
queries by minimally annotating existing representations. First experiments show that this architecture results in a scalable application.

1 Introduction

As the world of information technology grows, being able to quickly search data
of interest becomes one of the most important tasks in any kind of environment,
be it academic or not. This paper addresses the problem of searching mathematical formulae from a semantic point of view, i.e. to search for mathematical
formulae not via their presentation but their structure and meaning.

1.1 Semantic Search for Mathematical Formulae

Generally, searching for mathematical formulae is a non-trivial problem — especially if we want to be able to search occurrences of the query term as sub-
formulae:

1. *Mathematical notation is context-dependent.* For instance, binomial coefficients can come in a variety of notations depending on the context: $\binom{n}{k}$,
   $_nC^k$, C_k^n, and C_n^k all mean the same thing:[1] $\frac{n!}{k!(n-k)!}$. In a formula search we
   would like to retrieve all forms irrespective of the notations.
2. *Identical presentations can stand for multiple distinct mathematical objects,*
   e.g. an integral expression of the form $\int f(x)dx$ can mean a Riemann Integral, a Lebesgue Integral, or any other of the 10 to 15 known anti-derivative
   operators. We would like to be able to restrict the search to the particular
   integral type we are interested in at the moment.

---

[1] The third notation is the French standard, whereas the last one is the Russian one
(see [KK06] for a discussion of social context in mathematics). This poses a very
difficult problem for searching, since these two look the same, but mean different
things.

J. Calmet, T. Ida, and D. Wang (Eds.): AISC 2006, LNAI 4120, pp. 241–253, 2006.
© Springer-Verlag Berlin Heidelberg 2006

3. *Certain variations of notations are widely considered irrelevant*, for instance $\int f(x)dx$ means the same as $\int f(y)dy$ (modulo α-equivalence), so we would like to find both, even if we only query for one of them.

To solve this formula search problem, we concentrate on *content representations of mathematical formulae* (which solves the first two problems; see Section 1.3), since they are presentation-independent and disambiguate mathematical notions. Furthermore, we adapt term indexing techniques known from automatic theorem provers to obtain the necessary *efficiency and expressivity in query processing* (see Section 1.2) and to build in common equalities like α-equivalence.

Concretely, we present the web application MATHWEBSEARCH that is similar to a standard search engine like GOOGLE, except that it can retrieve content representations of mathematical formulae not just raw text. The system is released under the Gnu General Public License [FSF91] (see [Mat06] for details). A running prototype is available for testing at `http://search.mathweb.org`.

1.2 State of the Art in Math Search

There seem to be two general approaches to searching mathematical formulae. One generates string representations of mathematical formulae and uses conventional information retrieval methods, and the other leverages the structure inherent in content representations.

The first approach is utilized for the Digital Library of Mathematical Functions [MY03] and ACTIVEMATH system [LM06]: mathematical formulae are converted to text and indexed. The search string is similar to LaTeX commands and is converted to string before performing the search. This allows searching for normal text as well as mathematical content simultaneously but it cannot provide powerful mathematical search — for example searching for something like $a^2 + c = 2a$, where a must be the same expression both times, cannot be performed. An analogous idea to this would be to rely on an XML-based XQuery search engine. Both these methods have the important advantage that they rely on already existing technologies but they do not fully provide a mathematical formulae oriented search method.

The second approach is taken by the MBASE system [KF01], which applies the pattern matching of the underlying programming language to search for OMDOC-encoded [Koh06] mathematical documents in the knowledge base. The search engine for the HELM project indexes structural meta-data gleaned from Content MATHML representations for efficient retrieval [AS04]. The idea is that this metadata approximates the formula structure and can serve as a filter for very large term data bases. However, since the full structure of the formulae is lost, semantic equivalences like α-equivalence cannot be taken into account.

Another system that takes this second approach is described in [TSP06]. It uses term indexing for interfacing with Computer Algebra Systems while determining applicable algorithms in an automatically carried proof. This is closely related to what we present in this paper, the main difference being that we provide search for any formula in a predefined index, while in [TSP06] a predefined set of formulae characterizing an algorithm is automatically searched for in a changing index.

1.3 Content Representation for Mathematical Formulae

The two best-known open markup formats for representing mathematical formulae for the Web are MATHML [ABC+03] and OPENMATH [BCC+04][2] MATHML offers two sub-languages: Presentation MATHML for marking up the two-dimensional, visual appearance of mathematical formulae, and Content MATHML as a markup infrastructure for the functional structure of mathematical formulae. In Content MATHML, the formula $\int_0^a \sin(x)dx$ would be represented as the following expression:

Listing 1.1. Content Representation of an Integral

```
<apply><int/><bvar><ci>x</ci></bvar>
  <lowlimit><cn>0</cn></lowlimit><uplimit><cn>a</cn></uplimit>
  <apply><sin/><ci>x</ci></apply>
</apply>
```

The outer `apply` tags characterize this as as an application of an integral to the sin function, where x is the bound variable. The format differentiates numbers (`cn`) from identifiers (`ci`) and objects with a meaning fixed by the specification (represented by about 80 MATHML token elements like `int`, or `plus`). The OPENMATH format follows a similar approach, but replaces the fixed set of token elements for known concepts by an open-ended set of concepts that are defined in "content dictionaries": XML documents that specify their meaning in machine-readable form (see [BCC+04, Koh06] for details).

As content markup for mathematical formulae is rather tedious to read for humans, it is mainly used as a source to generate Presentation MATHML representations. Therefore content representations are often hidden in repositories, only their presentations are available on the web. In these cases, the content representations have to be harvested from the repositories themselves. For instance, we harvest the CONNEXIONS corpus, which is available under a Creative Commons License [Cre] for MATHWEBSEARCH. As we will see, this poses some problems in associating presentation (for the human reader) with the content representation. Other repositories include the ACTIVEMATH repository [MBG+03], or the MBASE system [KF01].

Fortunately, MATHML provides the possibility of "parallel markup", i.e. representations where content and presentation are combined in one tree[3] (see `http://functions.wolfram.com` for a widely known web-site that uses parallel markup).

---

[2] There are various other formats that are proprietary or based on specific mathematical software packages like Wolfram Research's MATHEMATICA [Wol02]. We currently support them if there is a converter to OPENMATH or MATHML.

[3] Modern presentation mechanisms will generate parallel markup, since that e.g. allows copy-and-paste into mathematical software systems [HRW02].

1.4   A Running Example: The Power of a Signal

A standard use case[4] for MATHWEBSEARCH is that of an engineer trying to solve a mathematical problem such as finding the power of a given signal $s(t)$. Of course our engineer is well-versed in signal processing and remembers that a signal's power has something to do with integrating its square, but has forgotten the details of how to compute the necessary integrals. He will therefore call up MATHWEBSEARCH to search for something that looks like $\int_?^? s^2(t)dt$ (for the concrete syntax of the query see Listing 1.3 in Section 3). MATHWEBSEARCH finds a document about Parseval's Theorem, more specifically $\frac{1}{T}\int_0^T s^2(t)dt = \sum_{k=-\infty}^{\infty}|c_k|^2$ where c_k are the Fourier coefficients of the signal. In short, our engineer found the exact formula he was looking for (he had missed the factor in front and the integration limits) and a theorem he may be able to use. So he would use MATHWEBSEARCH again to find out how to compute the Fourier transform of the concrete signal $s(t)$, eventually solving the problem completely.

2   Indexing Mathematical Formulae

For indexing mathematical formulae on the web, we will interpret them as first-order terms (see Subsection 4.1 for details). This allows us to use a technique from automated reasoning called *term indexing* [Gra96]. This is the process by which a set of terms is stored in a special purpose data structure (the **index**, normally stored in memory) where common parts of the terms are potentially shared, so as to minimize access time and storage. The indexing technique we work with is a form of tree-based indexing called *substitution-tree indexing*. A substitution tree, as the name suggests, is simply a tree

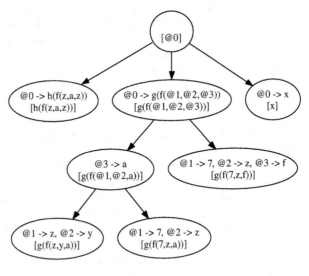

Fig. 1. An Index with Five Terms

where substitutions are the nodes. A term is constructed by successively applying substitutions along a path in the tree, the leaves represent the terms

---

[4] We use this simple example mainly for expository purposes here. Other applications include the retrieval of equations that allow to transform a formula, of Lemmata to simplify a proof goal, or to find mathematical theories that can be re-used in a given context (see [Nor06a] for a discussion of the latter).

stored in the index. Internal nodes of the tree are **generic terms** and represent similarities between terms.

The main advantage of substitution tree indexing is that we only store substitutions, not the actual terms, and this leads to a small memory footprint. Figure 1 shows a typical index for the terms $h(f(z, a, z))$, x, $g(f(z, y, a))$, $g(f(7, z, a))$, and $g(f(7, z, f))$. For clarity we present not only the substitutions in the node, but the term produced up to that node as well (between square brackets). The variables *@integer* are used to denote placeholder variables for parts that differ between terms. All placeholder variables are substituted before a leaf is reached.

Adding data to an existing index is simple and fast, querying the data structure is reduced to performing a walk down the tree. In contrast to automated reasoning our application does not need tree merging. Therefore we use substitutions only when building the index. Index building is done based on Algorithm 1. Once the index is built, we keep the actual term instead of the substitution at each node, so we do not have to recompute it with every search. Structure sharing methods conserve memory and make this tractable. To each of the indexed terms, some data is attached — an identifier that relates the term to its exact location. The identifier, location and other relevant data are stored in a database external to the search engine. We use XPointer [GMMW03] references to specify term locations (see Subsection 4.3 for more details).

Unfortunately, substitution tree indexing does not support subterm search in an elegant fashion, so when adding a term to the index, we add all its subterms as well. This simple trick works well: the increase in index size remains manageable (see Section 4.4) and it greatly simplifies the implementation. The rather small increase is caused by the fact that many of the subterms are shared among larger terms and they are only added once.

3   A Query Language for Content Mathematics

When designing a query language for mathematical formulae, we have to satisfy a couple of conflicting constraints. The language should be content-oriented and familiar, but it should not be specialized to a given content representation format. Our approach to this problem is to use a simple, generic extension mechanism for XML-based representation formats (referred to as base format) rather than a genuine query language itself.

The extension mechanism is represented by 4 tags and 4 attributes. The extension tags are `mq:query`, `mq:and`, `mq:or`, `mq:not`. The `mq:query` tag is used if one or more of the other extension tags are to be used and encloses the whole search expression. The tags `mq:and`, `mq:or`, `mq:not` may be nested and can contain tags from the base XML format, which may carry extension attributes (will be explained later). The `mq:and`, `mq:or`, `mq:not` tags are logical operators and may carry the `mq:target` attribute (the default value is **term**; only one other value allowed: **document**) which specifies the scope of the logical operator. Scope **term** is used to find *formulae* that contain the query terms as subformulae, while scope **document** does not restrict the occurrences of query terms.

There are 3 other attributes that may be used for any of the base format tags: mq:generic, mq:anyorder and mq:anycount. The first is used to specify that a term matches any subterm in the index; we call it a **generic term**. Note that generic terms with the same mq:generic value must be matched against identical target subterms. The mq:anyorder is used to specify that the order of the children can be disregarded. The mq:anycount attribute defines any number of occurrences of a certain base tag (if that base tag is known to be allowed multiple times). This is useful e.g. to define a variable number of bound variables (bvar MATHML).

Listing 1.2 shows a (somewhat contrived but illustrative) example query that searches for documents that contain at least one mathematical formula matching each of the math tags in the query. The first math tag will match any application of function f to three arguments, where at least two of the arguments are the same. The second math tag will match any formula containing at least two consecutive applications of the same function to some argument.

Listing 1.2. Example MATHML^Q Query

```
<mq:query xmlns:mq="http://mathweb.org/MathQuery">
  <mq:and mq:target="document">
    <math xmlns="http://www.w3.org/1998/Math/MathML">
      <apply><ci mq:anyorder="yes">f</ci>
        <ci mq:generic="same"/>
        <ci mq:generic="same"/>
        <ci mq:generic="other"/>
      </apply>
    </math>
    <math xmlns="http://www.w3.org/1998/Math/MathML">
      <apply><ci mq:generic="fun"/>
        <apply><ci mq:generic="fun"/><ci mq:generic="rest"/></apply>
      </apply>
    </math>
  </mq:and>
</mq:query>
```

Given the above, the MATHML^Q query of our running example has the form presented in Listing 1.3. Note that we do not know the integration limits or whether the formula is complete or not. Expressing this in MATHML^Q[5]

Listing 1.3. Query for Signal Power

```
<math xmlns="http://www.w3.org/1998/Math/MathML"
      xmlns:mq="http://mathweb.org/MathQuery">
  <apply><int/>
    <domainofapplication mq:generic="domain"/>
    <bvar> <ci mq:generic="time"/> </bvar>
    <apply><power/>
      <apply><ci mq:generic="fun"></ci><ci mq:generic="time"/></apply>
      <cn>2</cn>
    </apply>
  </apply>
</math>
```

---

[5] This is equivalent to the STRING representation #int(bvarset(bvar(@time)), @domain,power(@fun(@time),nr(2))).

4 The MathWebSearch Application

We have built a web search engine around the indexing technique explained above. Like commercial systems, MATHWEBSEARCH consists of three system components: a set of web crawlers[6] that periodically scan the Web, identify, and download suitable web content, a search server encapsulates the index, and a web server that communicates the results to the user. To ensure scalability, we have the system architecture in Figure 2, where individual search servers are replicated via a search meta-server that acts as a front-end.

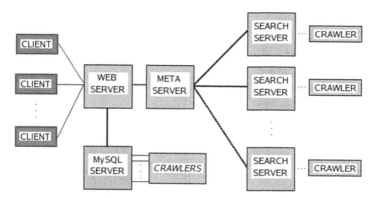

Fig. 2. The Architecture of the MATHWEBSEARCH Application

4.1 Input Processing

MATHWEBSEARCH can process any XML-based content mathematics. Currently, the system supports MATHML and OPENMATH (and MATHEMATICA notebooks via the system's MATHML converter). We will discuss input processing for the first here.

Given an XML document, we create an index term for each of its `math` (this is the case for MATHML) elements. Consider the example on the right: We have the standard mathematical notation of an equation (1), its Content MATHML representation (2), and

1) Mathematical expression:
$$f(x) = y$$

3) Term representation:
$$eq(f(x), y)$$

2) Content MATHML:
```
<apply><eq/>
  <apply>
    <ci>f</ci>
    <ci>x</ci>
  </apply>
  <ci>y</ci>
</apply>
```

the term we extract for indexing (3). As previously stated, any mathematical construct can be represented in a similar fashion.

When we process the Content MATHML formulae, we roughly create a term for every `apply` element tag, taking the first child of `apply` as the function and the rest of the children as arguments. Of course, cases like vectors or matrices

---

[6] At the moment, we are employing an OAI-based [OAI02] crawler for repositories like CONNEXIONS and a standard web-crawler for finding other MATHML repositories.

have to be treated specially. In some cases — e.g. for integrals — the same content can be encoded in multiple ways. Here, a simple standardization of both the indexed formulae and the queries leads to an improved recall of the search: for instance we can find an integral specified with `lowlimit` and `uplimit` tags (see Listing 1.1) using a query integral specified with the `interval` element[7], since we standardize argument order and integration domain representation for integrals.

Search modulo α-renaming becomes available via a very simple input processing trick: during input processing, we add a `mq:generic` attribute to every bound variable (but with distinct strings for different variables). Therefore in our running example the query variable t (`@time` in Listing 1.3) in the query $\int_?^? s^2(t)dt$ is made generic, therefore the query would also find the variant $\frac{1}{T}\int_0^T s^2(x)dx = \Sigma_{k=-\infty}^{\infty}|c_k|^2$: as t is generic it could principally match any term in the index, but given the MATHML constraints on the occurrences of bound variables, it will in reality

Fig. 3. Searching for Signal Power

only match variables (thus directly implementing α-equivalence).

Presentation MATHML in itself does not offer much semantic information, so it is not particularly well suited for our purposes. However, most of the available MATHML on the World Wide Web is Presentation MATHML. For this reason, we index it as well. The little semantic information we are offered, like when a number (`mn`), operator (`mo`) or identifier (`mi`) are defined, we use for recovering simple mathematical expressions which we then index as if the equivalent Content MATHML were found. This offers the advantage that when using a mixed index (both Presentation and Content MATHML) we have increased chances of finding a result.

4.2   Term Indexing

As the term retrieval algorithm for substitution trees is standard, we will concentrate on term insertion and memory management here. In a nutshell: we insert

---

[7] STRING representation: `#int(bvarset(bvar(id(x))), intervalclosed(lowlimit (nr(0)), uplimit(nr(a))),sin(id(x)))`.

a term in the first suitable place found. This will not yield minimal tree sizes, but (based on the experiments carried out in [Gra96]) the reduction in number of internal nodes is not significant and the extra computation time is large.

Algorithm 1. INSERT_TERM($node, term$)

      $found = true$
 2: **while** $found$ **do**
      $found = false$
 4:    **for all** sons of $node$ **do**
        if COMPLETE_MATCH($son.term, term$) **then**
 6:        $node = son, found = true$
          $break$
 8:      **end if**
      end for
10: **end while**
      $match = $ PARTIAL_MATCH($node.term, term$)
12: **for all** sons of $node$ **do**
      if PARTIAL_MATCH($son.term, term$) $> match$ **then**
14:    **return** INSERT_WITH_SEPARATION($node, son, term$)
      end if
16: **end for**
      return INSERT_AT($node, term$)

Concretely, an initial empty index contains a single node with the empty substitution. The term produced by that node is always the generic term @0. When a new term is to be inserted, we always try to insert from the root, using the algorithm INSERT_TERM, where

1. COMPLETE_MATCH checks if the second argument is an instance of the first argument. It uses a simple rule: a term is only an instance of itself and of any placeholder variable.
2. PARTIAL_MATCH returns an integer that represents the number of equal subterms.
3. INSERT_AT adds a new leaf to $node$ with a substitution from $node.term$ to $term$ unless that substitution is empty.
4. INSERT_WITH_SEPARATION creates a son of $node$ named n with a substitution to the shared parts of $son.term$ and $term$; it then adds proper substitutions to $son.term$ and $term$ from $n.term$ as sons of n.

4.3   Result Reporting

For a search engine for mathematical formulae we need to augment the set of result items (usually page title, description, and page link) reported to the user for each hit. As typical pages contain multiple formulae, we need to report the exact occurrence of the hit in the page. We do this by supplying an XPOINTER reference where possible. Concretely, we group all occurrences into one page item that can be expanded on demand and within this we order the groups by number of contained references. See Figure 4 for an example.

For any given result, a detailed view is available. This view shows the exact term that was matched and the used substitution (a mapping from the query variables specified by the `mq:generic` attributes to certain subterms) to match that specific term. A more serious problem comes from the fact that — as mentioned above — content representations are often the source from which presentations are generated. If MATH-WEBSEARCH can find out the correspondence between content and presentation documents, it will report both to the user. For instance for CONNEXIONS we present two links as results: one is the *source link*, a link to the document we actually index, and the *default link*, a link to the more aesthetically pleasing presentation document.

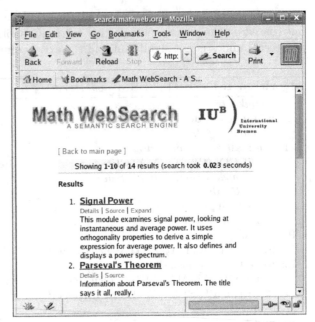

Fig. 4. Results for the Search in Fig. 3

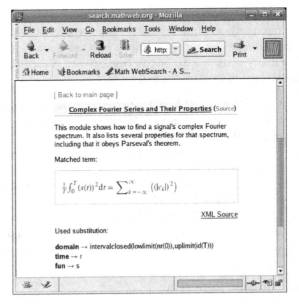

Fig. 5. Detailed Search Results

4.4 Case Studies and Results

We have tested our implementation on the content repository of the CONNEXIONS Project, available via the OAI protocol [OAI02]. This gives us a set of over 3,400 articles with mathematical expressions to work on. The number of terms represented in these documents is approximately 53,000 (77,000 including sub-terms). The average term depth is 3.6 and the maximal one is 14. Typical query execution times on this index are in the range of milliseconds. The search in our running example takes 23 ms for instance. There are, however, complex searches (e.g. using the `mq:anyorder` attribute) that internally call the searching routine multiple times and take up to 200 ms but for realistic examples execution time is below 50 ms. We also built an index of the 87,000 Content MathML formulae from `http://functions.wolfram.com`. Here, term depths are much larger (average term depth 8.9, maximally 53) resulting in a much larger index: 1.6 million formulae; total number of nodes in the index is 2.9 million, resulting in a memory footprint of 770MB. First experiments indicate that search times are largely unchanged by the increase in index size (for reasonably simple searches).

5   Conclusions and Future Work

We have presented a search engine for mathematical formulae on the Internet. In contrast to other approaches, MATHWEBSEARCH uses the full content structure of formulae, and is easily extensible to other content formats. A first proto-type is available for testing at `http://search.mathweb.org`. We will continue developing MATHWEBSEARCH into a production system.

A current weakness of the system is that it can only search for formulae that match the query terms up to α-equivalence. Many applications would benefit from similarity-based searches or stronger equalities. For instance, our search in Listing 1.3 might be used to find a useful identity for $\int_{\infty}^{0} f(x) \cdot g(x)dx$, if we know that $s(x) \cdot s(x) = s^2(x)$. MATHWEBSEARCH can be extended to a *E-Retrieval* engine (see [Nor06b]) without compromising efficiency by simply E-standardizing index and query terms.

We plan to index more content, particularly more OPENMATH. In the long run, it would be interesting to interface MATHWEBSEARCH with a regular web search engine and create a powerful, specialized, full-feature application. This would resolve the main disadvantage our implementation has – it cannot search for simple text. Finally we would like to allow specification of content queries us-ing more largely known formats, like LATEX: strings like `\frac{1}{x^2}` or `1/x^2` could be processed as well. This would make MATHWEBSEARCH accessible for a larger group of users.

References

[ABC+03]   Ron Ausbrooks, Stephen Buswell, David Carlisle, et al. Mathemati-cal Markup Language (MathML) version 2.0 (second edition). W3C recommendation, World Wide Web Consortium, 2003. Available at http://www.w3.org/TR/MathML2.

[AS04]   Andrea Asperti and Matteo Selmi. Efficient retrieval of mathematical statements. In Andrea Asperti, Grzegorz Bancerek, and Andrej Trybulec, editors, *Mathematical Knowledge Management, MKM'04*, number 3119 in LNCS, pages 1–4. Springer Verlag, 2004.

[BCC+04]   Stephen Buswell, Olga Caprotti, David P. Carlisle, Michael C. Dewar, Marc Gaetano, and Michael Kohlhase. The Open Math standard, version 2.0. Technical report, The Open Math Society, 2004. http://www.openmath.org/standard/om20.

[Cre]   Creative Commons. Web page at http://creativecommons.org.

[FSF91]   Free Software Foundation FSF. Gnu general public license. Software License available at http://www.gnu.org/copyleft/gpl.html, 1991.

[GMMW03]   Paul Grosso, Eve Maler, Jonathan Marsh, and Norman Walsh. Xpointer framework. W3c recommendation, World Wide Web Constortium W3C, 25 March 2003.

[Gra96]   Peter Graf. *Term Indexing*. Number 1053 in LNCS. Springer Verlag, 1996.

[HRW02]   Sandy Huerter, Igor Rodionov, and Stephen Watt. Content-faithful transformations for mathml. In *Second International Conference on MathML and Technologies for Math on the Web*, Chicago, USA, 2002. http://www.mathmlconference.org/2002/presentations/huerter/.

[ICW06]   Tetsuo Ida, Jacques Calmet, and Dongming Wang, editors. *Proceedings of Artificial Intelligence and Symbolic Computation, AISC'2006*, number 4120 in LNAI. Springer Verlag, 2006.

[KF01]   Michael Kohlhase and Andreas Franke. MBase: Representing knowledge and context for the integration of mathematical software systems. *Journal of Symbolic Computation; Special Issue on the Integration of Computer algebra and Deduction Systems*, 32(4):365–402, September 2001.

[KK06]   Andrea Kohlhase and Michael Kohlhase. Communities of practice in MKM: An extensional model. In Jon Borwein and William M. Farmer, editors, *Mathematical Knowledge Management, MKM'06*, number 4108 in LNAI. Springer Verlag, 2006.

[Koh06]   Michael Kohlhase. OMDoc *An open markup format for mathematical documents (Version 1.2)*. Number 4180 in LNAI. Springer Verlag, 2006. in press http://www.mathweb.org/omdoc/pubs/omdoc1.2.pdf.

[LM06]   Paul Libbrecht and Erica Melis. Methods for access and retrieval of mathematical content in ACTIVEMATH. In N. Takayama and A. Iglesias, editors, *Proceedings of ICMS-2006*, number 4151 in LNAI. Springer Verlag, 2006. forthcoming.

[Mat06]   Math web search. Web page at http://kwarc.eecs.iu-bremen.de/projects/mws/, seen July 2006.

[MBG+03]   Erica Melis, Jochen Büdenbender, George Goguadze, Paul Libbrecht, and Carsten Ullrich. Knowledge representation and management in ACTIVEMATH. *Annals of Mathematics and Artificial Intelligence*, 38:47–64, 2003. see http://www.activemath.org.

[MY03]   B. Miller and A. Youssef. Technical aspects of the digital library of mathematical functions. *Annals of Mathematics and Artificial Intelligence*, 38(1-3):121–136, 2003.

[Nor06a]   Immanuel Normann. Enhanced theorem reuse by partial theory inclusionss. In Ida et al. [ICW06].

[Nor06b]   Immanuel Normann. Extended normalization for e-retrieval of formulae. to appear in the proceedings of Communicating Mathematics in the Digital Era, 2006.

[OAI02]   The open archives initiative protocol for metadata harvesting, June 2002. Available at http://www.openarchives.org/OAI/openarchivesprotocol.html.

[TSP06]   Frank Theiß, Volker Sorge, and Martin Pollet. Interfacing to computer algebra via term indexing. In Silvio Ranise and Roberto Sebastiani, editors, CALCULEMUS-2006, 2006.

[Wol02]   Stephen Wolfram. *The Mathematica Book*. Cambridge University Press, 2002.

Hierarchical Representations with Signatures for Large Expression Management

Wenqin Zhou[1], J. Carette[2], D.J. Jeffrey[1], and M.B. Monagan[3]

[1] University of Western Ontario, London ON, Canada, N6A 5B7
{wzhou7, djeffrey}@uwo.ca
[2] McMaster University, Hamilton, ON, Canada, L8S 4L8
carette@mcmaster.ca
[3] Simon Fraser University, Burnaby, B.C. Canada, V5A 1S6
monagan@cecm.sfu.ca

Abstract. We describe a method for managing large expressions in symbolic computations which combines a hierarchical representation with signature calculations. As a case study, the problem of factoring matrices with non-polynomial entries is studied. Gaussian Elimination is used. Results on the complexity of the approach together with benchmark calculations are given.

Keywords: Hierarchical Representation, Veiling Strategy, Signature, Zero Test, Large Expression Management, Symbolic LU Decomposition, Time Complexity.

1 Introduction

One of the attractions of MAPLE is that it allows users to tackle large problems. However, when users undertake large-scale calculations, they often find that expression swell can limit the size of the problems they can solve [31]. Typically, users might meet two types of expression swell: one type we can call *inherent* expression swell, and the other *intermediate* expression swell.

A number of strategies have been proposed for coping with the large expressions generated during symbolic computation. We list a number of them here, but lack of space precludes an extensive discussion.

- *Avoid the calculation.* This strategy delays computation of a quantity whose symbolic expression is large until numerical data is given. For example, if the determinant of a matrix is needed in a computation, one uses an inert function until the point at which the elements of the matrix can be evaluated numerically, and then jumps to a numerical evaluation.
- *Use signatures.* See, for example, [5]. Signatures are one of the ideas used in this paper.
- *Use black-box calculations.* This is a strength of the Linbox project [7].
- *Approximate representations.* This is the growing area of symbolic-numeric computation.

J. Calmet, T. Ida, and D. Wang (Eds.): AISC 2006, LNAI 4120, pp. 254–268, 2006.

- *Use hierarchical representations.* These are studied in this paper, and the term will be abbreviated to HR.

Each of the above methods is successful for a different class of problems. This paper addresses a class of problems in which large expressions are built up from identifiable sub-expressions, and which as a result are suitable applications for hierarchical representations (HR). Hierarchical representations *per se* are not new in computer algebra. Similar ideas have appeared in the literature under a variety of names. Examples are as follows:

- *Maple DAGs.* Expressions in MAPLE are represented as DAGs with sub-expressions being reused and hence stored only once. For example, [13] uses this data structure to compute the determinant of polynomial matrices.
- *Straight-line programs.* The DAGWOOD [9] system computes with straight-line programs.
- *Common subexpression identification.* The MAPLE command `codegen[opti-mize]` searches a large expression for common subexpressions. (Also available as an option to commands in `CodeGeneration`) [16]
- *Computation sequences and* MAPLE*'s CompSeq.* An early example is given by Zippel in 1993 [11]. The function `CompSeq` in MAPLE is a placeholder for representing a computation sequence.
- *Large Expression Management (LEM).* This term was introduced in [10], and is the name of a MAPLE package.

The goal of this work is the combination of HR with signatures. We do this by modifying the `LargeExpressions` package in MAPLE and then applying it to a case study. The case study comes from DYNAFLEX [8], a system which computes the equations of motion for a mechanical device created from rigid or flexible bodies. It uses MAPLE for its computations and requires the factoring of matrices whose elements are multivariate polynomials or non-polynomial functions. In this paper, therefore, we consider the factoring of matrices with elements that are multivariate polynomials and exponential polynomials. We could have considered any application where the algorithm at hand only requires zero-recognition on the elements (as well as basic "arithmetic" operations); if obtaining other information, like degree or structural "shape" is absolutely necessary, this would need new ideas on top of the ones we present here.

2   Hierarchical Representation

The first point to establish is the need for a modified HR implementation. We begin by giving our definition of HR for this paper, with the purpose of distinguishing our implementation from similar definitions, such as straight-line programs.

Definition 1. *An exponential polynomial p over a domain $\mathbb{K}$ and a set of independent variables $\{x_1, ..., x_m\}$ is a polynomial $p \in \mathbb{K}[x_1, ..., x_m, y_1, ..., y_m]$ with $y_k = e^{x_k}, k = 1..m$.*

Definition 2. *A hierarchical representation (HR) over a domain $\mathbb{K}$ and a set of independent variables $\{x_1, ..., x_m\}$ is an ordered list $[S_1, S_2, ..., S_l]$ of symbols, together with an associated list $[D_1, D_2, ..., D_l]$ of definitions of the symbols. For each S_i with $i \geq 1$, there is a definition D_i of the form $S_i = f(\sigma_1, \sigma_2, ..., \sigma_k)$ where $f \in \mathbb{K}[\sigma_1, ..., \sigma_k]$, and each σ_j is either a symbol in $[S_1, S_2, ..., S_{i-1}]$ or an exponential polynomial in the independent variables.*

Hierarchical representation is a more general idea than the (algebraic) straight-line program defined in [14] and used in [12, 9, 23, 27]. A given expression can have different HR, i.e. different lists of definitions $[D_1, D_2, ...]$. The strategy used to assign the symbols during the generation of expressions will be something that can be varied by the implementation. The reason for inquiring an ordered list is to exclude implicit definitions. Details on how to build HR are in the section 4.

Remark 1. *An important part of the creation of HRs is the order in which assignments happen. For instance to use `codegen[optimize]`, an expression is completely generated first. Clearly, some expressions will be too large to be generated explicitly, in which case `codegen[optimize]` would have nothing to work with.*

Remark 2. *There are many types of computational procedures which naturally generate HR. One example is Gaussian elimination, which we study here. Another is the calculation described in [10]. Other computations that are known to generate large expressions, for example Gröbner basis calculations, do not have a obvious hierarchy, although [13] hints at one.*

Remark 3. *One can understand HRs as a compromise between full computations and no computations. Enough of the computation is performed to give a correct result, but not so much that a closed-form can be output. It is a compromise between immediately returning a placeholder and never returning a giant result.*

The key issue is control over expression simplification; this includes the identification of a zero expression. In an ordinary computer algebra system, the usual way this proceeds is by normalizing expressions, a step which frequently destroys the HR and causes the appearance of additional expression swell. For example, most systems will normalize[1] the expression

$$\frac{(2781 + 8565x - 4704x^2)^{23}(1407 + 1300x - 1067x^2)^{19} - \alpha}{(1809 + 9051x + 9312x^2)^{19}(2163 - 2162x + 539x^2)^{19} * (27 + 96x)^4(103 - 49x)^4}$$

by expanding it. The same strategy would be used by the system whether $\alpha = +1$ or $\alpha = -1$. However, in one case the result is zero, while in the other it is just a large expression, which now fills memory.

Consequently, the main purpose of creating user-controlled HR is to control normalization and to integrate more different (often more efficient) zero-testing

---

[1] Normalization is often confused with simplification, but [29] argues otherwise.

strategies into a computation in a convenient way. As well as creating a HR, one must give equal importance to the prevention of its destruction.

The original `LargeExpressions` package in MAPLE was created as a result of the investigations in [10]. The authors had external mathematical reasons for knowing that their expressions were nonzero, and hence no provision was made for more efficient testing. In the current implementation, we intend to apply the resulting code more widely, with the consequent need to test efficiently for zero. This we do by incorporating signature testing.

The basic action is the creation of a label for a sub-expression. The command for this was given the name `Veil` in the original `LargeExpressions` package, and so this will be used as the general verb here. Once an expression has been veiled, the system treats it as an inert object unless the user or the program issues an unveiling command, which reveals the expression associated with the label.

3 Signatures

The idea of using signatures is similar to the probabilistic identity testing of Zippel-Schwartz theorems [2, 1], and to the basis of `testeq` in MAPLE by Gonnet [3, 4], also studied in [5, 6]. The original polynomial results of Zippel-Schwartz were extended to other functions in [3, 4, 15].

Since we need to apply our method to matrices containing exponential polynomials, we first define a signature function that is appropriate for this class of functions.

Definition 3. *Given an expression e, an exponential polynomial, the signature $s(e)$ with characteristic prime p is defined in the following steps.*

- *If e is a variable, then its signature equals a random value of $\mathbb{Z}/p\mathbb{Z}$.*
- *If $e = e_1 + e_2$ then $s(e) = s(e_1) + s(e_2) \mod p$.*
- *If $e = e_1 * e_2$ then $s(e) = s(e_1) * s(e_2) \mod p$.*
- *If $e = e_1^n$, where n is a positive integer, then $s(e) = s(e_1)^n \mod p$.*
- *If $e = a^x$ is an exponential function a^x, where a could be the base of natural logarithms or any (non-zero) number less than p, then $s(e) = r^t \mod p$, where r is a primitive root of p, and $t = x \mod \phi(p)$. Here $\phi(p)$ is Euler's totient function.*

Note that unlike [3], we explicitly do not treat towers of exponentials, but only simple exponentials, which is frequently sufficient in applications.

Proposition 1. *For all non-zero $y \in \mathbb{Z}/p\mathbb{Z}$, there exists a unique $x \in \mathbb{Z}/\phi(p)\mathbb{Z}$, s.t. $s(a^x) = y$.*

PROOF: By the definition of a signature $s(a^x)$, $r = s(a)$ is a primitive root modulo p. By the definition of a primitive root of a prime [26], the multiplicative order of r modulo p is equal to $\phi(p) = p - 1$. So the powers r^i, $i = 1..p - 1$ range over all elements of $\mathbb{Z}/p\mathbb{Z} - \{0\}$. ∎

For the following theorems, we suppose that all random choices are made under the uniform probability distribution.

Theorem 1. *(Zippel-Schwartz theorem) Given $F \in \mathbb{Z}[x_1, ..., x_n]$, $F \mod p \neq 0$, and $\deg(F) \leq d$, the probability $Pr\{s(F) = 0 | F \neq 0\} \leq \frac{d}{p}$.*

A proof can be found in [2, 1].

Theorem 2. *Let $F \in \mathbb{Z}[y], y = a^x$, where a could be the base of the natural logarithms or any (non-zero) number less than p, $F \mod p \neq 0, \deg(F) = d$, the probability $Pr\{s(F) = 0 | F \neq 0\} \leq \frac{d}{p-1}$.*

PROOF: The polynomial $F \in \mathbb{Z}[y]$ has at most d roots in $\mathbb{Z}/p\mathbb{Z}$. For $z \in \mathbb{Z}/p\mathbb{Z}$ such that $F(z) = 0$, Proposition 1 gives that there exists unique $u_z \in \mathbb{Z}/\phi(p)\mathbb{Z}$, s.t. $s(a^{u_z}) = z$. Thus the number of values x such that $s(F(a^x)) = 0$ is at most d. Because the total number of choices for nonzero y is $p - 1$, the probability $Pr\{s(F) = 0 | F \neq 0\} \leq \frac{d}{p-1}$. ∎

Theorem 3. *Let $F \in \mathbb{Z}[x, y], y = a^x$, where a could be the base of natural logarithms or any non-zero integer less than p, $F \mod p \neq 0$, and $\deg(F) = d$, the probability $Pr\{s(F) = 0 | F \neq 0\} \leq \frac{d}{p-1}$.*

PROOF: The polynomial $F \in \mathbb{Z}[x, y]$ has at most dp roots in $\mathbb{Z}/p\mathbb{Z}$. For (x_i, y_i) such that $F(x_i, y_i) = 0$, based on Proposition 1, there exists unique x_u, s.t. $s(a^{x_u}) = y_i$. If $x_u = x_i$, then the solution (x_i, y_i) is the one to make $s(F(x_i, a^{x_i})) = 0$. Therefore the number of roots for x, such that $s(F(x, a^x)) = 0$ is at most dp.

As the total number of (independent) choices for (x, y) is $p(p - 1)$, the probability $Pr\{s(F) = 0 | F \neq 0\} \leq \frac{d.p}{p(p-1)} = \frac{d}{p-1}$. ∎

Signatures can be used to test if an expression is zero, as `testeq` does. However, `testeq` always starts fresh for each new zero-test. This is a source of inefficiency when the signature is part of a continuing computation, and will be seen in later benchmarks which use `testeq`.

The signature of the expression is computed before veiling an expression in HR. This value then becomes the signature of the veiling symbol. When that symbol itself appears in an expression to be veiled, the signature of the symbol is used in the calculation of the new signature. In particular, it is not necessary to unveil any symbol in order to compute its signature.

Other important references on this topic are [22, 21, 24]. Applications of this basic test is the determination of singularity and rank of a matrix [25] shows two applications of this basic technique: determining whether a matrix (of polynomials) is singular, and determining the rank of a polynomial matrix.

4   An Implementation of HR with Signatures

The simplest method for tracking HRs is to maintain an association list between an expression and its (new) label. This is easily implemented via (hash) tables; one table associates a "current" number to a symbol (used as an indexed name to generate fresh labels), and another table which associates to each indexed name to the underlying (unveiled) expression. The indexed names play the role

of the ordered list of symbols in definition 2. The main routine is `Veil[K](A)`.
Here K is the symbol and A is the expression to be veiled. This routine stores
the expression A in the slot associated to $K[c]$ where c is the "current" number,
increments c and returns the new symbol. For interactive use, a wrapper function
`subsVeil` can be used.

```
> subsVeil:=(e,A)->algsubs(e=Veil[op(procname)](e),A);
> A:= (x+y)^{10} + e^{x+y} + (x^2+1)^5 - 1:
> B:= subsVeil[K](x^2+1,A):
> C:= subsVeil[K](x+y, B);
```

$$k_2^{10} + e^{k_2} + k_1^5 - 1$$

Notice that there is no longer a danger of expanding the expression $(x^2 + 1)^5 - 1$
in a misguided attempt to simplify it. In order to retrieve the original expression,
one uses `Unveil`.

```
> Unveil[K](C) ;
```

$$(x + y)^{10} + e^{x+y} + (x^2 + 1)^5 - 1$$

At present, the expressions corresponding to K are stored in the memory space
of the implementation module[2]. After a computation is completed and the intermediate results are no longer needed. The memory occupied by K can be
cleared using the command `forgetVeil(K)`.

The signature must be remembered between calls to `Veil`, as commented
above. The signature could be attached more directly to K, or kept in a separate
array specified by the user. The above implementation seemed to provide the
best *information hiding*. Until we see, with more experience of case studies, which
method is best, we have for the present implementation used the MAPLE facility
`option remember` internally for handling some of the tables, for convenience
and efficiency. Thus after a call to the routine `SIG`, the signature of any veiled
expression is stored in an internal remember table and not re-computed.

The use of `Veil` to generate HRs together with the calculation of signatures
will be called Large Expression Management (LEM). In fact it is just expression
management, because the `Veil` tool can be used even on expressions which are
not large, for the convenience they give to understanding algebraic structure.

5   LU Factoring with LEM

A well-known method for solving matrix equations is LU factoring, in which a
matrix A is factored such that $PA = LU$, where L and U are triangular matrices
and P is a permutation matrix; see [18] for further details. The MAPLE command `LUDecomposition` uses large amounts of memory and is very slow for even
moderately sized matrices of polynomials. The large expression trees generated

---

[2] In other words, it is a stateful module, à la Parnas, which is also rather like a
singleton class in OO.

internally are part of the reason for this slowdown, but equally significant is the time taken to check for zero. For example,

```
> M :=Matrix(10,10,symbol=m);
> LinearAlgebra[LUDecomposition](M);
```

This LU factoring will not terminate. If the environment variable `Normalizer` is changed from its default of `normal` to the identity function, i.e. Normalizer:= $x- > x$, then the LU factoring can complete. This is why Large Expression Management requires both HR and signatures for its zero-test.

We modified the standard code for LU decomposition to include veiling and signature calculations. At the same time, we generalized the options for selecting pivots and added an option to specify a veiling strategy. One can see [30] for even more design points, and a general design strategy, for this class of algorithms.

Our LU factoring algorithm in high-level pseudo-code:

```
Get maximum_column, maximum_row for matrix A
  For current_column from 1 to maximum_column
    for current_row from current_column to maximum_row
      Check element for zero.
      Test element for being ''best'' pivot
      Veil pivot [invoke Veiling strategy]
      move pivot to diagonal, recording interchanges.
      row-reduce matrix A with veiling strategy
      store multipliers in L
    end do:
  end do:
return permutation_matrix, L, reduced matrix A
```

The function has been programmed with the following calling sequence.

```
    LULEM(A, K, p, Pivoting, Veiling, Zerotesting)
Parameters
    A           - square matrix
    K           - unassigned name to use as a label
    p           - prime
    Pivoting    - decide a pivot for a column
    Veiling     - decide to veil an expression or not
    Zerotesting - decide if the expression is zero.
```

5.1 Pivoting Strategy

The current MAPLE LUDecomposition function selects one of two pivoting strategies on behalf of the user, based on data type. Thus, at present, we have
```
> LUDecomposition(<<12345,1>|<1,1>>);
```

$$\begin{bmatrix} 1 & 0 \\ 0 & 1 \end{bmatrix}, \begin{bmatrix} 1 & 0 \\ 1/12345 & 1 \end{bmatrix}, \begin{bmatrix} 12345 & 1 \\ 0 & 12344/12345 \end{bmatrix}$$

even though

$$\begin{bmatrix} 0 & 1 \\ 1 & 0 \end{bmatrix}, \begin{bmatrix} 1 & 0 \\ 12345 & 1 \end{bmatrix}, \begin{bmatrix} 1 & 1 \\ 0 & -12344 \end{bmatrix}$$

is more attractive. If the matrix contains floating-point entries, partial pivoting is used.

```
> LUDecomposition(<<1,12345.>|<1,1>>);
```

$$\begin{bmatrix} 0 & 1 \\ 1 & 0 \end{bmatrix}, \begin{bmatrix} 1. & 0. \\ (8.1)10^{-5} & 1. \end{bmatrix}, \begin{bmatrix} 12345. & 1 \\ 0 & 0.99992 \end{bmatrix}$$

Since we wished to experiment with different pivoting strategies, we made it an option. Rather than make up names, such as 'partial pivoting' or 'non-zero pivoting', to describe strategies, we allow the user to supply a function which takes 2 arguments. The function returns true if the second argument is a preferred pivot to the first argument. For example, the preferred pivoting strategy for the example above (choose the smallest pivot) can be specified by the function `(p1,p2)->evalb(abs(p2)<abs(p1))`. In a symbolic and veiling context there are a number of conceivable strategies which one might wish to try. These can be based on operation count, size of expression or number of indeterminants. However, the definition of LU factors only allows pivoting on one column, so no form of full pivoting is offered.

5.2   Veiling Strategy

In the same spirit of experimentation, we have used a function to specify a veiling strategy. This function takes one argument and returns true if the expression should be veiled. The current **LargeExpressions** package, for example, follows a strategy of ignoring integers. Thus an integer, however large, cannot be veiled at present. Similarly, integer content is extracted from expressions before veiling. Rather than make these decisions in advance, we leave them to the declaration of a veiling-strategy function.

Of particular interest is the 'granularity' of the HR, namely whether one veils every pairwise operation, or whether one waits until an expression of a pre-determined size is allowed to accumulate. In the former case, the HR would look similar to a straight-line program as defined in [14]. For our experiments, we have based our strategies on the MAPLE **length** command, as being a convenient measure of expression complexity.

5.3   Zero Test Strategy

We need to do zero tests to find pivots. This can also help us simplify our expressions, if needed. During the LU factoring, we use signatures to perform this test quickly (more precisely, in random polynomial time). It is important to note that for LU factoring, we only need to find a provably non-zero pivot, so that a false positive (an entry which seems to be zero but in fact is not) rarely leads to a problem. And, in that case, we can always resort to a full zero-test.

We use the signatures computed along with the hierarchical representations to do the zero test for the expressions in HR. But a user could choose any MAPLE commands, like `Normalizer`, `testeq`, `simplify` or `evalb`, to do the zero test. Which one is best depends on the application at hand.

6 Time Complexity Analysis for LU with Veiling and Signatures

Since our case study compares current LU factoring and LU factoring with expression management, it is important to have some measure of the time complexity of each procedure. We therefore start with the time complexity of conventional Gaussian elimination (see [19, 20, 17] for early work). Although some cases of the following theorems are "well known", there seem to be no convenient published statement of them.

Here we consider the time complexity measure is the number of bit operations, which can be rigorously defined as the number of steps of a Turing or register machine or the number of gates of a Boolean circuit implementing the algorithm. [28] Throughout, we make the simplifying assumption that entries grow linearly, in both degree and in coefficient size. This is actually optimistic, as growth is usually worse than this if we apply the classical Gaussian elimination algorithm.

Theorem 4. *For an n by n matrix $A = [a_{i,j}]$, with $a_{i,j} \in \mathbb{Z}[x_1, \ldots, x_m]$, the time complexity of LU factoring for A is at least $\Omega(n^{2m+5})$ for naive arithmetic.*

PROOF: Let $d_r > \max_{i,j} \deg(a_{i,j}, x_r)$. Then $d_1 d_2 \times \ldots \times d_m$ bounds the number of non-zero terms of the polynomials $a_{i,j}$. Let l bound the length of the largest integer coefficient in the $a_{i,j}$. Suppose the degree of the polynomials $a_{i,j}$ in x_r and the size of their integer coefficients are growing linearly with each step of the LU factorization, i.e., at step k, $\deg(a_{i,j}, x_r) < k \, d_r$ and the largest integer coefficient is bounded by lk. When we do LU factoring, at the k'th step, we have $(n - k)^2$ entries to manipulate. For each new entry from step $k - 1$ to step k, we need to do at least one multiplication, one subtraction and one division. The cost will be at least $\Omega((k \, d_1 \; k \, d_2 \ldots k \, d_m \; l \; k)^2)$ for naive arithmetic.

The total cost for the LU factoring will be at least $\sum_{k=1}^{n-1}(n-k)^2 \times \Omega((kd_1 \times kd_2 \times \ldots \times kd_m \times kl)^2) = \Omega(d_1^2 d_2^2 \ldots d_m^2 l^2 n^{2m+5})$ (for naive arithmetic). ∎

With respect to the time complexity for LU factoring with veiling and signatures, we separate the time complexity analysis for LU factoring into two parts. Lemma 1 shows the time complexity for LU with veiling but without signature. Lemma 2 gives the time complexity for LU with signatures. The total cost will be the complexity for LU with veiling and signatures in Theorem 5.

This first lemma is valid for the following veiling strategy: we veil any expression with an integer coefficient of length larger than c_1, or whose degree in x_i is larger than c_2, where c_1, c_2 are positive constants. The cost for veiling an expression is $O(1)$. Then the length of each coefficient will be less than $c = c_1 * c_2^m$ and the degree in x_i will be less than c_2.

Lemma 1. *For an n by n matrix $A = [a_{i,j}]$ the time complexity of LU factoring with large expression management (and the above veiling strategy) is $O(n^3)$.*

PROOF: Let $a_{i,j} \in \mathbb{Z}[x_1, ..., x_m]$, $d_r > \max_{i,j} \deg(a_{i,j}, x_r)$, and the length of the integer coefficients of $a_{i,j}$ be at most l. At each step there are at most two multiplications, one division and one subtraction. The cost of each step will be less than $4 \times O((c_1.c_2^m)^2) + O(1)$ for naive arithmetic. At each step k, one performs arithmetic on $(n - k)^2$ matrix elements, for a total cost of $\sum_{k=0}^{n-1}(n - k)^2 \times O((c_1\, c_2^m)^2) = O(n^3)$. ∎

Remark 4. *To prevent the cost from growing exponentially with the number of variables, the above computation clearly shows that it is best to choose $c_2 = 1$.*

Lemma 2. *Let $A = [a_{i,j}]$ be an n by n matrix where $a_{i,j} \in \mathbb{Z}[x_1, ..., x_m]$. Let $d > \max_{i,j,r} \deg(a_{i,j}, x_r)$ and let l bound the length of the largest integer coefficient of the entries of the matrix. Let $T > ld^m$. So T bounds the size of matrix entries. Let p be the prime being used to compute signatures.*

Then the time complexity for computing all signatures modulo p in the LU factorization is $O((Tn^2 + n^3)(\log p)^2)$.

PROOF: The cost of dividing an integer coefficient of $a_{i,j}$ of length l by p is $O(l)$ arithmetic operations modulo p and there are less than d^m terms in $a_{i,j}$. Assuming nested horner form is used, the polynomial $a_{i,j}$ can be evaluated modulo p in less than d^m multiplications and d^m additions modulo p. Thus cT bounds the number of arithmetic operations modulo p needed to compute the signature of each input matrix entry for some positive integer c.

After the initial computation of signatures for the entries of A, we need at most four operations in $\mathbb{Z}/p\mathbb{Z}$ for computing the other entries' signatures at step k of the factorization. This costs $O((\log p)^2)$ for naive integer arithmetic. We compute all the signatures for the entries at each step, to greatly simplify zero-testing. So the total cost for computing signatures for the LU factoring is bounded by

$$\left[cTn^2 + \sum_{k=1}^{n-1} 4(n - k)^2\right] \times O((\log p)^2) = O((Tn^2 + n^3)(\log p)^2).$$ ∎

Theorem 5. *Let $A = [a_{i,j}]$ be an n by n matrix where $a_{i,j} \in \mathbb{Z}[x_1, ..., x_m]$ and let p be the prime used for signature arithmetic. Let $d > \max_{i,j,r} \deg(a_{i,j}, x_r)$ and l bound the length of the largest integer coefficient of the entries of the matrix. Let $T > ld^m$. So T bounds the size of matrix entries. The time complexity of LU factoring with the above veiling strategy and modulo p signature computation is $O((Tn^2 + n^3)(\log p)^2)$.*

PROOF: Immediate from the above two lemmas. ∎

From Theorem 4 and Theorem 5, we can see the more the variables and the bigger the size of the matrix, the bigger the difference between the algorithms which are with and without veiling and signatures. These results agree completely with our empirical results.

7   Empirical Results

We present some timing results. For the benchmarks described below, we use strategies based on MAPLE's `length` command. As these strategies are heuristics, any reasonable measure of the complexity of an entry is sufficient. The pivoting strategy searches for the element with the largest `length`. The `veiling` strategy depends on the type of matrix. For integer matrices, we veil all integers whose length is greater than 1000, while for polynomial matrices, the treshold is length 30. These constants reflect the underlying constants involved in the arithmetic for such objects.

For all benchmarks, three variations are compared: our own LU factoring algorithm with veiling and signatures, MAPLE's default `LinearAlgebra:-LUDecomposition`, and a version of `LinearAlgebra:-LUDecomposition` where `Normalizer` has been set to be the identity function and `Testzero` has been set to a version of `testeq`. We first had to "patch" MAPLE's implementation of `LUDecomposition` to use `Testzero` instead of an explicit call to `Normalizer(foo)` `<> 0`, and then had to further "patch" `testeq` to avoid a silly coding mistake that made the code extremely inefficient for large expressions[3]. All tests were first run with a time limit of 300 seconds. Then the first test that timed out at 300 seconds was re-run with a time limit of 1000 seconds, to see if that was sufficient for completion. Further tests in that column were attempted. Furthermore, the sizes of matrices used varies according to the results, to try and focus attention to the sizes where we could gather some meaningful results in (parts of) the three columns. All results are obtained using the TTY version of MAPLE10, running on an 1.8Ghz Intel P4 with 512Megs of memory running Windows XP SP2, and with garbage collection "frequency" set to 20 million bytes used, all results are for dense matrices. In each table, we report the times in seconds, and for the LEM column, the number in parentheses indicates how many[4] distinct labels (ie total number of veiled expressions) were needed by the computation, as an indication of memory requirements.

The reason for including the MapleFix column is to really separate out the effect of arithmetic and signature-based zero-testing from the effects of Large Expression Management; MapleFix measures the effect of not doing polynomial arithmetic and using signatures for zero-recognition, and is thus expected to be a middle ground between the other two extremes.

Table 1 shows the result for random matrices over the integers. Only for fairly large matrices (between 90x90 and 100x100) does the cost of arithmetic, due to coefficient growth, become so large that the overhead of veiling becomes worthwhile, as the LEM column shows. Since integer arithmetic is automatic in MAPLE, it is not surprising that the MapleFix column shows times that are the same as the Maple column. Here the veiling strategy really matters: for integers of length 500, veiling introduces so much overhead that for 110x110 matrices,

---

[3] Both of these deficiencies were reported to MAPLESOFT and will hopefully be fixed in later versions of MAPLE.

[4] and we use a postfix K or M to mean 10^3 and 10^6 as appropriate.

Table 1. Timings for LU factoring of random integer matrices generated by RandomMatrix(n,n,generator=$-10^{12}..10^{12}$). The entries are explained in the text.

Size	10	20	30	40	50	60	70	80	90	100	110
LEM	.03	.2	.8	2.3	6.1	12.5	17.8	27.6	42.4	56.4	75.4
	(0)	(0)	(0)	(0)	(148)	(902)	(2788)	(5948)	(12779)	(22396)	(36739)
Maplefix	.07	.2	.7	2.2	5.2	10.7	19.4	33.8	54.0	83.8	124.7
Maple	.04	.2	.7	2.2	5.2	10.5	19.2	32.6	52.8	85.8	123

Table 2. Timings for LU factoring of random matrices with univariate entries of degree 5, generated by RandomMatrix(n,n,generator=(() -> randpoly(x))). The entries are explained in the text.

Size	5	10	15	20	25	30	35	40	45	50
LEM	.12	.06	.18	.44	.87	1.9	3.0	4.5	7.8	9.1
	(26)	(237)	(872)	(2182)	(4417)	(7827)	(12K)	(19K)	(28K)	(39K)
MapleFix	.06	.07	.16	.30	.56	1.87	332	>1000	–	–
Maple	.53	1.5	9.3	39.2	110.4	269.8	431	845	>1000	–

Table 3. Timings for LU factoring of random matrices with trivariate entries, low degree, 8 terms RandomMatrix(n,n,generator=(() -> randpoly([x,y,z], terms = 8))). The entries are explained in the text.

Size	5	10	15	20	25	30	35	40	45	50
LEM	.05	.09	.23	.49	.99	1.7	2.8	4.2	6.0	8.8
	(26)	(237)	(872)	(2182)	(4417)	(7827)	(12K)	(19K)	(28K)	(39K)
MapleFix	.06	.09	.20	.39	.75	3.2	949	>1000	–	–
Maple	35.3	>1000	–	–	–	–	–	–	–	–

Table 4. Timings for LU factoring of fully symbolic matrix: Matrix(n,n,symbol=m). The entries are explained in the text.

Size	5	10	15	20	25	30	35
LEM	.047	.078	.20	.51	.88	1.7	2.95
	(22)	(218)	(858)	(2163)	(4393)	(7798)	(12K)
MapleFix	.03	.08	.14	.30	.58	3.8	>1000
Maple	1.56	>1000	–	–	–	–	–

Table 5. Timings for LU factoring of random matrix with entries over $\mathbb{Z}[x,3^x]$: RandomMatrix(n,n,generator=(()->eval(randpoly([x,y],terms=8),y=3^x))). The entries are explained in the text.

Size	5	10	15	20	25	30	35
LEM	.031	.094	.22	.50	.99	1.7	2.8
	(26)	(237)	(872)	(2182)	(4417)	(7827)	(12K)
MapleFix	xx	xx	xx	xx	xx	xx	xx
Maple	0.99	117	>1000	–	–	–	–

this overhead is still larger than pure arithmetic. For length 2000, no veiling at all occurs.

Table 2 shows the result for random univariate matrices, where the initial polynomials have degree 5 and small integer coefficients. The effect of LEM here is immediately apparent. What is not shown is that MapleFix uses very little memory (both allocated and "used"), while the Maple column involves a huge amount of memory "used", at all sizes, so that computation time was swamped by garbage collection time. Another item to notice is that while the times in the Maple column grow steadily, the ones in the MapleFix column are at first consistent with the LEM column, and then experience a massive explosion. Very careful profiling[5] was necessary to unearth the reason for this, and it seems to be somewhat subtle: for both LEM and MapleFix, very small DAGs are created, but for LEM we have full control of these, while for MapleFix, the DAGs are small but the underlying expression tree is enormous. All of Maple's operations on matrix elements first involve the element being *normalized* by the kernel (via the user-inaccessible `simpl` function), and then *evaluated*. While normalization follows the DAG, evaluation in a side-effecting language must follow the expression tree, and thus is extremely expensive. Along with the fact that no information is kept between calls to `testeq`, causes the time to explode for MapleFix for 35x35 (and larger) matrices. Since the veiling strategy used for the last 4 tables is the same, it is not very suprising that the number of veilings is essentially the same. The reason that the all-symbolic is a little lower is because we start with entries of degree 1 and coefficient size 1, and thus these entries do not get veiled immediately. However, one can observe a clear cubic growth in the number of veilings, as expected.

Table 3 shows the result for random trivariate matrices, where the initial polynomials have 8 terms and small integer coefficients. The results here clearly show the effect that multi-variate polynomial arithmetic has on the results. Table 4 shows the results for a matrix with all entries symbolic, further accentuating the results in the trivariate case. Again, MapleFix takes moderate amounts of memory (but a lot of CPU time at larger sizes), while Maple takes huge amounts, causing a lot of swapping and trashing already for 10x10 matrices.

Table 5 shows results for matrices with entries over $\mathbb{Z}[x, 3^x]$. Overall the behaviour is quite similar to bivariate polynomials, however the **xx** in the MapleFix entry indicate a weakness in MAPLE's `testeq` routine, where valid inputs (according to the theory in [3]) return FAIL instead. Our signature implementation can handle such an input domain without difficulty.

While we would have liked to present memory results as well, this was much more problematic, as MAPLEdoes not really provide adequate facilities to achieve this. One could look at **bytes used**, but this merely reflects the memory asked of the system, the vast majority of which is garbage and immediately reclaimed. This does measure the amount of overall memory *churn*, but does not give an indication of final memory use nor of the true *live set*. **bytes alloc** on the other

---

[5] Here we used a combination of procedure-level profiling via `CodeTools[Profiling]` and global profiling via `kernelopts(profile=true)`.

hand measure the actual amount of system memory allocated. Unfortunately, this number very quickly settles to something a little larger than **gcfreq**, in other words the amount of memory required to trigger another round of garbage collection, for all the tests reported here. This reflects the huge amount of memory used in these computations, but does not reflect the final amount of memory necessary to store the end result. Neither can we rely on MAPLE's **length** command to give an accurate representation of the memory needed for a result because, for some unfathomable reason, **length** returns the expression tree length rather than the DAG length! Thus, for matrices whose results are un-normalized polynomials, we have no easy way to measure their actual size. As a proxy, we can find out the total number of variables introduced by the veiling process.

Acknowledgements

We wish to thank Éric Schost for his suggestions and helps with this paper. We also thank the anonymous ISSAC referee for their suggestions to use the environment variable **Normalizer** when calling LUDecomposition.

References

1. Schwartz J.T. Fast Probabilistic Algorithms for Verification of Polynomial Identities. *J. ACM*, 27(4): 701-717, 1980.
2. Zippel R. Probabilistic Algorithms for Sparse Polynomials. *Proc. of Eurosam 79'*, Springer-Verlag LNCS, **72**, 216–226, 1979.
3. Gonnet G.H.. Determining Equivalence of Expressions in Random Polynomial Time, Extended Abstract, *Proc. of ACM on Theory of computing*, 334-341, 1984.
4. Gonnet G.H. New results for random determination of equivalence of expressions. *Proc. of ACM on Symbolic and algebraic comp.*, 127–131, 1986.
5. Monagan M.B. Signatures + Abstract Types = Computer Algebra − Intermediate Expression Swell. PhD Thesis, *University of Waterloo*, 1990.
6. Monagan M.B. Gauss: a Parameterized Domain of Computation System with Support for Signature Functions. *DISCO*, Springer-Verlag LNCS, **722**, 81-94, 1993.
7. http://www.linbox.org
8. http://www.maplesoft.com/products/thirdparty/dynaflexpro/
9. Freeman T.S., Imirzian G., Kaltofen E. and Yagati L. DAGWOOD: A system for manipulating polynomials given by straight-line programs. *ACM Trans. Math. Software*, 14(3):218-240, 1988.
10. Corless R.M., Jeffrey D.J., Monagan M.B. and Pratibha. Two Perturbation Calculation in Fluid Mechanics Using Large-Expression Management. *J. Symbolic Computation*, **11**, 1–17, 1996.
11. Zippel R. Effective Polynomial computation. *Kluwer*, 1993.
12. Kaltofen E. Greatest Common Divisors of Polynomials Given by Straight-Line Programs. *J. of the Association for Computing Machinery*, 35(1): 231-264, 1988.
13. Giusti M., Hägele K., Lecerf G., Marchand J. and Salvy B. The projective Noether Maple package: computing the dimension of a projective variety. *J. Symbolic Computation*, 30(3): 291–307, 2000.

14. Kaltofen E. Computing with polynomials given by straight-line programs I: greatest common divisors. *Proceedings of ACM on Theory of computing*, 131–142, 1985.
15. Monagan M.B. and Gonnet G.H. Signature Functions for Algebraic Numbers. *ISSAC*, 291–296, 1994.
16. Monagan M.B. and Monagan G. A toolbox for program manipulation and efficient code generation with an application to a problem in computer vision. *ISSAC*, 257–264, 1997.
17. Sasaki T. and Murao H. Efficient Gaussian elimination method for symbolic determinants and linear systems (Extended Abstract). *ISSAC*, 155–159, 1981.
18. W. Keith Nicholson. Linear Algebra with Applications, Fourth Edition. *McGraw-Hill Ryerson*, 2003.
19. Bareiss E.H. Sylvester's Identity and Multistep Integer-Preserving Gaussian Elimination. *Mathematics of Computation*, 22(103): 565-578, 1968.
20. Bareiss E.H. Computational Solutions of Matrix Problems Over an Integral Domain. *J. Inst. Maths Applics*, **10**, 68-104, 1972.
21. Heintz J. and Schnorr C.P. Testing polynomials which are easy to compute (Extended Abstract). *Proceedings of ACM on Theory of computing*, 262–272, 1980.
22. Martin W.A. Determining the equivalence of algebraic expressions by hash coding. *Proceedings of ACM on symbolic and algebraic manipulation*, 305–310, 1971.
23. Ibarra O.H. and Leininger B.S. On the Simplification and Equivalence Problems for Straight-Line Programs. *J. ACM*, 30(3): 641–656, 1983.
24. Ibarra O.H. and Moran S. Probabilistic Algorithms for Deciding Equivalence of Straight-Line Programs. *J. ACM*, 30(1): 217–228, 1983.
25. Ibarra O.H., Moran S. and Rosier L.E. Probabilistic Algorithms and Straight-Line Programs for Some Rank Decision Problems. *Infor. Proc. Lett.*, 12(5): 227–232, 1981.
26. Shoup V. A Computational Introduction to Number Theory and Algebra. *Cambridge University Press*, 2005.
27. Kaltofen E. On computing determinants of matrices without divisions. *ISSAC*, 342–349, 1992.
28. Gathen J. von zur and Gerhard J. Modern computer algebra *Cambridge : Cambridge University Press*, 1999.
29. Carette J. Understanding Expression Simplification. *ISSAC*, 72–79, 2004.
30. Carette J. and Kiselyov O. Multi-stage Programming with Functors and Monads: Eliminating Abstraction Overhead from Generic Code. *Generative Programming and Component Engineering*, 256–274, 2005.
31. S. Steinberg, P. Roach Symbolic manipulation and computational fluid dynamics. *Journal of Computational Physics*, 57, pp 251-284, 1985.

Author Index

Lecture Notes in Artificial Intelligence (LNAI)

Vol. 3960: R. Vieira, P. Quaresma, M.d.G.V. Nunes, N.J. Mamede, C. Oliveira, M.C. Dias (Eds.), Computational Processing of the Portuguese Language. XII, 274 pages. 2006.

Vol. 3955: G. Antoniou, G. Potamias, C. Spyropoulos, D. Plexousakis (Eds.), Advances in Artificial Intelligence. XVII, 611 pages. 2006.

Vol. 3949: F. A. Savacı (Ed.), Artificial Intelligence and Neural Networks. IX, 227 pages. 2006.

Vol. 3946: T.R. Roth-Berghofer, S. Schulz, D.B. Leake (Eds.), Modeling and Retrieval of Context. XI, 149 pages. 2006.

Vol. 3944: J. Quiñonero-Candela, I. Dagan, B. Magnini, F. d'Alché-Buc (Eds.), Machine Learning Challenges. XIII, 462 pages. 2006.

Vol. 3930: D.S. Yeung, Z.-Q. Liu, X.-Z. Wang, H. Yan (Eds.), Advances in Machine Learning and Cybernetics. XXI, 1110 pages. 2006.

Vol. 3918: W.K. Ng, M. Kitsuregawa, J. Li, K. Chang (Eds.), Advances in Knowledge Discovery and Data Mining. XXIV, 879 pages. 2006.

Vol. 3913: O. Boissier, J. Padget, V. Dignum, G. Lindemann, E. Matson, S. Ossowski, J.S. Sichman, J. Vázquez-Salceda (Eds.), Coordination, Organizations, Institutions, and Norms in Multi-Agent Systems. XII, 259 pages. 2006.

Vol. 3910: S.A. Brueckner, G.D.M. Serugendo, D. Hales, F. Zambonelli (Eds.), Engineering Self-Organising Systems. XII, 245 pages. 2006.

Vol. 3904: M. Baldoni, U. Endriss, A. Omicini, P. Torroni (Eds.), Declarative Agent Languages and Technologies III. XII, 245 pages. 2006.

Vol. 3900: F. Toni, P. Torroni (Eds.), Computational Logic in Multi-Agent Systems. XVII, 427 pages. 2006.

Vol. 3899: S. Frintrop, VOCUS: A Visual Attention System for Object Detection and Goal-Directed Search. XIV, 216 pages. 2006.

Vol. 3898: K. Tuyls, P.J. 't Hoen, K. Verbeeck, S. Sen (Eds.), Learning and Adaption in Multi-Agent Systems. X, 217 pages. 2006.

Vol. 3891: J.S. Sichman, L. Antunes (Eds.), Multi-Agent-Based Simulation VI. X, 191 pages. 2006.

Vol. 3890: S.G. Thompson, R. Ghanea-Hercock (Eds.), Defence Applications of Multi-Agent Systems. XII, 141 pages. 2006.

Vol. 3885: V. Torra, Y. Narukawa, A. Valls, J. Domingo-Ferrer (Eds.), Modeling Decisions for Artificial Intelligence. XII, 374 pages. 2006.

Vol. 3881: S. Gibet, N. Courty, J.-F. Kamp (Eds.), Gesture in Human-Computer Interaction and Simulation. XIII, 344 pages. 2006.

Vol. 3874: R. Missaoui, J. Schmidt (Eds.), Formal Concept Analysis. X, 309 pages. 2006.

Vol. 3873: L. Maicher, J. Park (Eds.), Charting the Topic Maps Research and Applications Landscape. VIII, 281 pages. 2006.

Vol. 3864: Y. Cai, J. Abascal (Eds.), Ambient Intelligence in Everyday Life. XII, 323 pages. 2006.

Vol. 3863: M. Kohlhase (Ed.), Mathematical Knowledge Management. XI, 405 pages. 2006.

Vol. 3862: R.H. Bordini, M. Dastani, J. Dix, A.E.F. Seghrouchni (Eds.), Programming Multi-Agent Systems. XIV, 267 pages. 2006.

Vol. 3849: I. Bloch, A. Petrosino, A.G.B. Tettamanzi (Eds.), Fuzzy Logic and Applications. XIV, 438 pages. 2006.

Vol. 3848: J.-F. Boulicaut, L. De Raedt, H. Mannila (Eds.), Constraint-Based Mining and Inductive Databases. X, 401 pages. 2006.

Vol. 3847: K.P. Jantke, A. Lunzer, N. Spyratos, Y. Tanaka (Eds.), Federation over the Web. X, 215 pages. 2006.

Vol. 3835: G. Sutcliffe, A. Voronkov (Eds.), Logic for Programming, Artificial Intelligence, and Reasoning. XIV, 744 pages. 2005.

Vol. 3830: D. Weyns, H. V.D. Parunak, F. Michel (Eds.), Environments for Multi-Agent Systems II. VIII, 291 pages. 2006.

Vol. 3817: M. Faundez-Zanuy, L. Janer, A. Esposito, A. Satue-Villar, J. Roure, V. Espinosa-Duro (Eds.), Nonlinear Analyses and Algorithms for Speech Processing. XII, 380 pages. 2006.

Vol. 3814: M. Maybury, O. Stock, W. Wahlster (Eds.), Intelligent Technologies for Interactive Entertainment. XV, 342 pages. 2005.

Vol. 3809: S. Zhang, R. Jarvis (Eds.), AI 2005: Advances in Artificial Intelligence. XXVII, 1344 pages. 2005.

Vol. 3808: C. Bento, A. Cardoso, G. Dias (Eds.), Progress in Artificial Intelligence. XVIII, 704 pages. 2005.

Vol. 3802: Y. Hao, J. Liu, Y.-P. Wang, Y.-m. Cheung, H. Yin, L. Jiao, J. Ma, Y.-C. Jiao (Eds.), Computational Intelligence and Security, Part II. XLII, 1166 pages. 2005.

Vol. 3801: Y. Hao, J. Liu, Y.-P. Wang, Y.-m. Cheung, H. Yin, L. Jiao, J. Ma, Y.-C. Jiao (Eds.), Computational Intelligence and Security, Part I. XLI, 1122 pages. 2005.

Vol. 3789: A. Gelbukh, Á. de Albornoz, H. Terashima-Marín (Eds.), MICAI 2005: Advances in Artificial Intelligence. XXVI, 1198 pages. 2005.

Vol. 3782: K.-D. Althoff, A. Dengel, R. Bergmann, M. Nick, T.R. Roth-Berghofer (Eds.), Professional Knowledge Management. XXIII, 739 pages. 2005.

Vol. 3763: H. Hong, D. Wang (Eds.), Automated Deduction in Geometry. X, 213 pages. 2006.

Vol. 3755: G.J. Williams, S.J. Simoff (Eds.), Data Mining. XI, 331 pages. 2006.

Vol. 3735: A. Hoffmann, H. Motoda, T. Scheffer (Eds.), Discovery Science. XVI, 400 pages. 2005.

Vol. 3734: S. Jain, H.U. Simon, E. Tomita (Eds.), Algorithmic Learning Theory. XII, 490 pages. 2005.

Vol. 3721: A.M. Jorge, L. Torgo, P.B. Brazdil, R. Camacho, J. Gama (Eds.), Knowledge Discovery in Databases: PKDD 2005. XXIII, 719 pages. 2005.

Vol. 3720: J. Gama, R. Camacho, P.B. Brazdil, A.M. Jorge, L. Torgo (Eds.), Machine Learning: ECML 2005. XXIII, 769 pages. 2005.

Vol. 3717: B. Gramlich (Ed.), Frontiers of Combining Systems. X, 321 pages. 2005.